Springer Series in Computational Physics

Editors:
H. Cabannes,   M. Holt,   H. B. Keller,   J. Killeen,   S. A. Orszag

Yu. I. Shokin

The Method of Differential Approximation

Translated by K. G. Roesner

With 75 Figures and 12 Tables

Springer-Verlag
Berlin Heidelberg New York Tokyo 1983

Dr. Yu. I. Shokin

Academy of Sciences of USSR,
Sibirian Branch,
Institute of Theoretical and
Applied Mechanics
SU-630090 Novosibirsk/USSR

Dr. K.G. Roesner

Institut für Mechanik, TH Darmstadt
Hochschulstraße 1
D-6100 Darmstadt
Federal Republic of Germany

Editors

Henri Cabannes

Mécanique Théoretique
Université Pierre et Marie Curie
Tour 66. 4, Place Jussieu
F-75005 Paris/France

Maurice Holt

College of Engineering and
Mechanical Engineering
University of California
Berkeley, California 94720/USA

H.B. Keller

Applied Mathematics 101-50
Firestone Laboratory
California Institute of Technology
Pasadena, California 91125/USA

John Killeen

Lawrence Livermore Laboratory
P.O. Box 808
Livermore, California 94551/USA

Stephen A. Orszag

Department of Mathematics
Massachusetts Institute of Technology
Cambridge, Massachusetts 02139/USA

ISBN-13: 978-3-642-68985-7   e-ISBN-13: 978-3-642-68983-3
DOI: 10.1007/ 978-3-642-68983-3

Library of Congress Cataloging in Publication Data.
Shokin, Yurii Ivanovich. The method of differential approximation.
(Springer series in computational physics)
Translation of: Metod differentsial'nogo priblizheniia.
Bibliography: p. Includes index. 1. Differential equations, Hyperbolic – Numerical solutions.
I. Title. II. Series.   QA377.S4913   1983   515.3'53   83-4681

Typesetting: Daten- und Lichtsatz-Service, Würzburg

2153/3020-543210

Preface

I am very glad that this book is now accessible to English-speaking scientists. During the three years following the publication of the original Russian edition, the method of differential approximation has been rapidly expanded and unfortunately I was unable to incorporate into the English edition all of the material which whould have reflected its present state of development. Nevertheless, a considerable amount of recently obtained results have been added and the bibliography has been enlarged accordingly, so that the English edition is one third longer than the Russian original.

Mathematical rigorousness is a basic feature of this monograph. The reader should therefore be familiar with the theory of partial differential equations and difference equations. Some knowledge of group theory as applied to problems in physics, especially the theory of Lie groups, would also be useful. The treatment of the approximation of gas dynamic equations focuses on the question of how to characterize the typical features of difference equations on the basis of the related differential approximation, which can be discussed using the fully developed theory of partial differential equations.

Part I is devoted to the basic concept of stability analysis of difference schemes. It gives a short introduction to the theory of linear differential and difference equations. The concept of differential approximation is also introduced, and the stability analysis of finite difference schemes is treated in relation to the first differential approximation. The dissipative character of difference schemes is investigated in great detail, and the construction of difference schemes with higher orders of approximation is treated at the end of Part I.

Part II acquaints the reader with the phenomenon of artificial viscosity of difference schemes and defines numerous other typical features of difference schemes, such as conservativity, dissipativity and dispersion. The monotonicity of difference schemes is defined in connection with the calculation of moving shocks. The results are applied to arbitrary curvilinear coordinate systems to show how they can be used in practical problems.

Part III introduces the concept of invariance of a difference scheme, which is important when calculating problems with a given symmetry character resulting from the physical problem formulation. This class of difference schemes is also discussed with respect to other properties such as conservativity, dissipativity, and stability. A large number of schemes are applied to an illustrative problem which demonstrates the different results obtained by invariant and noninvariant schemes. A group-theoretical investigation of difference schemes for gas dynamic problems is also included.

Part IV contains a detailed discussion of a variety of difference schemes used to solve the equations of propagation and one-dimensional gas dynamics. A complete list of the typical features of these difference schemes makes it possible to compare the results expected from an *a priori* analysis and the numerical results actually generated. Thus the book will be useful for readers mainly interested in applications of the method of first differential approximation.

I am deeply grateful to my colleague K. G. Roesner for his initiative in translating this book, his careful and accurate editing of the English version, and the efforts he devoted to the sometimes delicate problem of translating new and often rather exotic Russian mathematical terms.

The translator, K. G. Roesner, whishes to express his gratitude to his wife, Lotte, for typing the manuscript with such care and patience, to his colleague Maurice Holt for reading the manuscript and giving much helpful translating advice, and to Z. I. Fedotova, L. A. Kompaniecz, and A. I. Urusov of the Institute of Theoretical and Applied Mechanics in Akademgorodok for their valuable assistance in checking for misprints and errors. The translator is also particularly indebted to Mrs. I. Schmidt of the Institut für Mechanik in Darmstadt for her meticulous execution of the tables and figures.

I would like to express my especial gratitude to Professor W. Beiglböck and Springer-Verlag for publishing this translation in so ambitious a form.

Finally, I wish to thank my wife, Eleonora, for her support during the working at this book.

Akademgorodok, March 1983   Yu. I. Shokin

Preface to the Russian Edition

A fundamental aim of applied mathematics is to find new mathematical methods to solve important and complicated problems which arise in connection with the rapid progress of science and technology. The use of fast computers has substantially increased the number of applied problems which can be handled by a mathematical approach and has provided many guidelines for the modern branches of mathematical research.

Of the numerical methods used to solve applied problems, the most important is the method of finite differences. This is because of its universal applicability to wide classes of differential equations, including linear and nonlinear equations as well as ordinary and partial differential equations with steady and unsteady coefficients and different initial and boundary conditions. The successful application of difference methods in various scientific and technical fields spurred the intensive development of the theory of finite-difference methods which has covered the period of the last 25–30 years, although some fundamental facts in this field were known much earlier. The literature devoted to the theory of difference methods includes hundreds of names (see the monographs [1–15], which contain fairly complete bibliographies and describe the essential results of the theory).

It is apparent to those working on various problems of applied mathematics that unless new methods in numerical mathematics are created, it will be practically impossible to obtain solutions in several important cases. In the USSR and other countries considerable effort is being devoted to the construction of effective finite-difference methods and to methods of analyzing existing and newly discovered difference schemes. The results of investigations of the problems mentioned above are contained in publications by such renowned Soviet mathematicians as A. N. Tikhonov, G. I. Marchuk, A. A. Samarskii, N. N. Yanenko, O. M. Belotserkovskii, K. I. Babenko, S. K. Godunov, and V. V. Rusanov.

The present book is devoted to the method of differential schemes for equations of hyperbolic character, in particular the equations of gas dynamics [16–161]. In the fifties, A. I. Zhukov [38–40] showed for the first time that it is possible to apply the differential approximation to the investigation of difference schemes. In the ensuing years, this idea was followed up by mathematicians from the Soviet Union and other countries, and a theory was developed. With the method of differential approximation it is possible to construct new difference schemes with well-defined properties, analyze the properties of already existing and newly developed difference schemes, and classify difference schemes according to defined properties.

I should like to acknowledge my especial indebtedness to N. N. Yanenko for many fruitful discussions and much assistance in finishing this book. I also wish to extend my thanks to V. V. Kobkov, A. I. Urusov, and Z. I. Fedotova for their help in preparing the manuscript for the printer.

Akademgorodok, 1979   Yu. I. Shokin

Contents

Part I

Stability Analysis of Difference Schemes by the Method of Differential Approximation

1. Certain Properties of the Theory of Linear Differential Equations and Difference Schemes

1.1 Cauchy's Problem

We introduce some properties of the theory of linear differential equations and of the theory of difference equations which are necessary for a better understanding of the following chapters. For a more detailed information the interested reader is referred to the monographs [1–15, 162].

We will discuss *Cauchy's problem* for linear hyperbolic systems of partial differential equations of first order:

$$\frac{\partial u}{\partial t} = \sum_{k=1}^{s} A_k(x, t) \frac{\partial u}{\partial x_k} := P\left(x, t, \frac{\partial}{\partial x}\right) u, \quad t \geq 0, \ -\infty < x_k < \infty. \tag{1.1}$$

$u = u(x, t)$ is a m-dimensional vector-valued function of the real independent variables $x = \{x_1, \ldots, x_s\} \in \mathbb{R}_s$ and t; $A_k(x, t)$ are real $m \times m$ matrices with elements depending on x and t.

Cauchy's problem for the system of equations (1.1) consists in finding a solution $u(x, t)$ which coincides with the given initial values $u(x, 0) = u_0(x)$ on a non-characteristic initial hyperplane $t = 0$.

Definition. The system of equations (1.1) is called hyperbolic in the point (x^0, t^0) $= \{x_1^0, \ldots, x_s^0, t^0\}$ if for arbitrary real numbers $\omega_1, \ldots, \omega_s$ with $\sum_{i=1}^{s} \omega_i^2 = 1$ the eigenvalues of the matrix $P(x^0, t^0, i\omega)$ are purely imaginary and if the matrix itself can be transformed into diagonal form for all real $\omega = \{\omega_1, \ldots, \omega_s\}$ [8, 163].

From this it follows that the equation

$$\det \left\| \lambda I - \sum_{k=1}^{s} \omega_k A_k(x^0, t^0) \right\| = 0$$

has real roots, where I is the unit matrix.

Definition. Cauchy's problem for the system of equations (1.1) is called correct in a space $\mathbb{B}$ if its solution exists, is unique and if it depends continuously on the initial data with respect to a norm in the space $\mathbb{B}$.

Especially if A_k are symmetric matrices ($A_k' = A_k$, A_k' is the transposed matrix) and $u_0(x) \in \mathbb{L}_2(-\infty, \infty)$, then Cauchy's problem for the system of equations (1.1) is correct in $\mathbb{L}_2$. The scalar product and the norm are defined in the usual way:

$$(u, v) = \int\limits_{\mathbb{R}_s} \sum_{k=1}^{m} u_k v_k \, dx; \qquad \|u\|^2 = (u, u).$$

1.2 One-dimensional Time-dependent Case

We consider the system of equations (1.1) for one space variable

$$\frac{\partial u}{\partial t} = A \frac{\partial u}{\partial x}, \tag{1.2}$$

where the matrix A has different real eigenvalues $\xi_1(x, t), \ldots, \xi_m(x, t)$, and is continuous with respect to the independent variables in the half-plane $t \geq 0$, $-\infty < x < \infty$, and satisfies a *Lipshitz's condition*.

Definition. If in each point of a curve l_j the relation

$$\frac{dx}{dt} = \xi_j(x, t)$$

holds the curve l_j is called a characteristic of the system of equations (1.2).

Through each point (x^0, t^0) m different characteristic curves $l_1, \ldots, l_m$ of the system of equations (1.2) can be drawn.

For the system of equations (1.2) we consider Cauchy's problem with the initial data $u(x, 0) = u_0(x)$. Then its solution in the point (x, t) depends only on the initial distribution in the interval $[a, b]$ of the axis $t = 0$ which is defined by the intersection points of the characteristic curves l_1 and l_m with the axis. Therefore, a jump of the initial data in the point a leads to a jump of the solution of the system of equations (1.2) only along the characteristics which start from the point a.

Definition. The interval $[a, b]$ is called the region of dependence of the point (x, t) for the system of equations (1.2). Generally we will call Q the region of dependence of the point (x, t) for the system of equations (1.1) if Q consists of the smallest bounded set of points on the hyperplane $t = 0$ such that if the initial data are zero in an open set of points containing Q the corresponding solution in the point (x, t) is also zero.

Definition. The set of points which is bounded by the curves l_1 and l_m starting from the point $(x^0, 0)$ is called the region of influence of the point $(x^0, 0)$ for the system of equations (1.2).

1.3 Systems of Second-order Equations

In the following we will deal with systems of equations of second order

$$\frac{\partial u}{\partial t} = \sum_{j,k=1}^{s} C_{jk} \frac{\partial^2 u}{\partial x_j \partial x_k} + \sum_{j=1}^{s} D_j(x,t) \frac{\partial u}{\partial x_j}. \tag{1.3}$$

Definition. The system of equations (1.3) is called not totally parabolic in the point (x^0, t^0), if for all real $\omega_1, \ldots, \omega_s$ with $\sum_{i=1}^{s} \omega_i^2 = 1$ the roots of the equation

$$\det \left\| - \sum_{j,k=1}^{s} \omega_j \omega_k C_{jk}(x^0, t^0) - \lambda I \right\| = 0 \tag{1.4}$$

have non-positive real parts

$$\mathrm{Re}\,\{\lambda_j\} \leq 0, \quad (j = 1, \ldots, m).$$

Definition. If $\mathrm{Re}\,\{\lambda_j\} \leq \delta < 0$, $(j = 1, \ldots, m)$, the system of equations (1.3) is called parabolic in the sense of Petrovskii [163].

It can be proved easily that from the not total parabolicity of the system of equations (1.3) it follows that the one-dimensional system of equations

$$\frac{\partial u}{\partial t} = C_{jj}(x,t) \frac{\partial^2 u}{\partial x_j^2} + D_j(x,t) \frac{\partial u}{\partial x_j}$$

is also not totally parabolic.

The hyperbolic character of the following one-dimensional system of equations

$$\frac{\partial u}{\partial t} = A_j(x,t) \frac{\partial u}{\partial x_j}$$

can be derived in a manner similar to that of the system (1.1) to establish the same property.

1.4 Basic Concepts of the Theory of Difference Schemes

We will discuss basic concepts of the theory of difference schemes for the scheme

$$u^{n+1}(x) = S_h(T) u^n(x), \tag{1.5}$$

which approximates the system of equations (1.1) (for a more detailed discussion see [1–15]). Here S_h – the step-operator for the difference scheme (1.5) – is given by the following expression:

$$S_h := \sum_\alpha B_\alpha T^{\kappa \lambda_\alpha} = \sum_\alpha B_\alpha T_1^{\kappa_1 \lambda_\alpha^1} \ldots T_s^{\kappa_s \lambda_\alpha^s};$$

B_α are real $(m \times m)$-matrices; $T := \{T_1, \ldots, T_s\}$; T_j is the shift-operator along the x_j-axis, $(j = 1, \ldots, s)$:

$$T_j f(x_1, \ldots, x_j, \ldots, x_s) := f(x_1, \ldots, x_{j-1}, x_j + h_j, x_{j+1}, \ldots, x_s),$$

h_j is the step-size in x_j-direction; $\kappa := \tau/h$; $\kappa_j := \tau/h_j = \text{const}$; $h = \{h_1, \ldots, h_s\}$; $t = n\tau$; $\lambda_\alpha = \{\lambda_\alpha^1, \ldots, \lambda_\alpha^s\}$.

The values of α are taken from an index set.

Definition. If the set of parameter values α is finite, the difference scheme is called explicit, otherwice implicit.

In the following we are faced with the shift-operator in t-direction

$$T_0 f(x, t) = f(x, t + \tau)$$

and with the following difference operators:

$$\Delta_j = T_j - E, \qquad \Delta_{-j} = E - T_{-j},$$
$$\Delta_0 = T_0 - E, \qquad Ef(x, t) = f(x, t).$$

In the case of one space variable we will use the notation $T_0, T_1, E, \Delta_0, \Delta_1, \Delta_{-1}$.

Definition. The largest number p, for which all solutions of the system of equations (1.1) with continuous derivatives of $(p + 1)$ order satisfy the equation (1.5) with an accuracy up to $O(\tau^{p+1})$ is called the order of approximation of the difference scheme (1.5).

If the difference scheme (1.5) is approximating the system of equations (1.1) with first order the following compatibility conditions are satisfied:

$$\sum_\alpha B_\alpha = I; \qquad \sum_\alpha \lambda_\alpha^k B_\alpha = A_k \qquad (k = 1, \ldots, s).$$

Definition. The difference scheme (1.5) is called stable, if the norm of the step-operator S_h satisfies the inequality

$$\|S_h\| \leq 1 + O(\tau).$$

Here and in the following chapters, when not specified especially, the stability will be discussed with respect to the norm in the space $\mathbb{L}_2$.

Definition. The difference scheme (1.5) is called convergent if its solution tends to the solution of the system of differential equations (1.1) if τ and h tend to zero.

The equivalence theorem [8] states that, if the original differential problem is correct, a necessary and sufficient condition for convergence of the approximating difference scheme is, that the scheme is stable.

We introduce in the the usual way the *Fourier transform*

$$\hat{u}(y) = (2\pi)^{-s/2} \int\limits_{-\infty}^{\infty} e^{-ixy} u(x)\, dx,$$

where $\hat{u}(y)$ is the Fourier transform of the function $u(x)$. Here

$$xy = \sum_{k=1}^{s} x_k y_k$$

is the scalar product. It follows that

$$u(x) = (2\pi)^{-s/2} \int_{-\infty}^{\infty} e^{ixy} \hat{u}(y)\,dy.$$

In the case of constant coefficients performing the Fourier transformation in the difference scheme (1.5) we get

$$\hat{u}^{n+1}(y) = G\,\hat{u}^n(y),$$

where the matrix

$$G := \sum_{\alpha} B_{\alpha} e^{i\tau\lambda_{\alpha} y}$$

is called the amplification matrix of the system.

Definition. The smallest closed set of points in the hyperplane $t = 0$ with the property that, if the initial distribution $u_0(x)$ is equal to 0 in an open set containing Q' the corresponding solution of the difference scheme in the point (x, t) is also zero for all values of τ, is called the region of dependence Q' of the point (x, t).

In the case of $s = 1$ the difference scheme (1.5) has the form:

$$u^{n+1}(x) = \sum_{\alpha} B_{\alpha} u^n(x + \tau\lambda_{\alpha}), \qquad (1.6)$$

where λ_{α} are given numbers. Let $\lambda_{\alpha} := \alpha h/\tau$, $\tau/h = \text{const}$, $h > 0$, $\alpha = -q_1, \ldots,$ q_2, $q_1 > 0$, $q_2 > 0$ then the solution of the system in the point $x = \Theta h$, $t = n\tau$ will depend on the values of the function $u^0(x) = u_0(x)$ on the interval $[(\Theta - nq_1)h, (\Theta + nq_2)h]$ of the x-axis. This interval is the region of dependence of the point $(\Theta h, n\tau)$ of the difference scheme (1.6). The value of the function $u_0(x)$ in the point $(x, 0)$ influences the solution of the difference scheme (1.6) in a region which is bounded by straight lines with the slopes:

$$\frac{dx}{dt} = -q_2 \frac{h}{\tau} \quad \text{and} \quad \frac{dx}{dt} = q_1 \frac{h}{\tau}.$$

The given region will be called the region of influence of the point $(x, 0)$ of the difference scheme (1.6).

Courant et al. [164] have shown that for convergence of a difference scheme it is necessary, that the region of dependence of the original system of differential equations is contained in the region of dependence of the corresponding difference scheme. Assuming some additional limitations this condition is also a sufficient stability condition of the difference scheme and, consequently, according to the equivalence theorem also a sufficient condition for the convergence of the scheme.

2. The Concept of the Differential Approximation of a Difference Scheme

2.1 Γ-form and Π-form of the Differential Representation of a Difference Scheme

Let

$$\Lambda_1 u^{n+1}(x) = \Lambda_0 u^n(x) \tag{2.1}$$

be a difference scheme which approximates the differential equation

$$\frac{\partial u}{\partial t} = Lu. \tag{2.2}$$

Here the operator Λ_1 has an inverse,

$$\Lambda_k := \Lambda_k(t, x, \tau, h, T); \qquad T := \{T_1, \ldots, T_s\}; \qquad (k = 0, 1);$$

$$L := L(t, x, D); \qquad D := \frac{\partial}{\partial x} = \left\{\frac{\partial}{\partial x_1}, \ldots, \frac{\partial}{\partial x_s}\right\}.$$

Approximating the differential equation by a difference scheme means that we change from an infinite-dimensional function space of a continuous argument to a finite-dimensional space of grid functions and that we change the equations for functions of continuous arguments to algebraic equations.

Such a treatment which is very convenient in practice, leads to difficulties for the theoretical investigation of properties of difference schemes, because of the fact that a grid function and a function of a continuous argument are defined on different spaces. In this case it is more convenient to investigate difference operators in the same function space, to which the differential operators belong, which are approximated by them [8].

In this connection it is of importance that the solutions of the difference schemes are functions of discrete arguments in each point of the considered region. In the following we will always keep to this treatment.

The following operator representations are well-known:

$$T_0 = e^{\tau D_0} = e^{\tau(\partial/\partial\tau)} = \sum_{l=0}^{\infty} \frac{\tau^l}{l!} D_0^l;$$

$$T_j = e^{h_j D_j} = e^{h_j(\partial/\partial x_j)} = \sum_{l=0}^{\infty} \frac{h_j^l}{l!} D_j^l; \quad j = 1, \ldots, s;$$

$$\ln(e^{\tau D_0}) = \tau D_0.$$

The difference scheme (2.1) can be written in the form:

$$\Lambda_1(t, x, \tau, h, e^{hD}) e^{\tau(\partial/\partial t)} u = \Lambda_0(t, x, \tau, h, e^{hD}) u$$

or

$$e^{\tau(\partial/\partial t)} u = \Lambda_1^{-1}(t, x, \tau, h, e^{hD}) \Lambda_0(t, x, \tau, h, e^{hD}) u, \tag{2.3}$$

where

$$e^{hD} = \{e^{h_1(\partial/\partial x_1)}, \ldots, e^{h_s(\partial/\partial x_s)}\}.$$

From this it follows that

$$\sum_{l=1}^{\infty} \frac{\tau^{l-1}}{l!} \frac{\partial^l u}{\partial t^l} = \frac{1}{\tau} [\Lambda_1^{-1}(t, x, \tau, h, e^{hD}) \Lambda_0(t, x, \tau, h, e^{hD}) - E] u$$

and, consequently, because of the approximation we get finally

$$\sum_{l=1}^{\infty} \frac{\tau^{l-1}}{l!} \frac{\partial^l u}{\partial t^l} = Lu + \sum_{l=1}^{\infty} \sum_{l_1 + \ldots + l_s = 1}^{l} \alpha_{l_1 \ldots l_s} \frac{\partial^l u}{\partial x_1^{l_1} \ldots \partial x_s^{l_s}}, \tag{2.4}$$

where $\alpha_{l_1 \ldots l_s}$ are coefficients which depend on t, x, τ, and h.

Definition. Equation (2.4) is called the Γ-form of the differential representation of the difference scheme (2.1).

Example. Let us consider the difference scheme

$$\frac{u^{n+1}(x) - u^n(x)}{\tau} = a \frac{u^n(x+h) - u^n(x)}{h}; \quad \frac{\tau}{h} = \kappa = \text{const} \tag{2.5}$$

of the differential equation

$$\frac{\partial u}{\partial t} = a \frac{\partial u}{\partial x}, \quad a = \text{const}. \tag{2.6}$$

It is easy to proof that the Γ-form of the differential representation of the difference scheme (2.5) has the form

$$\sum_{l=1}^{\infty} \frac{\tau^{l-1}}{l!} \frac{\partial^l u}{\partial t^l} = a \frac{\partial u}{\partial x} + a \sum_{l=2}^{\infty} \frac{h^{l-1}}{l!} \frac{\partial^l u}{\partial x^l} \tag{2.7}$$

or

$$\frac{\partial u}{\partial t} = a \frac{\partial u}{\partial x} + \sum_{l=2}^{\infty} \frac{h^{l-1}}{l!} \left(a \frac{\partial^l u}{\partial x^l} - \kappa^{l-1} \frac{\partial^l u}{\partial t^l} \right). \tag{2.8}$$

In the following we will consider the differential representation of the difference scheme normally in the Π-form which we can derive from the Γ-form by replacing the derivatives $\partial^l u/\partial t^l$, ($l \geq 2$), by the derivatives with respect to x using the Γ-form of the differential representation.

From the equations (2.4) or (2.3) we get

$$\ln \left(e^{\tau(\partial/\partial t)}\right) u = \ln \left[A_1^{-1}(t, x, \tau, h, e^{hD}) A_0 (t, x, \tau, h, e^{hD})\right] u$$

$$= \ln \left[E + \tau \left(L + \sum_{l=1}^{\infty} \sum_{l_1 + \ldots + l_s = 1}^{l} \alpha_{l_1 \ldots l_s} \frac{\partial^l}{\partial x_1^{l_1} \ldots \partial x_s^{l_s}} \right) \right] u$$

and, consequently,

$$\frac{\partial u}{\partial t} = \frac{1}{\tau} \ln \left[A_1^{-1}(t, x, \tau, h, e^{hD}) A_0 (t, x, \tau, h, e^{hD})\right] u$$

$$= \frac{1}{\tau} \ln \left[E + \tau \left(L + \sum_{l=1}^{\infty} \sum_{l_1 + \ldots + l_s = 1}^{l} \alpha_{l_1 \ldots l_s} \frac{\partial^l}{\partial x_1^{l_1} \ldots \partial x_s^{l_s}} \right) \right] u$$

$$= Lu + \sum_{l=1}^{\infty} \sum_{l_1 + \ldots + l_s = 1}^{l} c_{l_1 \ldots l_s} \frac{\partial^l u}{\partial x_1^{l_1} \ldots \partial x_s^{l_s}},$$

where $c_{l_1 \ldots l_s}$ are coefficients which depend on t, x, τ, and h.

Definition. The equation

$$\frac{\partial u}{\partial t} = Lu + \sum_{l=1}^{\infty} \sum_{l_1 + \ldots + l_s = 1}^{l} c_{l_1 \ldots l_s} \frac{\partial^l u}{\partial x_1^{l_1} \ldots \partial x_s^{l_s}} \tag{2.9}$$

is called the Π-form of the differential representation of the difference scheme (2.1).

Example. The Π-form of the differential representation of the difference scheme (2.5) can be written

$$\frac{\partial u}{\partial t} = a \frac{\partial u}{\partial x} + \sum_{l=2}^{\infty} c_l \frac{\partial^l u}{\partial x^l},$$

where $c_2 = ah(1 - \kappa a)/2$; $c_k = O(h^{k-1})$; $k \geq 3$, are the coefficients.

It can easily be proved that in formula (2.9)

$$c_l = O(\tau^{\gamma_1}, h^{\gamma_2})$$

if the difference scheme (2.1) has the order of approximation γ_1 and γ_2 with respect to t and x, respectively.

The formal way to construct the Γ-form and Π-form of the differential representation of the difference scheme as shown in the previous discussion can be carried out in practice in the following way: We demonstrate this using as an example the following difference scheme:

$$u^{n+1}(x) = A(t, x, \tau, h, T_0, T) u^n(x), \tag{2.10}$$

which approximates the equation (2.2) with the order γ assuming that $h = h(\tau)$. In the general case the process will be carried out in an analogous way. The condition $h = h(\tau)$ is essential for many difference schemes which are practically used.

If we are developing in the difference scheme (2.10) functions in the form

$$u^{n+\alpha}(x_1 + \beta_1 h_1, \ldots, x_s + \beta_s h_s) = T_0^\alpha T_1^{\beta_1} \ldots T_s^{\beta_s} u^n(x)$$
$$= \exp(\alpha\tau D_0 + \beta_1 h_1 D_1 + \ldots + \beta_s h_s D_s) u^n(x)$$

with given numbers $\alpha, \beta_1, \ldots, \beta_s$ into a series of powers of the parameters τ and h_j, we get

$$\frac{\partial u}{\partial t} = \bar{L}(t, x, \tau, h, D_0, D) u = L^1(t, x, \tau, h, D_0, D) u + R, \tag{2.11}$$

where

$$L^1 u = L(t, x, D) u + \tau P_1(t, x, D_0, D) u + \ldots + \tau^\gamma P_\gamma(t, x, D_0, D) u,$$

$$R := \sum_{v > \gamma} \tau^v P_v(t, x, D_0, D) u.$$

The equation

$$\frac{\partial u}{\partial t} = \bar{L}(t, x, \tau, h, D_0, D) u \tag{2.12}$$

is the Γ-form of the differential representation of the difference scheme (2.10). The Π-form of the differential representation

$$\frac{\partial u}{\partial t} = \tilde{L}(t, x, \tau, h, D) u \tag{2.13}$$

of the difference scheme (2.10) is derived from the Γ-form (2.12) by replacing in the right side of equation (2.12) the partial derivatives with respect to t and the mixed derivatives with respect to x and t by derivatives with respect to x alone using the Γ-form of the differential representation.

The algorithm for getting the Π-form of the differential representation of the difference scheme is demonstrated using the difference scheme (2.5) as an example. We rewrite the Γ-form of the differential representation (2.8) of the difference scheme (2.5) in the following form:

$$\frac{\partial u}{\partial t} - a \frac{\partial u}{\partial x} - a \sum_{l=2}^\infty \frac{h^{l-1}}{l!} \frac{\partial^l u}{\partial x^l} + \sum_{l=2}^\infty \frac{\tau^{l-1}}{l!} \frac{\partial^l u}{\partial t^l} = 0. \tag{2.14}$$

To eliminate the term $(\tau/2) \partial^2 u/\partial t^2$ we let the operator $-(\tau/2)\partial/\partial t$ act on the equation (2.14), and the equation which is found is added to the equation (2.10). We get

$$\frac{\partial u}{\partial t} - a \frac{\partial u}{\partial x} + \frac{\tau a}{2} \frac{\partial^2 u}{\partial x \partial t} + \sum_{l=3}^\infty \frac{\tau^{l-1}}{l!} \frac{\partial^l u}{\partial t^l} - a \sum_{l=2}^\infty \frac{h^{l-1}}{l!} \frac{\partial^l u}{\partial x^l}$$

$$+ \frac{\tau a}{2} \sum_{l=2}^\infty \frac{h^{l-1}}{l!} \frac{\partial^{l+1} u}{\partial t \partial x^l} - \frac{1}{2} \sum_{l=2}^\infty \frac{\tau^l}{l!} \frac{\partial^{l+1} u}{\partial t^{l+1}} = 0. \tag{2.14a}$$

If we let the operator $-(\tau a/2)\partial/\partial x$ act on the equation (2.14) again and if we add the resulting equation to (2.14a), the term $(\tau a/2)\partial^2 u/\partial t\,\partial x$ can be eliminated from equation (2.14a). Then we get

$$\frac{\partial u}{\partial t} - a\frac{\partial u}{\partial x} - \frac{ah}{2}\left(1 - \frac{a\tau}{h}\right)\frac{\partial^2 u}{\partial x^2}$$
$$+ \sum_{l=3}^{\infty}\left[\frac{ah^{l-1}}{l!}\left(1 - \frac{l}{2}\kappa a\right)\frac{\partial^l u}{\partial x^l} + \frac{\tau^{l-1}}{l!}\left(1 - \frac{l}{2}\right)\frac{\partial^l u}{\partial t^l}\right.$$
$$\left.+ \frac{a}{2}\frac{\tau h^{l-2}}{(l-1)!}\left(1 - \frac{l-1}{2}\kappa a\right)\frac{\partial^l u}{\partial t\,\partial x^{l-1}} - \frac{a}{2}\frac{\tau^l}{l!}\left(1 - \frac{l}{2}\right)\frac{\partial^{l+1} u}{\partial x\,\partial t^l}\right] = 0.$$

In the same way all the other derivatives with respect to t and the mixed derivatives with respect to t and x are eliminated. As a result we get the following equation:

$$\frac{\partial u}{\partial t} = a\frac{\partial u}{\partial x} + \sum_{l=2}^{\infty} c_l\frac{\partial^l u}{\partial x^l} \tag{2.15}$$

with coefficients $c_l = O(\tau^{l-1}, h^{l-1})$, which depend on a, τ and h, e.g. $c_2 = ah(1 - \kappa a)/2$. The equation (2.15) is the Π-form of the differential representation of the difference scheme (2.5).

It should be noted that the process just described of elimination of the derivates with respect to t and of the mixed derivatives with respect to t and x from the Γ-form of the differential representation of the difference scheme can be given in form of a scheme which can easily be programmed on a computer which can handle algebraic operations.

2.2 General Form of the Π-form

For difference schemes which are used in practice to approximate equations of the form (2.6) a more general form of the Γ-form of the differential representation is the following equation:

$$\frac{\partial u}{\partial t} - \frac{\partial u}{\partial x} - \sum_{j=2}^{\infty}\sum_{l=1}^{j+1} \alpha_l^j\frac{\partial^j u}{\partial t^{j-l+1}\partial x^{l-1}} = 0, \tag{2.16}$$

where

$$\alpha_l^j = O(\tau^{j-1}).$$

The Π-form of the differential representation can be written as follows:

$$\frac{\partial u}{\partial t} = a\frac{\partial u}{\partial x} + \sum_{j=2}^{\infty} c_j\frac{\partial^j u}{\partial x^j}, \tag{2.17}$$

where

$$c_j = \bar{c}_j + (\alpha_2^2 + 2 a\alpha_1^2)\bar{c}_{j-1} + (\alpha_3^{3(1)} + 2 a\alpha_2^{3(1)} + 3 a^2 \alpha_1^{3(1)})\bar{c}_{j-2} + \dots$$
$$+ [\alpha_{j-2}^{j-2(j-4)} + 2 a\alpha_{j-3}^{j-2(j-4)} + \dots + (j-2) a^{j-3} \alpha_1^{j-2(j-4)}]\bar{c}_3$$
$$+ [\alpha_{j-1}^{j-1(j-3)} + 2 a\alpha_{j-2}^{j-1(j-3)} + \dots + (j-1) a^{j-2} \alpha_1^{j-1(j-3)}]\bar{c}_2; \quad (2.18)$$

$$c_2 = \bar{c}_2; \quad \bar{c}_j = \sum_{l=1}^{j+1} a^{j-i+1} \alpha_i^j; \quad (2.19)$$

$$\alpha_k^{j(s)} = \alpha_k^{j(s-1)} + \alpha_1^{s+1(s-1)} \alpha_k^{j-s} + [\alpha_2^{s+1(s-1)} + a\alpha_1^{s+1(s-1)}]\alpha_{k-1}^{j-s}$$
$$+ [\alpha_3^{s+1(s-1)} + a\alpha_2^{s+1(s-1)} + a^2 \alpha_1^{s+1(s-1)}]\alpha_{k-2}^{j-s} + \dots$$
$$+ [\alpha_{s+1}^{s+1(s-1)} + a\alpha_s^{s \mid 1(s-1)} + \dots + a^s \alpha_1^{s+1(s-1)}]\alpha_{k-s}^{j-s}; \quad (2.20)$$

$$s = 1, 2, \dots; \quad j = 2, 3, \dots; \quad k = 1, 2, \dots, j+1;$$

and

$$\alpha_k^{j(0)} = \alpha_k^j.$$

The formulas (2.18–20) can be proved by the method of transfinite induction.

2.3 Γ- and Π-form of the First Differential Approximation

We assume that the difference scheme (2.1) has orders of approximation γ_1 and γ_2 with respect to t and x, respectively. Then omitting terms of order $O(\tau^{\gamma_1+1}, h^{\gamma_2+1})$ in equation (2.4) we get

$$\frac{\partial u}{\partial t} + \frac{\tau}{2} \frac{\partial^2 u}{\partial t^2} + \dots + \frac{\tau^{\gamma_1}}{(\gamma_1+1)!} \frac{\partial^{\gamma_1+1} u}{\partial t^{\gamma_1+1}} = Lu + L_1(D)u, \quad (2.21)$$

where $L_1(D)$ is a differential operator with coefficients of the order $O(\tau, \tau^2, \dots, \tau^{\gamma_1}, h, h^2, \dots, h^{\gamma_2})$.

Definition. Equation (2.21) is called the Γ-form of the first differential approximation of the difference scheme (2.1).

Examples. The Γ-form of the first differential approximation of the difference scheme (2.5) has the form

$$\frac{\partial u}{\partial t} + \frac{\tau}{2} \frac{\partial^2 u}{\partial t^2} = a \frac{\partial u}{\partial x} + \frac{ah}{2} \frac{\partial^2 u}{\partial x^2}. \quad (2.22)$$

In the case of the difference scheme (2.10) the Γ-form of the first differential approximation is the following

$$\frac{\partial u}{\partial t} = L^1 u. \quad (2.23)$$

If we omit in the Π-form of the differential representation (2.9) terms of order $O\left(\tau^{\gamma_1+1}, h^{\gamma_2+1}\right)$ we get the equation

$$\frac{\partial u}{\partial t} = Lu + \tilde{L}_1(D)u, \tag{2.24}$$

where $\tilde{L}_1(D)$ is a differential operator with coefficients of the order $O\left(\tau^{\gamma_1}, h^{\gamma_2}\right)$.

Definition. We will call the equation (2.24) the Π-form of the first differential approximation of the difference scheme (2.1).

Examples. The Π-form of the difference scheme (2.5) has the form

$$\frac{\partial u}{\partial t} = a\frac{\partial u}{\partial x} + \frac{ah}{2}(1 - \kappa a)\frac{\partial^2 u}{\partial x^2}. \tag{2.25}$$

In the case of the difference scheme (2.10) the Π-form of the first differential approximation is the following:

$$\frac{\partial u}{\partial t} = \bar{L}^1(t, x, \tau, h, D)u, \tag{2.26}$$

where

$$\bar{L}^1 u = Lu + \tau^\gamma \bar{P}_\gamma(t, x, D)u.$$

It is not difficult to prove that the Π-form of the first differential approximation (2.24) can be derived from the Γ-form of the first differential approximation (2.21) by means of an approximation of the formal replacing of the derivatives with respect to t up to the order γ_1 by derivatives with respect to x using the basic differential equation (2.1).

Example. The Π-form of the first differential approximation (2.25) of the difference scheme (2.5) can be derived from the Γ-form (2.22) by replacing in the equation (2.22) the derivative $\partial^2 u/\partial t^2$ by a derivative with respect to x using the relation

$$\frac{\partial^2 u}{\partial t^2} = a^2\frac{\partial^2 u}{\partial x^2},$$

which holds for any solution of the differential equation (2.6).

Omitting in equation (2.9) terms of order $O\left(\tau^{\gamma_1+2}, h^{\gamma_2+2}\right)$ and $O\left(\tau^{\gamma_1+3}, h^{\gamma_2+3}\right)$ etc. we get the Π-form of the second, third etc. differential approximation. In a similar way the Γ-forms of differential approximations are defined.

Adding, for example, to equation (2.25) the terms $c_3\,\partial^3 u/\partial x^3, c_4\,\partial^4 u/\partial x^4$, etc. we get the Π-form of the second, third etc. differential approximation of the difference scheme (2.5).

In the following, if it is not specifically stated to the contrary, we understand by the first differential approximation of a difference scheme the Π-form of the first differential approximation. We will normally obtain the latter by a formal replacement of the corresponding partial derivatives with respect to t and the mixed partial derivatives with respect to t and x in the Γ-form of the first differential approximation using the given differential equation.

2.4 Remarks on Nonlinear Differential Equations

In the case of variable coefficients or nonlinear differential equations the form of the differential approximation is more complicated. If, for example, in the equation (2.6) $a = a(x, t)$, the Γ-form of the first differential approximation for the difference scheme (2.5) has the same form [see (2.22)], but the Π-form of the first differential approximation can be written in the following way:

$$\frac{\partial u}{\partial t} = a\frac{\partial u}{\partial x} + \frac{ah}{2}(1 - \kappa a)\frac{\partial^2 u}{\partial x^2} - \frac{\tau}{2}\left(\frac{\partial a}{\partial t} + a\frac{\partial a}{\partial x}\right)\frac{\partial u}{\partial x},$$

as we get from equation (2.6)

$$\frac{\partial^2 u}{\partial t^2} = a^2\frac{\partial^2 u}{\partial x^2} + \left(\frac{\partial a}{\partial t} + a\frac{\partial a}{\partial x}\right)\frac{\partial u}{\partial x}.$$

In the following some examples will be given of the first and the following differential approximations of difference schemes for different nonlinear differential equations and especially for the system of equations of gas dynamics.

2.5 The Role of the First Differential Approximation

The differential representation of a difference scheme is a transformation of the given scheme in terms of differential operators and contains full information about the difference scheme.

The differential representation of a difference scheme is a differential equation which has an intermediate position between the original differential equation and the difference scheme which approximates it. The differential approximation contains the information of the original differential equation and also of the difference scheme. The question arises whether it is possible on the basis of the first differential approximation (or the following differential approximations) to judge on the features of the difference scheme (e.g., on the stability, dissipation etc.) and further to get a deeper knowledge about the scheme which could not be found using the traditional methods of analysis of difference schemes.

In the following the usefulness of the application of the method of differential approximations will be demonstrated concerning the investigation of characteristics of difference schemes.

2.6 On the Correctness of Giving the Π-form as an Infinite Differential Equation

We will investigate in detail the question of the correctness to rewrite the Π-form of a differential representation of a difference scheme in the form of an infinite differential equation.

Let

$$\frac{u^{n+1}(x) - u^n(x)}{\tau} = \Lambda_0 u^n(x) \tag{2.27}$$

be an explicit difference scheme which approximates in a class of sufficiently smooth functions the system of differential equations

$$\frac{\partial u}{\partial t} = L(D)u. \tag{2.28}$$

Here the operator Λ_0 is bounded and does not depend on $x \in \mathbb{R}_1$; it is an analytical function of the parameters $\tau > 0, h$ and the shift-operator T_1. For the system of differential equations (2.28) and the difference scheme (2.27) we pose initial conditions

$$u^0(x) = u_0(x). \tag{2.29}$$

Then the solution of Cauchy's problem (2.27, 29)

$$u^n(x) = \tilde{u}(t, x) = (E + \tau \Lambda_0)^n u_0(x) = (E + \tau \Lambda_0)^{t/\tau} u_0(x) \tag{2.30}$$

is analytically continued for arbitrary t and τ. If (2.30) is differentiated with respect to t we get

$$\frac{\partial \tilde{u}(t, x)}{\partial t} = \frac{1}{\tau} \ln(E + \tau \Lambda_0) \tilde{u}(t, x), \tag{2.31}$$

which is a system of differential equations describing the behavior of the solution of the difference scheme (2.27) by the function $\tilde{u}(t, x)$ of the discrete argument x and which is the Π-form of the differential representation of the difference scheme (2.27).

The function $\tilde{u}(t, x)$ defined as a solution of the system of equations (2.31) with the initial condition $\tilde{u}(0, x) = u_0(x)$ following from (2.29) coincides with $u^n(x)$ for $t = n\tau$.

The formal rewriting of the system of equations of the Π-form of the differential representation (2.31) of the difference scheme (2.27) as a power series in τ

$$\frac{\partial \tilde{u}(t, x)}{\partial t} = \left(\Lambda_0 + \sum_{j=1}^{\infty} (-1)^j \frac{\tau^j}{j+1} \Lambda_0^{j+1}\right) \tilde{u}(t, x) \tag{2.32}$$

follows from expansion of the right side of (2.31) into a power series with respect to the parameter τ.

To rewrite equation (2.32) in another form one can use the property of the approximation $\Lambda_0(T_1) \sim L(D)$. If $h = h(\tau)$ and

$$\Lambda_0(h) = L(D) + \sum_{j=1}^{\infty} h^j L_j(D) \tag{2.33}$$

after replacing this expansion of $\Lambda_0(h)$ on the right side of (2.32) we get the Π-form of the differential representation of the difference scheme (2.27) in the form of a differential equation, the right side of it consists of an operator series with respect to powers of τ and h

$$\frac{\partial \tilde{u}(t, x)}{\partial t} = L(D)\tilde{u} + \sum_{j=1}^{\infty} \left\{ h^j L_j(D) + (-1)^j \frac{\tau^j}{j+1} \left[L(D) + \sum_{i=1}^{\infty} h^i L_i(D) \right]^{j+1} \right\} \tilde{u}. \tag{2.34}$$

Omitting in the right side of the system of equations (2.32) some of the highest terms with respect to $\tau \to 0$ we get differential approximations of the difference scheme (2.27) of different order where it is necessary to take into account the dependence of Λ_0 on the parameter τ

$$\frac{\partial \tilde{u}}{\partial t} = \left[\Lambda_0 + \sum_{j=1}^{k} (-1)^j \frac{\tau^j}{j+1} \Lambda_0^j(h(\tau)) \right] \tilde{u}. \tag{2.35}$$

In order that the derived differential approximations describe the behavior of the difference solution $\tilde{u}(t, x)$ for $\tau \to 0$ it is necessary that the series on the right side of equation (2.32) converges in a vicinity of $\tau = 0$. This series converges for

$$\tau \| \Lambda_0(h(\tau)) \| < 1 \tag{2.36}$$

and diverges for

$$\tau \| \Lambda_o(h(\tau)) \| > 1.$$

Here the norm is the *Hermitean norm* of the operator Λ_0 in the space of the vector components of $u = (u_1, \ldots, u_m)$ in those points x which are connected with the difference scheme (2.27).

The higher the order k of the differential approximation the more accurately its solution describes the solution of the difference scheme (2.27) for arbitrary τ, which satisfies the condition (2.36).

We remark that the derived condition is similar to the usual condition for stability of the difference scheme (2.27) but in the general case it is not identical with it. Normally the condition (2.36) is stricter than the stability condition of the scheme

$$\| E + \tau \Lambda_0 \| \leq 1 + O(\tau).$$

The equation (2.34) is a formal restatement of the difference scheme (2.27) which holds under the condition (2.36). The latter bounds in fact the law for the limit, because we get from it the inequality

$$\tau < \frac{1}{\|A_0(h)\|},$$

a condition under which the difference scheme (2.27) is asymptotically equivalent for $\tau \to 0$ to the equation of infinite order with respect to x (2.34). From the condition (2.36) it follows which terms in (2.33) must be taken into account when (2.33) is substituted in (2.35), when we confine ourselves to powers τ^k. Thus we get the first differential approximation in the form of an equation with partial derivatives if we put $k = 1$ in (2.35) and take into account the corresponding term in h corresponding to (2.33).

In a similar way differential approximations are constructed also for implicit difference schemes

$$\frac{u^{n+1}(x) - u^n(x)}{\tau} = A_0 u^{n+1}(x)$$

or

$$\frac{u^{n+1}(x) - u^n(x)}{\tau} = (E - \tau A_0)^{-1} A_0 u^n(x).$$

We discuss in more detail the question of the correctness of the formulas (2.31, 32). Let $u^n \in \mathbb{B}$ ($\mathbb{B}$ is a *Banach space* on the field of complex numbers $\mathbb{C}$) and $A_0: \mathbb{B} \to \mathbb{B}$. By $\sigma(A)$ we denote the spectrum of the bounded operator A: $\mathbb{B} \to \mathbb{B}$, which means the set of complex numbers λ for which the operator $A - \lambda E$ has no inverse.

Let us consider the step-operator of the difference scheme (2.27)

$$S_h = E + \tau A_0.$$

The spectrum of the operator S_h depending on τ will be denoted by $\sigma_\tau(S_h)$ and for the set of all those τ for which the null belongs to the unbounded component $\mathbb{C} \setminus \sigma_\tau(S_h)$ – where $\mathbb{C}$ is the field of complex numbers – (component is the maximal connected set [165]) we write $\Theta(A_0)$. Then for each $\tau \in \Theta(A_0)$ the logarithm of the operator $E + \tau A_0$ exists, which is a bounded operator from $\mathbb{B}$ in $\mathbb{B}$ whereas the operator $E + \tau A_0$ possesses roots of all orders (see [165]). Therefore for each $\tau \in \Theta(A_0)$ operators $(E + \tau A_0)^{t/\tau}: \mathbb{B} \to \mathbb{B}$ and $\ln(E + \tau A_0)$: $\mathbb{B} \to \mathbb{B}$ are defined for which the following representations hold:

$$(E + \tau A_0)^{t/\tau} = \frac{1}{2\pi i} \int_\Gamma \lambda^{t/\tau} [(\lambda - 1)E - \tau A_0]^{-1} d\lambda;$$

$$(2.37)$$

$$\ln(E + \tau A_0) = \frac{1}{2\pi i} \int \ln \lambda [(\lambda - 1)E - \tau A_0]^{-1} d\lambda,$$

where the integration is carried out along an arbitrary contour Γ which is surrounding a region Q with

$$\sigma_\tau(S_h) \subset Q; \quad 0 \notin Q, \quad \tau \in \Theta(\Lambda_0).$$

Consequently, it is possible to consider a function (2.30) which is well defined for arbitrary values of $\tau \in \Theta(\Lambda_0)$. Differentiating (2.30) with respect to t we get the equation (2.31) which holds for all $\tau \in \Theta(\Lambda_0)$ and which is the Π-form of the differential representation of the difference scheme (2.27). If the conditions (2.36) are satisfied $\ln(E + \tau\Lambda_0)$ can be defined by

$$\ln(E + \tau\Lambda_0) = \sum_{j=0}^{\infty} (-1)^j \frac{(\tau\Lambda_0)^{j+1}}{j+1}. \tag{2.38}$$

On the basis of the Theorem 10.27 from the monograph [165] it is easy to show, that the representations (2.37) and (2.38) are identical. In an analogous way also the case of an implicit difference scheme can be considered.

2.7 Differential Representations of Difference Schemes in Spaces of Generalized Functions

In the following we need some properties derived from the theory of generalized functions [165, 166]. For simplicity we confine ourselves to the case of one space variable while all results can be extended easily to the multidimensional case. We discuss now the question of the differential representations of difference schemes in spaces of generalized functions.

1) As basic space we consider a linear topological space which consists of functions $\phi(x)$ which are defined on a set Ω. The set Ω normally will be the real axis $\mathbb{R}$ or the complex plane $\mathbb{C}$. Functions which belong to the basic space will be called basic functions.

We need the following spaces of basic functions:

    I) The space $\mathbb{S}$ of rapidly decaying functions of the real argument,

    II) The space $\mathbb{D}(\Omega)$ of finite functions of the real argument the support of which is dense in the region Ω,

    III) The space $\mathbb{Z}$ of analytical functions of complex argument which satisfy the following condition: For each function $\phi(z) \in \mathbb{Z}$ and for arbitrary integer $k \geq 0$, there exist constants a, c_k such that the inequality holds:

$$|z^k \phi(z)| \leq c_k e^{a|\text{Im}\,(z)|}. \tag{2.39}$$

The spaces mentioned above are linear topological spaces.

Definition. A linear continuous functional on a basic space is called a generalized function.

Thus the generalized functions are not defined by themselves but in correspondence to a basic space. The spaces of generalized functions on the spaces $\mathbf{S}$, $\mathbf{D}(\Omega)$, $\mathbf{Z}$ will be denoted by $\mathbf{S}'$, $\mathbf{D}'(\Omega)$, $\mathbf{Z}'$, respectively. If f is a usual function, then

$$(f, \phi) = \int\limits_{-\infty}^{+\infty} f(x)\phi(x)\,dx. \tag{2.40}$$

In the spaces of generalized functions which we will use the following operators are defined:

$$F[\phi](\xi) := \int\limits_{-\infty}^{+\infty} e^{ix\xi}\phi(x)\,dx; \quad F^{-1}[\phi](\xi) := \frac{1}{2\pi}\int\limits_{-\infty}^{\infty} e^{-ix\xi}\phi(x)\,dx;$$

$$\check{\phi}(x) := \phi(-x);$$

$$D^m\phi(x) := \frac{\partial^m\phi}{\partial x^m}; \tag{2.41}$$

$$T_y\phi(x) := \phi(x+y);$$

$$(\phi * \psi)(x) := \int\limits_{-\infty}^{\infty} \phi(y)\psi(x-y)\,dy;$$

these operators can be defined in the spaces of generalized functions in such a way that in the case when the generalized function is a basic function the corresponding operators act on this function in the same way as they would in the space of basic functions. In other words the definitions of the operators in the space of generalized functions coincide with the definitions of the corresponding operators in the space of basic functions. In the following the definitions of operators (2.41) are given in the space of generalized functions (by f and g we will denote generalized functions, by ϕ a basic function):

$$(F[f], \phi) = (f, F[\phi]); \quad (F^{-1}[f], \phi) = (f, F^{-1}[\phi]);$$

$$(\check{f}, \phi) = (f, \check{\phi});$$

$$(D^m f, \phi) = (-1)^m(f, D^m\phi);$$

$$(T_y f, \phi) = (f, T_{-y}\phi);$$

$$(f * \phi)(x) = (f, T_{-x}\check{\phi});$$

$$(f * g, \phi) = (g, \check{f} * \phi).$$

We note that the operation $*$ between two generalized functions f and g is not always defined. But if one of them has a compact support $f * g$ is defined whereas $f * g = g * f$. Generally $f * g$ is defined if $\check{f} * g$ is a basic function for an arbitrary basic function ϕ or if the functional g has a unique continuation in the space to which the functions $(\check{f} * \phi)(x)$ belong where ϕ represents the whole space of basic functions.

By δ_y we denote a generalized function which acts according to the formula:

$$(\delta_y, \phi) = \phi(y).$$

For any generalized function f the $*$ product $\delta_y * f$ is defined. One can prove that the following relations hold:

$$\delta_y * f = T_{-y} f,$$
$$F[\delta_y] = e^{iy\xi},$$
$$F^{-1}[\delta_y] = \frac{1}{2\pi} e^{-iy\xi}.$$

Further we need the following formulas:

$$F[D^m f](x) = (-ix)^m F[f](x),$$
$$F^{-1}[D^m f(x) = (ix)^m F^{-1}[f](x),$$
$$D^m F[f](x) = F[(i\xi)^m f(\xi)](x),$$
$$D^m F^{-1}[f](x) = F^{-1}[(-i\xi)^m f(\xi)](x), \tag{2.42}$$
$$F[f*g] = F[f]F[g],$$
$$F^{-1}[f*g] = F^{-1}[f]F^{-1}[g],$$
$$T_x \delta_y = \delta_{y-x}.$$

In the following we will often make use of the fact that the Fourier representation defines an isomorphism of the following topological spaces:

$$\mathbf{S} \text{ and } \mathbf{S}; \quad \mathbf{S}' \text{ and } \mathbf{S}'; \quad \mathbb{D}(\mathbb{R}) \text{ and } \mathbf{Z}; \quad \mathbb{D}'(\mathbb{R}) \text{ and } \mathbf{Z}'.$$

2) We consider now the difference scheme

$$\sum_{|\alpha| \leq q_1} b_\alpha^1 T_{\alpha h} f^{n+1} = \sum_{|\beta| \leq q_0} b_\beta^0 T_{\beta h} f^n \tag{2.43}$$

in the space of generalized functions $\mathbb{D}'(\mathbb{R})$. Here $0 \leq q_0, q_1 < \infty$, b_α^1 and b_β^0 are some real constants, h is a parameter. It is apparent that the operators

$$\Lambda_1 := \sum_{|\alpha| \leq q_1} b_\alpha^1 T_{\alpha h}, \quad \Lambda_0 := \sum_{|\beta| \leq q_0} b_\beta^0 T_{\beta h}$$

are linear continuous operators from $\mathbb{D}'(\mathbb{R})$ in $\mathbb{D}'(\mathbb{R})$. The difference scheme (2.43) can be written in the form:

$$w_1 * f^{n+1} = w_0 * f^n, \tag{2.44}$$

where

$$w_1 = \Lambda_1 \delta_0 = \sum_{|\alpha| \leq q_1} b_\alpha^1 \delta_{-\alpha h},$$
$$w_0 = \Lambda_0 \delta_0 = \sum_{|\beta| \leq q_0} b_b^0 \delta_{-\beta h}.$$

Because of the fact that w_1 and w_0 are generalized functions with a compact support the $*$ products $w_1 * f$ and $w_0 * f$ are defined for an arbitrary generalized function $f \in \mathbb{D}'(\mathbb{R})$. As the $*$ product of a generalized function with a compact support with a function from the space $\mathbf{S}$ is a function from $\mathbf{S}$ [166] the operators Λ_1 and Λ_0 are also continuous operators from $\mathbf{S}'$ in $\mathbf{S}'$.

Lemma. Let

$$w = \sum_{|j| \leq J} c_j \delta_{-jh},$$

c_j are real constants and

$$F[w](x) = \sum_{|j| \leq J} c_j e^{-ixjh} \neq 0 \qquad \text{for all } x \in \mathbb{R}.$$

Then the function $\ln F[w](x)$ is a multiplier for $\mathbf{S}$, i.e. for each $\phi \in \mathbf{S}$, $\ln F[w]$ $\cdot \phi \in \mathbf{S}$ and the operator of multiplication by $\ln F[w](x)$ is continuous from $\mathbf{S}$ in $\mathbf{S}$.

Proof. The logarithm of a complex argument is assumed to be a single valued function and therefore it is sufficient to give a precise definition for the meaning of $\ln F[w](x)$. The function $F[w](x)$ is an integral analytical function which does not tend to zero on the real axis. As the representation of the real axis under the mapping $F[x]$ is a closed curve which in general contains $z = 0$ we can not choose any branch of the function $\ln F[w](z)$ but using the analytical continuation one can construct an analytical function $\ln F[w](z)$ (in some neighborhood of the real axis) such that

$$\ln F[w](z)|_{z=0} = \ln\left(\sum_{|j| \leq J} c_j\right).$$

We will have in mind this function when we use the symbol

$$\ln F[w](x).$$

Under the conditions of the lemma the function $F[w](x)$ is periodic and therefore there exist such constants K_1, K_2, K_3, and K_4 that

$$0 < K_1 \leq |F[w](x)| \leq K_2 < \infty,$$
$$|\arg F[w](x)| \leq K_3(1 + |x|),$$
$$|\ln F[w](x)| \leq K_4(1 + |\arg F[w](x)|).$$

Then there exists a constant c_0 such that for some $N < \infty$

$$|\ln F[w](x)| \leq c_0(1 + |x|)^N.$$

In addition using Cauchy's theorem on the representation of an analytical function in a point by its values on a contour containing that point, one can get easily the inequality:

$$\left|\frac{\partial^m}{\partial x^m} \ln F[w](x)\right| \leq c_m(1 + |x|)^{Nm}.$$

From these estimates it follows that

$$\sup_{x \in \mathbb{R}} (1 + |x|^2)^M \left|\frac{\partial^m}{\partial x^m} [\ln F[w](x) \phi(x)]\right| < \infty$$

for all $\phi(x) \in \mathbf{S}$, and also the continuity of the operator of multiplication by $\ln F[w](x)$ on the topological space $\mathbf{S}$ can be shown. Thus the lemma is proved. ☐

Using the Fourier transform from (2.44) we get:

$$F[w_1]g^{n+1} = F[w_0]g^n,$$

where $g^n = F[f^n]$ or, if $F[w_1](x) \neq 0$ for all $x \in \mathbb{R}$

$$g^{n+1} = \frac{F[w_0]}{F[w_1]} g^n. \tag{2.45}$$

Equation (2.45) can be written in the form:

$$g(t) = \left(\frac{F[w_0]}{F[w_1]}\right)^{t/\tau} g(0), \tag{2.46}$$

where $t = n\tau$. We consider the relation (2.46) for all $t \geq 0$ and differentiate it with respect to the parameter t. Under the assumption that $F[w_0](x) \neq 0$ for all $x \in \mathbb{R}$ we have

$$\frac{dg(t)}{dt} = \frac{1}{\tau}[\ln F[w_0] - \ln F[w_1]]g(t). \tag{2.47}$$

According to the lemma the right side of equation (2.47) is defined for any generalized function $g(t) \in \mathbf{S}'$. It is apparent that the equation (2.47) can be considered in the space $\mathbb{D}'(\Omega)$. The following statement holds [166]:

A) Let Φ a basic space, $\Psi = F[\Phi]$ is the dual space (with respect to the Fourier transform). If a locally summable function g is a multiplicator in the space Ψ' (i.e. for each $f \in \Psi$ $gf \in \Psi$ and the operator of multiplication by the function g is a linear, continuous operator from Ψ' in Ψ') then the functional $f = F^{-1}[g]$ is such that:

    I) For each $\phi \in \Phi$ the $*$ product $f * \phi \in \Phi$,
   II) The operation $*$ with a function f is a linear continuous operator from Φ in Φ,
 III) For all $u \in \Phi'$ the equation holds:

$$F[f * u] = F[f] \cdot F[u].$$

If a generalized function w satisfies the condition of the lemma then by virtue of the statement (A) it is easy to show that

$$F[F^{-1}[\ln F[w]] * f] = \ln F[w] \cdot F[f],$$

for any generalized function $f \in S'$. Then if we let act the inverse Fourier transform [assuming that $g(t) \in \mathbf{S}'$] on (2.47) we get the equation

$$\frac{df}{dt} = \frac{1}{\tau} F^{-1}[\ln F[w_0] - \ln F[w_1]] * f, \tag{2.48}$$

which is equivalent to equation (2.47) in the space $\mathbf{S}'$ and which we will call the differential representation of the difference scheme (2.43). Thus the right side of the equation (2.48) is defined for all $f \in \mathbf{S}'$.

Remark. By virtue of the lemma and the statement (A) the right side of equation (2.48) is defined for all $f \in \mathbf{S}$. Consequently, the equation (2.48) can be considered also in the space $\mathbf{S}$.

We consider Cauchy's problem for the equation (2.43, 48) with the initial conditions

$$f^0 = f(0) = f_0. \qquad (2.49)$$

The following theorem holds:

. . . .

Theorem 2.1. Let $f_0 \in \mathbf{S}'$ (or $\mathbf{S}$) then Cauchy's problems (2.43, 49) and (2.48, 49) have a unique solution in the space $\mathbf{S}'$ (correspondingly in the space $\mathbf{S}$). In addition we get
a) if $f(t)$ is the solution of Cauchy's problem (2.48, 49) then $\{f^n\}_{n=\infty}^{\infty}$ is the solution of Cauchy's problem (2.43, 49) where $f^n = f(n\tau)$;
b) if $\{f^n\}_{n=\infty}^{\infty}$ is the solution of Cauchy's problem (2.43, 49) then a one-parameter class of generalized functions $f(t)$ exists which is the solution of Cauchy's problem (2.48, 49) such that $f(n\tau) = f^n$.

. . . .

Proof. It is apparent that the solution of Cauchy's problem (2.43) and (2.49) exists and is unique in the spaces $\mathbf{S}'$ and $\mathbf{S}$.

We will show that this statement is true also for the problem (2.48, 49). For this we will make use of the following statement which is proved in [162].

B) Let A be a linear continuous operator from $\mathbf{S}$ in $\mathbf{S}$ and A^* the adjoint operator. If Cauchy's problem

$$\frac{d\phi(t)}{dt} = A\,\phi(t), \quad \phi(t_0) = \phi_0 \in \mathbf{S}, \quad 0 \leq t \leq T, \quad 0 \leq t_0 \leq T$$

is always solvable (i.e. for all $t_0, 0 \leq t_0 \leq T$ and all $\phi_0 \in \mathbf{S}$ a solution $\phi(t)$ exists which is defined in $0 \leq t \leq T$ and which tends to ϕ_0 for $t = t_0$, and this solution $\phi(t)$ depends linearly and continuously on ϕ_0) then Cauchy's problem

$$\frac{du(t)}{dt} = -A^* u(t),$$

$$u(0) = u_0 \in \mathbf{S}', \quad 0 \leq t \leq T,$$

in the adjoint space $\mathbf{S}'$ has a solution for an arbitrary initial distribution u_0; this solution is unique in $\mathbf{S}'$ and depends continuously on the functional u_0 in the sense of the topological space $\mathbf{S}'$.

Let $b = F[w_0]/F[w_1]$. Using the statement (A) it is easy to prove that the operator $Q_{t_0}^t$ which is defined by

$$Q_{t_0}^t \phi = F^{-1}[b^{t_0-t}] * \phi$$

is a continuous operator from $\mathbf{S}$ in $\mathbf{S}$. As the function $\phi(t) = Q_{t_0}^t \phi$ is a solution of the problem

$$\frac{d\phi}{dt} = -F^{-1}[\ln b] * \phi,$$

$$\phi|_{t=t_0} = \phi_0,$$

Cauchy's problem

$$\frac{df}{dt} = F^{-1}[\ln b] * f,$$

$$f|_{t=0} = f_0 \in \mathbf{S}'$$

has always a unique solution in the space $\mathbf{S}'$ by virtue of the statement (B). Consequently, Cauchy's problem (2.48, 49) is always solvable in a unique way in the space $\mathbf{S}'$. An analogous proof can be given for functions from the space $\mathbf{S}$.

The statement a) in Theorem 2.1 follows from the fact that (2.48) is in $\mathbf{S}'(\mathbf{S})$ equivalent to (2.47), and (2.47) is equivalent to (2.43) for $t = n\tau$. The statement b) follows from the uniqueness of the solution of the problem (2.48, 49) in $\mathbf{S}'(\mathbf{S})$; therefore (2.46) gives the solution (in terms of a Fourier transform). The theorem is proved. $\quad\square$

3) Now we will discuss the question of the representation of the expression $F^{-1}[\ln F[w]] * f$ in the form of a series

$$\sum_k \eta_k h^k (iD)^k f.$$

As the function $\ln F[w](x)$ is analytical in some neighborhood of the point $x_0 \in \mathbb{R}$ it can be expanded in a *Taylor series:*

$$\ln F[w](x) = \sum_{k=0}^{\infty} \eta_k(x_0) h^k (x - x_0)^k \tag{2.50}$$

which converges in some region $\Omega \ni x_0$. Further we will use the following notations:

$$P_m(x_0; x) := \sum_{k=0}^{\infty} \eta_k(x_0) h^k (x - x_0)^k,$$

$$\eta_k(0) := \eta_k,$$

$$P_m(0; x) := P_m(x)$$

and for simplicity we will consider an explicit difference scheme

$$f^{n+1} = \sum_{|j| \le J} c_j f^n = w * f^n \tag{2.51}$$

although all the following statements will be true also for difference schemes of the form (2.43). As before we will assume that

$$F[w](x) \neq 0$$

for all $x \in \mathbb{R}$. Because of

$$F^{-1}[(x - x_0)^k g(x)](\xi) = F^{-1}[T_{-x_0}(x^k T_{x_0} g(x))](\xi)$$
$$= e^{-ix_0\xi} F^{-1}[x^k T_{x_0} g(x)](\xi) = e^{-ix_0\xi}(iD_\xi)^k F^{-1}[T_{x_0} g(x)](\xi),$$

we get

$$F^{-1}[P_m(x_0; x)] * f = \sum_{k=0}^m \eta_k(x_0) h^k F^{-1}[(x - x_0)^k F[f]]$$

$$= e^{-ix_0\xi} \sum_{k=0}^m \eta_k(x_0) h^k (iD_\xi)^k F^{-1}[T_{x_0} F[f]](\xi)$$

$$= \sum_{k=0}^m \eta_k(x_0) h^k \sum_{l=0}^k (-x_0)^{k-l} \frac{k!}{l!(k-l)!} (iD)^l f$$

and, especially,

$$F^{-1}[P_m(x)] * f = \sum_{k=0}^m \eta_k h^k (iD_\xi)^k f(\xi).$$

In the following we will prove some results for polynomials $P_m(x)$, although all these statements remain true also for the general case of polynomials of the form $P_m(x_0; x)$.

. . . .

Theorem 2.2. Let the series $\sum_{k=0}^\infty \eta_k h^k x^k$ converge to the function $\ln F[w](x)$ in the region $\Omega \ni 0$. Then for any integral function f for which $F[f] \in \mathbb{D}(\Omega)$ the series

$$\sum_{k=0}^\infty \eta_k h^k (iD)^k f$$

converges in the topology of the space $\mathbb{Z}$ to the function

$$F^{-1}[\ln F[w]] * f,$$

and, consequently, in an arbitrary bounded region of $\mathbb{C}$ the series

$$\sum_{k=0}^\infty \eta_k h^k (iD)^k f$$

converges uniformly to the function

$$F^{-1}[\ln F[w]] * f$$

together with all its derivatives.

. . . .

Proof. First of all we remark that the function $F^{-1}[\ln F[w]]*f$ is an element of the space $\mathbb{Z}$ (if the conditions of the theorem for the function f are satisfied).

Indeed as $F[f] \in \mathbb{D}(\Omega)$ we have

$$\ln F[w] \cdot F[f] \in \mathbb{D}(\Omega)$$

and, consequently, (by virtue of Theorem 7.22 from [165])

$$F^{-1}[\ln F[w]]*f \in \mathbb{Z}.$$

But

$$P_m F[f] \to \ln F[w] \cdot F[f] \quad \text{in } \mathbb{D}(\Omega) \text{ for } m \to \infty$$

and therefore by virtue of the continuity of the Fourier transform and its inverse we get

$$F^{-1}[P_m]*f \to F^{-1}[\ln F[w]]*f \quad \text{in } \mathbb{Z} \text{ for } m \to \infty.$$

Thus the Theorem 2.2 is proved.   □

We remark that the statement of Theorem 2.2 can not be applied to all integral functions, i.e. such integral analytical functions f exist that the series

$$\sum_{k=0}^{\infty} \eta_k h^k (\mathrm{i}D)^k f$$

converges in the sense of the topology of the space $\mathbb{Z}$. In addition, this series even can converge pointwise to some integral function. Indeed, let the series

$$\sum_{k=0}^{\infty} \eta_k h^k x^k$$

be non-convergent in the point x_0. We consider the integral function $f(x) = \exp(-\mathrm{i}x_0 x)$, as $(\mathrm{i}D)^k f = x_0^k f$ also in an arbitrary point $x \in \mathbb{R}$ the series

$$\sum_{k=0}^{\infty} \eta_k h^k (\mathrm{i}D)^k f$$

in non-convergent.

. . . .

Theorem 2.3. Let the series $\sum_{k=0}^{\infty} \eta_k h^k x^k$ converge in an arbitrary point to the function $\ln F[w](x)$. Then for all $f \in \mathbb{S}'$ we get

$$\frac{df}{dt} - \frac{1}{\tau} F^{-1}[\ln F[w]]*f = \lim_{m \to \infty} \left[\frac{df}{dt} - \frac{1}{\tau} \sum_{k=0}^{m} \eta_k h^k (\mathrm{i}D)^k f \right],$$

where the limes is understood in the sense of the topology of the space $\mathbb{Z}'$.

. . . .

Proof. We remark that the statement of the theorem is equivalent the statement

$$F^{-1}[P_m]*f \to F^{-1}[\ln F[w]]*f \quad \text{for } m \to \infty \text{ and all } f \in S'.$$

Because of $\mathbb{D}(\mathbb{R}) \subset S$ and $F: S \to S$ it follows that $\mathbb{Z} = F^{-1}[\mathbb{D}(\mathbb{R})] \subset S$. If $\phi_\nu \in \mathbb{Z}$ and $\phi_\nu \to 0$ in $\mathbb{Z}$ then $\psi_\nu = F[\phi_\nu] \in \mathbb{D}(\mathbb{R})$ and $\psi_\nu \to 0$ in $\mathbb{D}(\mathbb{R})$, but as $\psi_\nu \to 0$ in S and, consequently, $F^{-1}[\psi_\nu] \to 0$ in S then $\phi_\nu \to 0$ in S. From this it follows that $S' \subset \mathbb{Z}'$. Therefore one can consider a generalized function $f \in S'$ also as a generalized function space $\mathbb{Z}'$. Further, the sequence P_m converges uniformly to the function $\ln F[w]$ on an arbitrary compactum, therefore

$$P_m F[f] \to \ln F[w] \cdot F[f]$$

in $\mathbb{D}'(\mathbb{R})$ for $m \to \infty$. But then

$$F^{-1}[P_m]*f \to F^{-1}[\ln F[w]]*f \quad \text{for } m \to \infty \text{ in } \mathbb{Z}'. \qquad \square$$

4) We will formulate some statements which correspond to the more general case especially when $P_m(x_0; x)$ converges to $\ln F[w](x)$ at least in a region Ω of the space $\mathbb{R}$.

. . . .

Theorem 2.4. For all $\phi \in \mathbb{D}(\mathbb{R})$ the following equation holds

$$\lim_{m \to \infty} (F^{-1}[P_m]*f, F[\phi]) = (F^{-1}[\ln F[w]]*f, F[\phi]), \quad f \in S'.$$

. . . .

Proof. As $P_m(x) \to \ln F[w](x)$ for all $x \in \Omega$, for all $\phi \in \mathbb{D}(\Omega)$

$$\phi(x) P_m(x) \to \phi(x) \ln F[w](x) \quad \text{for } m \to \infty$$

in the topology of the space S. This follows from *Leibniz's formula* and from the fact that absolutely and uniformly convergent series can be differentiated term by term. Therefore by virtue of the continuity the Fourier transforms in S, $F[\phi \cdot P_m] \to F[\phi \cdot \ln F[w]]$ for $m \to \infty$. But then for all $f \in S'$

$$(f, F[\phi \cdot P_m]) \to (f, F[\phi \ln F[w]]),$$

and from this

$$(F^{-1}[P_m]*f, F[\phi]) \to (F^{-1}[\ln F[w]]*f, F[\phi]),$$

which proves the statement. $\qquad \square$

Let Φ_Ω be the space of Fourier transforms of functions from $\mathbb{D}(\Omega)$ with a topology in which the transforms $F: \mathbb{D}(\Omega) \to \Phi_\Omega$ and $F^{-1}: \Phi_\Omega \to \mathbb{D}(\Omega)$ are continuous, Φ_Ω is the basic space. It is apparent that functionals from S' can be considered as functionals from Φ'_Ω, and Theorem 2.4 can be formulated as follows.

Theorem 2.5. For all $f \in S'$ in the topology of the space Φ'_Ω we have

$$\lim_{m \to \infty} \left[\frac{df}{dt} - \frac{1}{\tau} \sum_{k=0}^{m} \eta_k(x_0) h^k \sum_{l=0}^{k} (-x_0)^{k-l} \frac{k!}{l!(k-l)!} (iD)^l f \right]$$

$$= \frac{df}{dt} - \frac{1}{\tau} F^{-1} [\ln F[w]] * f.$$

The following theorem is applicable even in the case of more than one independent variables. Therefore we will assume further that $x \in \mathbb{R}_n$, Ω is some region in $\mathbb{R}_n$ in which the polynomials P_m converge to functions $\ln F[w]$.

Theorem 2.6. For all $f \in Z'$ for which supp $F[f] \subset \Omega \subset \mathbb{R}_n$ in the topology of the space Z' we get

$$\lim_{m \to \infty} \{ F^{-1} [P_m] * f \} = F^{-1} [\ln F[w]] * f.$$

Proof. Let $g = F[f]$ then $g \in D'(\mathbb{R}_n)$. We will show that $gP_m \to g \ln F[w]$ for $m \to \infty$ in $D'(\mathbb{R}_n)$. Let ϕ be an arbitrary function from $D(\mathbb{R}_n)$, χ_ϕ an infinitely often differentiable function such, that $\chi_\phi(x) = 1$ for

$$x \in \text{supp } \{g\} \cap \text{supp } \{\phi\} \quad \text{and} \quad \chi_\phi(x) = 0 \quad \text{for } x \in \Omega.$$

(The existence of such a function follows, for example, from the Theorem 6.20 [165]). As supp $\{g\} \in \Omega$, we get

$$(g(P_m - \ln F[w]), \phi) = (\chi_\phi g(P_m - \ln F[w]), \phi).$$

The function $\chi_\phi(x) \cdot \phi(x) \in D(\mathbb{R}_n)$ has a compact support in the region Ω and therefore the sequence $\chi_\phi(P_m - \ln F[w])\phi$ converges to zero in the topology of $D(\mathbb{R}_n)$. But $g \in D'(\mathbb{R}_n)$ and

$$(g, \chi_\phi(P_m - \ln F[w])\phi) \to 0 \quad \text{for } m \to \infty.$$

This means that $g P_m \to g \ln F[w]$ for $m \to \infty$ in $D'(\mathbb{R}_n)$. As the mapping F^{-1}: $D'(\mathbb{R}_n) \to Z'$ is continuous we get

$$F^{-1}[P_m g] \to F^{-1}[\ln F[w]g] \quad \text{for } m \to \infty \text{ in } Z'.$$

By virtue of the statement (A) and because of the fact that $\ln F[w]$ is a multiplier in the space $D'(\mathbb{R}_n)$ for all $f \in Z'$ the functional $F^{-1}[\ln F[w]] * f \in Z'$ and the equation

$$F[F^{-1}[\ln F[w]] * f] = \ln F[w] \cdot F[f].$$

holds. Consequently,

$$F^{-1}[\ln F[w]] * f = F^{-1}[\ln F[w]g].$$

In an analogous way we get:

$$F^{-1}[P_m]*f = F^{-1}[P_m g].$$

From these results we get the statement of the theorem

$$F^{-1}[P_m]*f \to F^{-1}[\ln F[w]]*f \quad \text{for } m \to \infty \text{ in } \mathbf{Z}'. \qquad \square$$

Remark. If Ω is a bounded region in $\mathbb{R}_n$ then according to the theorem of Wiener-Schwarz f is an integral analytic function the order of which is not larger than 1. Therefore in this case the Theorem 2.6 ensures the convergence of the differential series in the weak topology at least for some integral functions (i.e. the statement of Theorem 2.6 is weaker than the statement of Theorem 2.2). If the region Ω is not bounded in $\mathbb{R}_n$ then the statement of Theorem 2.6 is more essential. Thus we have shown that the equation

$$F^{-1}[\ln F[w]]*f = \int \sum_{k=0}^{\infty} \eta_k h^k (iD)^k f \qquad (2.52)$$

holds in a weaker topology compared with the topology in the space $\mathbf{S}'$.

2.8 Asymptotic Expansion of the Solution of a Difference Scheme

We will show that the right side of (2.52) can be interpreted as an asymptotic expansion of the generalized function

$$F^{-1}[\ln F[w]]*f \quad \text{for } h \to 0.$$

We remember that the formal series $\sum_{n=0}^{\infty} a_n h^n$ is called an asymptotic expansion of the function $\phi(h)$ for $h \to 0$ if for all integers N

$$\lim_{h \to 0} \left(\frac{\phi(h) - \sum_{k=0}^{N} a_k h^k}{h^N} \right) = 0.$$

This can be written in the following form:

$$\phi(h) \sim \sum_{n=0}^{\infty} a_n h^n, \quad h \to 0.$$

In an analogous way an asymptotic expansion of generalized functions can be defined which depend on a parameter.

Definition. The series $\sum_{n=0}^{\infty} h^n f_n$ is called an asymptotic expansion of the one-parameter family of generalized functions $f(h)$ ($f(h) \in \mathbf{S}'$, $f_n \in \mathbf{S}'$ for all n) for

$h \to 0$ if for any basic function $\phi \in \mathbf{S}$ the relation

$$(f(h), \phi) \sim \sum_{n=0}^{\infty} h^n (f_n, \phi), \quad h \to 0$$

holds.

It is easy to prove that if

$$f(h) \sim \sum_{n=0}^{\infty} f_n h^n, \quad h \to 0,$$

then

$$F[f] \sim \sum_{n=0}^{\infty} F[f_n] h^n, \quad h \to 0.$$

In addition, for each fixed $x \in \mathbb{R}$ we get:

$$\ln F[w](x) \sim \sum_{k=0}^{\infty} \eta_k x^k h^k, \quad h \to 0.$$

We will show that for any basic function $\phi \in \mathbf{S}$ the product

$$\psi_N(h) = \frac{1}{h^N} \left[\ln F[w](x) - \sum_{n=0}^{N} \eta_n x^n h^n \right] \phi(x)$$

converges to zero for $h \to 0$ in the topology of the space $\mathbf{S}$.

As the series $\sum_{n=0}^{\infty} \eta_n x^n h^n$ converges to the function $\ln F[w](x)$ for $|xh| \leq r_0$ for some $r_0 > 0$, we get

$$\left| \ln F[w](x) - \sum_{n=0}^{N} \eta_n x^n h^n \right| \leq c_0(h) h^N |x|^N \quad \text{for} \quad |hx| \leq r_0,$$

where $c_0(h) \to 0$ for $h \to 0$. Further, for $x \in \mathbb{R}$ we get

$$|\ln F[w](x)| \leq c_1 (1 + |hx|)^{M_1}$$

for some constants c_1 and M_1 and, consequently,

$$|\ln F[w](x)| \leq c_2 |hx|^{M_2} \quad \text{for} \quad |xh| \geq r_0, \quad M_2 > N,$$

from which follows that

$$\left| \ln F[w](x) - \sum_{n=0}^{N} \eta_n x^n h^n \right| \leq \begin{cases} c_0(h) h^N |x|^N & \text{for} \quad |hx| \leq r_0, \\ c h^M |x|^M & \text{for} \quad |hx| \geq r_0. \end{cases}$$

Therefore, for all $\phi(x) \in \mathbf{S}$ we get:

$$\sup_{x \in \mathbb{R}} \left\{ \left| \frac{1}{h^N} \left(\ln F[w](x) - \sum_{n=0}^{N} \eta_n x^n h^n \right) \phi(x) \right| \right\} \to 0 \quad \text{for} \quad h \to 0.$$

In an analogous way we can prove that

$$\sup_{x \in R} \left\{ (1 + |x|^2)^M \left| \frac{\partial^m}{\partial x^m} (\psi_N \phi) \right| \right\} \to 0 \quad \text{for} \quad h \to 0.$$

Thus, for an arbitrary basic function $\phi(x) \in S$ the one-parameter family of functions $\psi_N(h)$ converges to zero in S for $h \to 0$. As for every generalized function $f \in S'$

$$(\psi_N f, \phi) = (f, \psi_N \phi) \to 0 \quad \text{for} \quad h \to 0$$

the following statement is proved.

. . . .

Theorem 2.7. For all $f \in S'$ the following relation holds in S'

$$F^{-1}[\ln F[w]] * f \sim \sum_{n=0}^{\infty} \eta_n h^n (iD)^n f, \quad h \to 0.$$

. . . .

We consider Cauchy's problem of the following kind:

$$\frac{df}{dt} = \frac{1}{\tau} F^{-1}[\ln F[w]] * f, \tag{2.53}$$

$$f|_{t=0} = f_0 \in S',$$

and

$$\frac{df}{dt} = \frac{1}{\tau} F^{-1}[P_m] * f, \tag{2.54}$$

$$f|_{t=0} = f_0 \in S'.$$

Here

$$P_m = P_m(hx) = \sum_{k=0}^{m} \eta_k h^k x^k.$$

We will assume as above that the series $\sum_{k=0}^{\infty} \eta_k h^k x^k$ converges to the function $\ln F[w](hx)$ for $hx \in \Omega \subset \mathbb{R}$.

Let $R_{\tau, h}(t)$ and $R_{m, \tau, h}(t)$ be the solution-operators of the problems (2.53, 54), respectively. It is easy to see that

$$R_{\tau, h}(t)f = F^{-1}[(F[w])^{t/\tau}] * f, \quad R_{\tau, h}(t): S' \to S, \tag{2.55}$$

$$R_{m, \tau, h}(t)f = F^{-1}[e^{tP_m/\tau}] * f, \quad R_{m, \tau, h}(t): S' \to \mathbb{Z}'. \tag{2.56}$$

. . . .

Theorem 2.8. Let $f_{\tau, h}(t)$ be the solution of problem (2.53), and $f_{m, \tau, h}$ the solution of the problem (2.54). Then

$$f_{\tau, h}(t) - f_{m, \tau, h}(t) = o(h^m), \quad h \to 0 \text{ in } \mathbb{Z}'.$$

. . . .

Proof. It is necessary to show that for all $\psi \in \mathbf{Z}$

$$\lim_{h \to 0} \left[\frac{1}{h^m} (f_{\tau, h}(t) - f_{m, \tau, h}(t), \psi) \right] = 0. \tag{2.57}$$

As

$$f_{\tau, h}(t) = R_{\tau, h}(t) f_0, \quad f_{m, \tau, h}(t) = R_{m, \tau, h}(t) f_0,$$

from (2.55, 56) it follows that the relation (2.57) is equivalent to the equation

$$\lim_{h \to 0} \left[\frac{1}{h^m} ((F[w]^{t/\tau} - \exp(t P_m/\tau)) g_0, \phi) \right] = 0, \quad \text{for all } \phi \in \mathbf{D}(\mathbb{R}), \tag{2.58}$$

where $g_0 = F[f_0]$. Let be

$$g(h; x) = h^{-m} (\exp(\ln F[w] - P_m) - 1).$$

It is easy to see that for every nonnegative integer k and for an arbitrary compactum $\mathscr{K}$, $\partial^k g(h; x)/\partial x^k \to 0$ uniformly in $\mathscr{K}$ for $h \to 0$. But then for all $\phi \in \mathbf{D}(\mathbb{R})$

$$g(h; x) \phi(x) \to 0 \quad \text{in } \mathbf{D}(\mathbb{R}) \text{ for } h \to 0.$$

As $g_0 \in \mathbf{S}'$ we get $(g g_0, \phi) \to 0$ for $h \to 0$ and all $\phi \in \mathbf{D}(\mathbb{R})$. The latter means that

$$\lim_{h \to 0} \left[\frac{1}{h^m} (\exp(P_m)(\exp(\ln F[w] - P_m) - 1) g_0, \phi) \right]$$

$$= \lim_{h \to 0} \left[\frac{1}{h^m} ((F[w] - \exp(P_m)) g_0, \phi) \right] = 0.$$

Thus the equation (2.58) is proved and, consequently, also the statement of the theorem. $\qquad \square$

Up to now we have considered a two-parameter family of difference schemes (2.43) and we have investigated the behaviour of the solution for $h \to 0$. In practise usually a coupling between the parameters τ and h is given. E.g. for the difference schemes of the form (2.43) which approximate equations of hyperbolic type a natural coupling is defined by $\tau = \kappa h$; $\kappa = \text{const.}$

In the contributions [46, 201] the asymptotic behaviour of solutions of a one-parameter family of difference schemes of the form (2.43) ($\tau = \kappa h^\alpha$, $\alpha > 0$) was investigated and it was shown that in some linear topological space $\mathbb{F}$ (for sufficiently strong limitations on the scheme (2.43) and the differential approximation) the solutions of the differential approximations lead to an asymptotic representation of the solution of the difference scheme (see [46, 201]).

We will show that in the topological space $\mathbf{Z}'$ an analogous statement is true without the essential limitations on the difference schemes and the differential approximations. Thus let $\tau = \kappa h^\alpha$, $0 < \alpha \leq l$ and in (2.50) $\eta_0 = \eta_1 = \ldots = \eta_{l-1} = 0$. Then

$$\frac{1}{\tau} \ln F[w](hx) = \frac{1}{\kappa} x^\alpha \sum_{k=l}^{\infty} \eta_k (hx)^{k-\alpha}.$$

As in $\mathbb{D}(\mathbb{R})$ for $h \to 0$

$$\frac{1}{h^{m-\alpha}} \left[\frac{1}{\kappa h^{\alpha}} \ln F[w](hx) - \frac{1}{\kappa} x^{\alpha} \sum_{k=l}^{m} \eta_k (hx)^{x-\alpha} \right] \to 0,$$

the following statement holds.

. . . .

Theorem 2.9. Let be $\tau = \kappa h^{\alpha}$, $0 < \alpha \leq l$, $\alpha = \text{const}$, $\kappa = \text{const}$, $\eta_0 = \ldots$
$= \eta_{l-1} = 0, f_h(t)$ a solution of problem (2.53), $f_{m,h}(t)$ a solution of the problem

$$\frac{df}{dt} = \frac{1}{\kappa} F^{-1} \left[x^{\alpha} \sum_{k=l}^{m} \eta_k h^{k-\alpha} x^{k-\alpha} \right] * f, \quad m \geq l, \ f|_{t=0} = f_0.$$

Then

$$f_h(t) - f_{m,h}(t) = o(h^{m-\alpha}) \qquad \text{for } h \to 0 \text{ in } \mathbb{Z}'.$$

. . . .

The proof of this theorem is analogous to the proof of the Theorem 2.8.

2.9 On the Injective Character of the Mapping of Difference Schemes in the Set of Differential Representations

We consider two difference schemes of the form (2.43)

$$\Lambda_1 f^{n+1} = \Lambda_0 f^n \tag{2.59}$$

and

$$\bar{\Lambda}_1 f^{n+1} = \bar{\Lambda}_0 f^n, \tag{2.60}$$

where

$$\Lambda_s = \sum_{j=p_s}^{q_s} b_j^s T_{jh}, \quad b_{p_s}^s \neq 0, \quad b_{q_s}^s \neq 0;$$
$$\qquad\qquad\qquad\qquad\qquad\qquad\qquad s = 0, 1.$$
$$\bar{\Lambda}_s = \sum_{j=\bar{p}_s}^{\bar{q}_s} \bar{b}_j^s T_{jh}, \quad \bar{b}_{\bar{p}_s}^s \neq 0, \quad \bar{b}_{\bar{q}_s}^s \neq 0,$$

Let $w_s = \Lambda_s \delta_0, \bar{w}_s = \bar{\Lambda}_s \delta_0$ $(s = 0, 1)$. Then the schemes (2.59, 60) can be written in the form

$$w_1 * f^{n+1} = w_0 * f^n, \tag{2.61}$$
$$\bar{w}_1 * f^{n+1} = \bar{w}_0 * f^n. \tag{2.62}$$

As above we will assume that $F[w_s](x) \neq 0$ for all $x \in \mathbb{R}$, $F[\bar{w}_s](x) \neq 0$ for all $x \in \mathbb{R}$, $s = 0, 1$.

. . . .

Theorem 2.10. The differential representations of the difference schemes (2.61) and (2.62) coincide if and only if there exist generalized functions w, $\bar{w}$, w'_0 w'_1 with a compact support such that

$$w_s = w * w'_s, \qquad \bar{w}_s = \bar{w} * w'_s \qquad (s = 0, 1). \tag{2.63}$$

. . . .

Proof. If for the schemes (2.61, 62) the equations (2.63) hold, then

$$\ln F[w_0] - \ln F[w_1] = \ln F[\bar{w}_0] - \ln F[\bar{w}_1]$$

and, consequently, the differential representations of the schemes (2.61, 62) coincide.

On the other side let the differential representations of the schemes (2.61, 62) coincide. Then by virtue of the fact that the coefficients b^s_j and $\bar{b}^s_j$ are real $F[w_0]F[\bar{w}_1] = F[\bar{w}_0]F[w_1]$. As it was shown in the paper [202] the scheme (2.59) can be represented in the form

$$\Lambda\Lambda'_1 f^{n+1} = \Lambda\Lambda'_0 f^n,$$

where the characteristic polynomials of the operators Λ'_0 and Λ'_1 do not have common roots. As

$$\frac{F[\Lambda\Lambda'_0\delta_0]}{F[\Lambda\Lambda'_1\delta_0]} = \frac{F[\Lambda'_0\delta_0]}{F[\Lambda'_1\delta_0]},$$

one can assume without limitation of generality that the characteristic polynomials of the operators Λ_0 and Λ_1 (and also of the operators $\bar{\Lambda}_0$ and $\bar{\Lambda}_1$) do not have common roots. Let be

$$P_s(x) = \sum_{j=p_s}^{q_s} b^s_j x^{j-p_s}, \qquad \bar{P}_s(x) = \sum_{j=\bar{p}_s}^{\bar{q}_s} \bar{b}^s_j x^{j-\bar{p}_s}$$

the characteristic polynomials of the operators Λ_s and $\bar{\Lambda}_s$, respectively,

$$x_{j,s} \quad (1 \le j \le q_s - p_s), \qquad \bar{x}_{k,s} \quad (1 \le k \le \bar{q}_s - \bar{p}_s)$$

the roots of the equations $P_s(x) = 0$ and $\bar{P}_s(x) = 0$, respectively. Then

$$P_s(x) = a_s \prod_{j=1}^{q_s-p_s} (x - x_{j,s}), \qquad \bar{P}_s(x) = \bar{a}_s \prod_{j=1}^{\bar{q}_s-\bar{p}_s} (x - \bar{x}_{j,s}).$$

As $F[w_0]F[\bar{w}_1] = F[\bar{w}_0]F[w_1]$ it follows that $P_0\bar{P}_1 = \bar{P}_0 P_1$. According to the assumption P_1 and P_0, $\bar{P}_1$ and $\bar{P}_0$ have no common roots and, therefore, from the equation

$$a_0\bar{a}_1 \prod_{j=1}^{q_0-p_0} (x - x_{j,0}) \prod_{k=1}^{\bar{q}_1-\bar{p}_1} (x - \bar{x}_{k,1}) = \bar{a}_0 a_1 \prod_{k=1}^{\bar{q}_0-\bar{p}_0} (x - \bar{x}_{k,0}) \prod_{j=1}^{q_1-p_1} (x - x_{j,1})$$

$$\tag{2.64}$$

it follows that the polynomials P_1 and $\bar{P}_1$ (and also the polynomials P_0 and $\bar{P}_0$)

have the same roots. Consequently,

$$\prod_{j=1}^{q_1-p_1} (x - x_{j,1}) = \prod_{k=1}^{\bar{q}_1-\bar{p}_1} (x - \bar{x}_{k,1}),$$

$$\prod_{j=1}^{q_0-p_0} (x - x_{j,0}) = \prod_{k=1}^{\bar{q}_0-\bar{p}_0} (x - \bar{x}_{k,0}), \quad \text{or}$$

$$\bar{P}_1 = \frac{\bar{a}_0}{a_0} P_1, \quad \bar{P}_0 = \frac{\bar{a}_1}{a_1} P_0.$$

As from (2.64) we get $a_0 \bar{a}_1 = \bar{a}_0 a_1$ it follows that $\bar{a}_0/a_0 = \bar{a}_1/a_1 = c$. Thus

$$q_1 - p_1 = \bar{q}_1 - \bar{p}_1 = r_1,$$
$$\bar{q}_0 - \bar{p}_0 = q_0 - p_0 = r_0,$$
$$\bar{b}_j^s = c b_{k_s+j}^s; \quad s = 0, 1;$$
$$j = \bar{p}_s, \bar{p}_s + 1, \ldots, \bar{q}_s; \quad k_s = q_s - \bar{q}_s = p_s - \bar{p}_s.$$

Therefore

$$\bar{A}_s = c T_{k_s h} A_s, \quad s = 0, 1. \tag{2.65}$$

But

$$\frac{F[\bar{A}_0 \delta_0]}{F[\bar{A}_1 \delta_0]} = \frac{F[c T_{k_0 h} \delta_0] F[A_0 \delta_0]}{F[c T_{k_1 h} \delta_0] F[A_1 \delta_0]} = \frac{F[A_0 \delta_0]}{F[A_1 \delta_1]},$$

i.e.

$$F[T_{k_0 h} \delta_0] = F[T_{k_1 h} \delta_0],$$

from which we get $k_0 = k_1 = k$. Finally, we get

$$\bar{A}_s = c T_{kh} A_s, \quad s = 0, 1 \tag{2.66}$$

and, consequently,

$$\bar{w}_s = w * w_s, \quad s = 0, 1.$$

Thus the theorem is proved. □

This theorem can be interpreted as follows. Let Sh be the set of difference schemes of the form (2.59) satisfying the conditions $F[w_0](x) \neq 0$ for all $x \in \mathbb{R}$ and $F[w_1](x) \neq 0$ for all $x \in \mathbb{R}$.

We denote by $\mathscr{P}$ the set of differential representations of schemes from Sh. Let $Q: Sh \to \mathscr{P}$ be a mapping which relates to each scheme from Sh its differential representation. Two schemes S_1 and S_2 from Sh of the form (2.61, 62) will be called equivalent ($S_1 \sim S_2$) if such generalized functions with compact support $w, \bar{w}, w_0', w_1'$ exist which are finite combinations of *Dirac's δ-functions* that the equations (2.63) hold. It is easy to see that this relation is indeed an equivalence relation on the set Sh.

We denote by $\tilde{S}h$ the set of equivalence classes. As it follows from Theorem 2.10, if $S_1 \sim S_2$ then $QS_1 = QS_2$ and therefore the mapping Q induces a mapping $\tilde{Q}: \tilde{S}h \to \mathscr{P}$, which relates the class of equivalent difference schemes with the differential representation of any representative of this equivalence class.

Now we can formulate the Theorem 2.10 in the following way: The mapping $\tilde{Q}$ is a one-to-one mapping of the set $\tilde{S}h$ on the set $\mathscr{P}$. Especially from this it follows that the mapping $\tilde{Q}: \tilde{S}h \to \mathscr{P}$ is injective, i.e. different differential representations correspond to different equivalence classes.

3. Stability Analysis of Difference Schemes with Constant Coefficients by Means of the Differential Representation

3.1 Absolute and Conditional Approximation

Let us consider equation (2.2) for constant coefficients. In the general case the Π-form of the differential representation of the difference scheme (2.1) for the equation (2.2) has the form

$$\frac{\partial u}{\partial t} = L(D)u + \sum_l \sum_{l_1 + \ldots + l_s = 1}^{l} c_{l_1 \ldots l_s} \frac{\partial^l u}{\partial x_1^{l_1} \ldots \partial x_s^{l_s}}, \tag{3.1}$$

and represents a differential equation which can be solved by a finite-difference algorithm and which contains the full information about the difference scheme.

Definition. The order of approximation is defined by the lowest order of powers of τ and h which are met among the coefficients $c_{l_1 \ldots l_s}$.

It is clear that the order of approximation introduced coincides with the order of approximation in the usual sense, which means how well the solution of the given differential equation satisfies the difference equation, while the order of approximation defined by the differential approximation shows how well the original differential equation is approximated by the differential representation.

Definition. The difference scheme (2.1) approximates absolutely the differential equation (2.2) if

$$R(D)u = \sum_l \sum_{l_1 + \ldots + l_s = 1}^{l} c_{l_1 \ldots l_s} \frac{\partial^l u}{\partial x_1^{l_1} \ldots \partial x_s^{l_s}}$$

converges to zero for τ and h which may converge to zero in an arbitrary way. Otherwise the difference scheme approximates the differential equation conditionally, then for $h = h(\tau)$, τ and h converging to zero, the differential representation leads immediately to the differential equation which is approximated by the difference scheme.

For the scheme (2.5), for example, the differential representation has the form (2.15) with $c_l = O(\tau^{l-1}, h^{l-1})$ and the given scheme is absolutely approximating the differential equation. On the other hand the *Lax' scheme*

$$\frac{u^{n+1}(x) - \bar{u}^n(x)}{\tau} = a\frac{u^n(x+h) - u^n(x-h)}{2h},$$

$$\bar{u}^n(x) = \tfrac{1}{2}[u^n(x+h) + u^n(x-h)]$$

(3.2)

for equation (2.6) has the differential representation (2.15) with $c_2 = h^2(1 - \varkappa^2 a^2)/2\tau$, $c_l = O(\tau^{l-1}, h^{l-1}, h^l/\tau); l \geq 3$; and, consequently, is conditionally approximating the differential equation, as for $\tau/h = \kappa = $ const the scheme approximates the equation (2.6) and for $\tau/h^2 = $ const the equation

$$\frac{\partial u}{\partial t} = a\frac{\partial u}{\partial x} + \frac{h^2}{2\tau} \cdot \frac{\partial^2 u}{\partial x^2}.$$

3.2 Lax' Equivalence Theorem

We will formulate some statements of the stability theory for difference schemes in terms of differential representations.

Let us find a one-parameter family of elements $u(t) \in \mathbb{B}$, where $\mathbb{B}$ is a Banach space, which satisfy the equation:

$$\frac{du(t)}{dt} = A u(t); \quad 0 \leq t \leq t_0; \quad u(0) = u_0;$$

(3.3)

A is a linear operator with constant coefficients. We assume that the problem (3.3) is correct; especially for the solution-operator S we have

$$u(t) = S(t)u_0,$$

and the inequality

$$\|S(t)\| \leq K_1 = \text{const}$$

holds for $0 \leq t \leq t_0$ according to a norm of the space $\mathbb{B}$.

We approximate the equation (3.3) by the difference scheme

$$\Lambda_1 u^{n+1} = \Lambda_0 u^n,$$

(3.4)

where Λ_1 and Λ_0 are difference operators with constant coefficients while the operator Λ_1 is assumed to be nonsingular. Then the step-operator Λ of the difference scheme (3.4) is defined by the equation

$$u^{n+1} = \Lambda u^n.$$

(3.5)

The Π-form of the differential representation of the difference scheme (3.4) has the form

$$\frac{du(t)}{dt} = A u(t) + R u(t), \tag{3.6}$$

where R is a linear operator which depends on τ. The lowest power of τ which appears in R defines the order of approximation of the difference scheme. For the equation (3.6) we consider Cauchy's problem with $u(0) = u_0$.

Definition. The difference scheme (3.4) is called stable, if Cauchy's problem for the equation (3.6) is correct; especially for the solution operator $\Omega = \Lambda^n$ of equation (3.6) the inequality

$$\|\Omega(t)\| \leq K_2 = \text{const} \quad 0 \leq t \leq t_0,$$

holds.

Then Lax' equivalence theorem can be formulated as follows:

Equivalence theorem. For the convergence of the difference scheme (3.4) which approximates the correct problem (3.3) it is necessary and sufficient that Cauchy's problem for the differential representation (3.6) of the difference scheme is correct.

The proof of this theorem follows directly the proof of Lax' equivalence theorem [7].

The necessary stability condition of von Neumann is formulated as follows:

$$|\varrho_j| \leq 1 + O(\tau); \quad j = 1, \ldots, m, \tag{3.7}$$

where ϱ_j are the eigenvalues of the amplification matrix G of the difference scheme or, equivalently, of the spectral counterpart of the step-operator of the differential representation (3.6) of the difference scheme

$$G = \exp[\tau \tilde{A}(ik) + \tau \tilde{R}(ik)]. \tag{3.8}$$

Here $\tilde{A}(ik)$ and $\tilde{R}(ik)$ are the spectral counterparts of the operators A und R, respectively; k is the dual variable.

The condition (3.7) will be also a sufficient stability condition, if the matrix G is normal (especially for $m = 1$).

3.3 On the Necessary Stability Conditions for Difference Schemes

In the general case a two-level difference scheme (3.4) for the equation (2.6) has a differential representation in the form:

$$\frac{\partial u}{\partial t} = a \frac{\partial u}{\partial x} + \sum_l c_l \frac{\partial^l u}{\partial x^l}. \tag{3.9}$$

A necessary and sufficient condition for stability of the difference scheme in this case is the inequality

$$|\varrho| \leq 1, \tag{3.10}$$

where

$$\varrho = \exp\left[i\kappa a\xi + \sum_l \frac{\tau}{h^l} i^l c^l \xi^l\right], \quad \kappa := \frac{\tau}{h}; \quad \xi := kh.$$

Then

$$|\varrho| = \exp\left[\sum_l (-1)^l \frac{\tau}{h^{2l}} c_{2l} \xi^{2l}\right]. \tag{3.11}$$

For small ξ we get the following necessary stability condition

$$(-1)^{q+1} c_{2q} > 0, \tag{3.12}$$

where c_{2q} is the first coefficient different from zero from all the coefficients c_{2l}.

In particular, let the difference scheme (3.4) have the form

$$u^{n+1}(x) = \sum_\alpha b_\alpha u^n(x + \alpha h). \tag{3.13}$$

Here α runs through a set of values (finite number in the case of an explicit scheme and infinite number in the case of an implicit scheme) and approximates the equation (2.6) with the order p. Then the compatibility conditions

$$\sum_\alpha \alpha^j b_\alpha = \kappa^j a^j; \quad j = 0, 1, \ldots, p$$

are satisfied and the differential representation can be written in the form:

$$\frac{\partial u}{\partial t} = a \frac{\partial u}{\partial x} + \sum_{l=p+1}^\infty c_l \frac{\partial^l u}{\partial x^l}, \tag{3.14}$$

where

$$c_{p+1} = \frac{h^{p+1}}{\tau(p+1)!}\left[\sum_\alpha \alpha^{p+1} b_\alpha - \kappa^{p+1} a^{p+1}\right];$$

$$c_{p+2} = \frac{h^{p+2}}{\tau(p+2)!}\left[\sum_\alpha \alpha^{p+2} b_\alpha - \kappa^{p+2} a^{p+2}\right] - \tau a c_{p+1}$$

$$= \frac{h^{p+2}}{\tau(p+2)!}[\tilde{c}_{p+2} - (p+2)\kappa a \tilde{c}_{p+1}]; \quad \tilde{c}_j := \sum_\alpha \alpha^j b_\alpha - \kappa^j a^j.$$

The necessary stability condition (3.12) in this case has the form

$$\begin{aligned} (-1)^{N+1} c_{2N} > 0; \quad & p = 2N - 1; \\ (-1)^N c_{2N+2} > 0; \quad & p = 2N, \end{aligned} \tag{3.15}$$

or

$$(-1)^{N+1} \tilde{c}_{2N} > 0; \quad p = 2N - 1;$$

$$(-1)^N [\tilde{c}_{2N+2} - (2N+2)\kappa a \tilde{c}_{2N+1}] > 0; \quad p = 2N.$$

Thus the following lemma holds.

Lemma 3.1. A necessary stability condition for the difference scheme (3.13) is the inequality (3.15).

We remark that we have derived a necessary stability condition in terms of the first differential approximation of a difference scheme (for a scheme of odd order of approximation) and in terms of the second differential approximation (for a scheme of even order of approximation). In the latter case we do not assume explicitly that $c_{2N+2} \neq 0$. Otherwise we arrive at limitations for the following even coefficients of the differential representation.

Let us consider e.g. the difference scheme (2.5) of first order of approximation ($\kappa a \neq 1$, $a > 0$). Here $c_2 = a h (1 - \kappa a)/2 \neq 0$, and, consequently, the necessary stability condition has the form $\kappa a < 1$. As it is known the given inequality is also sufficient for stability.

Let us also consider the *Lax-Wendroff difference scheme* [7] of second order of approximation for the equation (2.6):

$$\frac{\Delta_0 u^n(x)}{\tau} = a \frac{\Delta_1 + \Delta_{-1}}{2h} u^n(x) + \frac{\tau a^2}{2} \frac{\Delta_1 \Delta_{-1}}{h^2} u^n(x).$$

The Π-form of the differential representation of this difference equation is given by

$$\frac{\partial u}{\partial t} = a \frac{\partial u}{\partial x} + \frac{a h^2}{6} (1 - \kappa^2 a^2) \frac{\partial^3 u}{\partial x^3} - \frac{\tau h^2 a^2}{8} (1 - \kappa^2 a^2) \frac{\partial^4 u}{\partial x^4} + \dots.$$

As $p = 2$ the necessary stability condition can be written in the form:

$$-c_4 = \frac{\tau h^2 a^2}{8} (1 - \kappa^2 a^2) > 0,$$

i.e. $\kappa^2 a^2 < 1$. This inequality is also sufficient for stability.

In the case of constant coefficients and one space variable the Π-form of the differential representation of a difference scheme for the equation (2.2) in the general case has the form

$$\frac{\partial u}{\partial t} = \sum_l c_l \frac{\partial^l u}{\partial x^l} \tag{3.16}$$

or

$$\frac{\partial u}{\partial t} = \sum_{l=0}^{\infty} c_{2l+1} \frac{\partial^{2l+1} u}{\partial x^{2l+1}} + \sum_{l=1}^{\infty} c_{2l} \frac{\partial^{2l} u}{\partial x^{2l}},$$

where the operator L is enclosed in some of the terms of the sum. Then

$$\varrho = \exp \left[i \sum_{l=0}^{\infty} (-1)^l \tilde{c}_{2l+1} \xi^{2l+1} + \sum_{l=1}^{\infty} (-1)^l \tilde{c}_{2l} \xi^{2l} \right], \tag{3.17}$$

where

$$\tilde{c}_l := \frac{\tau}{h^l} c_l \quad (l = 1, 2, \dots),$$

and, consequently,

$$|\varrho| = \exp\left[\sum_{l=1}^{\infty} (-1)^l \tilde{c}_{2l} \xi^{2l}\right].$$ (3.18)

For small ξ we get the following necessary stability condition

$$(-1)^{q+1} c_{2q} > 0,$$ (3.19)

where c_{2q} is the first coefficient different from zero from the whole set of coefficients c_{2l}. The case of a hyperbolic equation will be discussed in Chap. 3.
   We consider the example of a difference scheme

$$\frac{\Delta_0 u^n(x)}{\tau} = \sigma \frac{\Delta_1 \Delta_{-1}}{h^2} u^n(x),$$ (3.20)

for the parabolic equation

$$\frac{\partial u}{\partial t} = \sigma \frac{\partial^2 u}{\partial x^2}.$$ (3.21)

   It is known that the scheme (3.20) is stable for $\sigma \tau/h^2 \leq \frac{1}{2}$. The Π-form of the differential representation can be written as

$$\frac{\partial u}{\partial t} = \sigma \frac{\partial^2 u}{\partial x^2} + \frac{h^2 \sigma}{12}\left(1 - 6\frac{\sigma \tau}{h^2}\right)\frac{\partial^4 u}{\partial x^4} + \ldots.$$

A necessary condition for stability will be the inequality $c_2 = \sigma > 0$. Here we have the situation which shows that the derived criterion is necessary but not sufficient. This should not be regarded as a surprising fact because our criterion is derived for small ξ which corresponds to small wave numbers k and, consequently, is not sufficient in the general case. The maximal wave number for a numerical calculation is $k = \pi/h$. The stability condition in this limiting case is important, because the Fourier components, which correspond to the maximal wave number ($k \sim h^{-1}$), are not at all stable. The differential approximation of a difference scheme can not always be used for finding stability conditions. Thus, for example, for a difference scheme which is an approximation of equation (2.6), $c_{2l} \sim O(h^{2l-1})$ if $\tau/h = \kappa = \text{const}$, but if $\tilde{c}_{2l} \sim O(1)$; $\xi \sim O(1)$ and, consequently, it is necessary to consider the whole series of terms in the differential representation and not only the differential approximation. But in a number of cases which are discussed later the derived criterion is also sufficient.
   Finally, we emphasize the fact which underlines the right to apply the differential approximation to analyse the properties of difference schemes and which follows from the results of the following publications [8, 167, 168]. We consider especially a difference scheme of first order of approximation

$$u^{n+1}(x) = \sum_{\alpha} b_\alpha u^n(x + \alpha h),$$

for the equation (2.6) and we use the notation: $u(x, t)$ is the solution of equation (2.6); $v(x, t)$ is the solution of the difference scheme; $w(x, t)$ is the solution of the first differential approximation of the difference scheme

$$\frac{\partial u}{\partial t} = a \frac{\partial u}{\partial x} + c_2 \frac{\partial^2 u}{\partial x^2},$$

where

$$c_2 := \frac{h^2}{2\tau} \left(\sum_\alpha \alpha^2 b_\alpha - \kappa^2 a^2 \right).$$

Let the difference scheme be stable and, consequently, $c_2 > 0$. Then for discontinuous initial conditions (of the step-type) the following bounds in the norm of the space $\mathbb{L}_2$ hold:

$$\| v(x, t) - u(x, t) \| \leq \text{const} \cdot \tau^{1/4};$$
$$\| v(x, t) - w(x, t) \| \leq \text{const} \cdot \tau^{3/4}.$$

In the case of a smooth initial distribution it is easy to show that

$$\| v(x, t) - u(x, t) \| \leq \text{const} \cdot \tau;$$
$$\| v(x, t) - w(x, t) \| \leq \text{const} \cdot \tau^2.$$

Thus even for discontinuities of the step-type the solution of the first differential approximation is closer to the solution of the difference scheme than the solution of the original differential equation to the solution of the difference scheme.

4. Connection Between the Stability of Difference Schemes and the Properties of Their First Differential Approximations

4.1 Simple Difference Schemes

We consider Cauchy's problem for equation (1.2) where the matrix A has different real eigenvalues $a_1(x, t) < \ldots < a_m(x, t)$.

We approximate the system of equations (1.2) by a simple difference scheme of first order of approximation [40, 61, 169]:

$$u^{n+1}(x) = \sum_{\alpha=1}^{2} B_\alpha u^n (x + \tau \lambda_\alpha). \tag{4.1}$$

Here

$$\sum_{\alpha=1}^{2} B_\alpha = I, \quad \sum_{\alpha=1}^{2} \lambda_\alpha B_\alpha = A. \tag{4.2}$$

The following theorem holds.

. . . .

Theorem 4.1. In the case of constant coefficients it is necessary for the stability of the simple difference scheme (4.1) that its first differential approximation is a not totally parabolic system of equations. If the matrix A is symmetric then the condition of not total parabolicity of the first differential approximation is also sufficient for stability.

. . . .

Proof. The Π-form of the difference scheme (4.1) has the form

$$\frac{\partial u}{\partial t} = D \frac{\partial u}{\partial x} + C_2 \frac{\partial^2 u}{\partial x^2}, \tag{4.3}$$

where

$$D = A - \frac{\tau}{2} \left[\frac{\partial A}{\partial t} + A \frac{\partial A}{\partial x} \right]; \quad C_2 = \frac{\tau}{2} \left[\sum_{\alpha=1}^{2} \lambda_\alpha^2 B_\alpha - A^2 \right].$$

If the coefficients are constant then $D = A$.

In the considered case we have $B_1 B_2 = B_2 B_1$. Indeed, from the compatibility conditions (4.2) we find that

$$B_1 = (\lambda_1 - \lambda_2)^{-1}(A - \lambda_2 I);$$
$$B_2 = (\lambda_1 - \lambda_2)^{-1}(-A + \lambda_1 I),$$

and, consequently, the matrices B_1 and B_2 commute. We get

$$C_2 = \frac{\tau}{2}(\lambda_1 - \lambda_2)^2 B_1 B_2$$

$$= \frac{\tau}{2}(\lambda_1 - \lambda_2)^2 B_1 (I - B_1)$$

$$= \frac{\tau}{2}(\lambda_1 - \lambda_2)^2 B_2 (I - B_2). \tag{4.4}$$

Let the difference scheme (4.1) be stable. From this by virtue of Lemma 3.1 we have $C_2 \geq 0$, i.e. the eigenvalues of the matrix C_2 are nonnegative and, consequently, the system of equations (4.3) is not totally parabolic.

On the contrary, let A be a symmetric matrix and the system of equations (4.3) not totally parabolic i.e. $C_2 \geq 0$. Then from (4.4) it follows that $B_\alpha \geq 0 (\alpha = 1, 2)$, and the scheme is stable according to *Friedrichs' theorem* [8]. □

. . . .

Theorem 4.2. If A is a Lipshitz-continuous symmetric matrix then the not total parabolicity of the system of equations (4.3) is a necessary and sufficient stability condition for the simple difference scheme (4.1).

. . . .

Proof. As it is known [7, 8, 170] a necessary stability condition of the scheme (4.1) is the stability of all schemes

$$u^{n+1}(x) = \sum_{\alpha=1}^{2} B_a(x_0, t_0) u^n(x + \tau \lambda_\alpha)$$

with constant coefficients which coincide with the values of the coefficients of the scheme (4.1) in the points (x_0, t_0). In the same way as in the proof of the previous theorem it can be shown that in each point (x_0, t_0) the matrices B_1 and B_2 are nonnegative and therefore on the basis of (4.4) $C_2(x_0, t_0) \geq 0$, i.e. the first differential approximation is a not totally parabolic system of equations.

On the contrary, if the system of equations (4.3) is not totally parabolic then we have $C_2 \geq 0$. From the equation (4.4) and the compatibility equation (4.2) it follows that the matrices B_1 and B_2 are nonnegative, symmetric and Lipshitz-continuous. Then according to Friedrich's theorem the scheme is stable. Thus the theorem is proved. □

4.2 Majorant Difference Schemes

Let us assume that the system of equations (1.2) is approximated by a three-point difference scheme

$$u^{n+1}(x) = \sum_{\alpha=-1}^{1} B_\alpha u^n(x + \alpha h),$$ (4.5)

where

$$\sum_{\alpha=-1}^{1} B_\alpha = I,$$

$$\sum_{\alpha=-1}^{1} \alpha B_\alpha = \kappa A, \quad \kappa = \frac{\tau}{h} = \text{const.}$$ (4.6)

The Π-form of the first differential approximation of the difference scheme (4.5) has the form (4.3), where D is defined as earlier and

$$C_2 = \frac{h^2}{2\tau} \left[\sum_{\alpha=-1}^{1} \alpha^2 B_\alpha - \kappa^2 A^2 \right].$$ (4.7)

Definition. The difference scheme (4.5) is called majorant [90, 100, 171] if

$$B_1 = \kappa A^+; \quad B_{-1} = -\kappa A^-;$$

$$A^+ \geq 0; \quad A^- \leq 0;$$

$$A = A^+ + A^-.$$

The matrices A^+, A^- are chosen such that if one would transform them into diagonal form $\tilde{A}^+, \tilde{A}^-$ then in the diagonal of the matrix $\tilde{A}^+$ only positive eigenvalues and zeros are encountered and in the diagonal of the matrix $\tilde{A}^-$ zeros and negative eigenvalues of the matrix A would appear.

....

Theorem 4.3. For the stability of the majorant scheme (4.5) it is necessary that its first differential approximation is a not totally parabolic system of equations. If $A' = A$ (A' is the transposed matrix) then the given condition is also sufficient.

....

The proof of this theorem is analogous to the proof of Theorem 4.1. In the present case from the compatibility condition it follows

$$B_0 = I - \kappa(A^+ - A^-) = I - \kappa|A|;$$

$$C_2 = \frac{h^2}{2\tau}(I - B_0)B_0.$$ (4.8)

....

Theorem 4.4. If A is a symmetric matrix, and if A^+, A^- are Lipshitz-continuous matrices, then the not total parabolicity of the first differential approximation of a majorant scheme is a necessary and sufficient stability condition of the scheme.

....

Proof. A necessary stability condition of a majorant scheme is the stability of all those schemes

$$u^{n+1}(x) = \kappa A^+(x_0, t_0) u^n(x + h) + [I - \kappa|A(x_0, t_0)|]u^n(x)$$
$$- \kappa A^-(x_0, t_0) u^n(x - h)$$

with constant coefficients. From this it follows according to Theorem 4.3 that in each point (x_0, t_0) the matrix $C_2(x_0, t_0) \geq 0$, i.e. the first differential approximation of a majorant scheme is a not totally parabolic system of equations.

On the contrary, from $C_2 \geq 0$ it follows with respect to (4.8) that the matrix B_0 is nonnegative, and therefore according to Friedrichs' theorem the scheme is stable. The theorem is proved.   □

. . . .

Theorem 4.5. Let A be a symmetric matrix. For the stable three-point difference scheme (4.5) to be majorant it is necessary and sufficient that the following conditions for the coefficients of the scheme hold:

$$B'_\alpha = B_\alpha;$$
$$B_\alpha B_\beta = B_\beta B_\alpha; \quad (\alpha, \beta = -1, 0, 1);$$
$$B_1 B_{-1} = 0.$$

. . . .

Proof. That the conditions are necessary follows from the definition of a majorant scheme and from the fact that in this case $A^+ A^- = A^- A^+ = 0$ and A^+ and A^- are symmetric matrices. Indeed, as A is a symmetric matrix which has different real eigenvalues it is orthogonally similar to the diagonal matrix $\tilde{A} = Q^{-1} A Q$, where Q is an orthogonal matrix. As long as $A = A^+ + A^-$ ($A^+ \geq 0$, $A^- \leq 0$) the matrix $A^+ = Q^{-1} \tilde{A}^+ Q$ and $A^- = Q^{-1} \tilde{A}^- Q$, where $\tilde{A}^+ \geq 0$, $\tilde{A}^- \leq 0$. Then

$$A^+ A^- = Q^{-1} \tilde{A}^+ \tilde{A}^- Q = Q^{-1} \tilde{A}^- \tilde{A}^+ Q = A^- A^+ = 0.$$

Proof. Under the assumptions made the matrix $C_2 \geq 0$ and, consequently,

$$0 \leq B_0 \leq I. \tag{4.9}$$

As the matrices B_α commute and as they are symmetric there exists an orthogonal matrix Q which transforms simultaneously all these matrices in diagonal form: $B_\alpha = Q^{-1} \tilde{B}_\alpha Q$ ($\alpha = -1, 0, 1$), where $\tilde{B}_\alpha$ is a diagonal matrix. From the compatibility condition and the inequality (4.9) it follows that $B_1 + B_{-1} = I - B_0 \geq 0$, i.e. $\tilde{B}_1 + \tilde{B}_{-1} = I - \tilde{B}_0 \geq 0$. As $\tilde{B}_1 \tilde{B}_{-1} = 0$ then $\tilde{B}_1 \geq 0$, $\tilde{B}_{-1} \geq 0$ and $B_1 \geq 0$, $B_{-1} \geq 0$. As $B_1 - B_{-1} = \kappa A$ then we have $\tilde{B}_1 - \tilde{B}_{-1} = \kappa \tilde{A} = \kappa \tilde{A}^+ + \kappa \tilde{A}^-$. Then $\tilde{B}_1 = \kappa \tilde{A}^+$, $\tilde{B}_{-1} = -\kappa \tilde{A}^-$ and, consequently, $B_1 = \kappa A^+$, $B_{-1} = -\kappa A^-$, i.e. the difference scheme is majorant.   □

Let $A = A(x)$. We consider the difference scheme (4.5) for $B_0 = 0$ (Friedrichs' scheme)

$$u^{n+1}(x) = \tfrac{1}{2}[u^n(x + h) + u^n(x - h)] + \tfrac{1}{2}\kappa A[u^n(x + h) - u^n(x - h)]. \tag{4.10}$$

Theorem 4.6. If the system of equations of the first differential approximation of the difference scheme (4.10) is strongly parabolic then the difference scheme is stable.

Proof. The Π-form of the first differential approximation of the difference scheme (4.10) has the form (4.3) where D is the same but

$$C_2 = \frac{h^2}{2\tau}(I - \kappa^2 A^2).$$

The condition of strong parabolicity in the case under consideration means that $C_2 > 0$, i.e. $\kappa^2 A^2 < I$ and, consequently, $|\kappa a_j| < 1$ for all κ and $j = 1, \ldots, m$. Especially $|\kappa a_0| < 1$, where $a_0 = \max_{j, x}\{|a_j(x)|\}$. As shown in [172] the latter inequality is sufficient for the stability of the difference scheme (4.10).

Thus the theorem is proved. $\square$

4.3 Fractional-step Method

For the system of equations (1.1) we consider the difference scheme with splitting

$$u^{n+1}(x) = \Omega_s \Omega_{s-1} \ldots \Omega_1 u^n(x), \tag{4.11}$$

which is equivalent to the following scheme

$$u^{n+(j/s)}(x) = \Omega_j u^{n+(j-1)/s}(x); \quad j = 1, \ldots, s, \tag{4.12}$$

where $u^{n+(j/s)}(x)$ $(j = 1, \ldots, s-1)$ are auxiliary functions. Thus in the jth intermediate step the difference scheme (4.12) is approximating the system of equations.

$$\frac{\partial u}{\partial t} = A_j(x, t)\frac{\partial u}{\partial x_j}; \quad j = 1, \ldots, s. \tag{4.13}$$

Theorem 4.7. Let

$$\Omega_j = \sum_{\alpha=1}^{2} B_\alpha^j T_j^{\tau \lambda_\alpha^j}; \quad j = 1, \ldots, s, \tag{4.14}$$

i.e. let us consider a simple scheme with splitting. If A_j are Lipshitz-continuous symmetric matrices and the first differential approximation of the simple scheme with splitting is a parabolic system of equations, then the scheme is stable.

Proof. The first differential approximation of the simple scheme with splitting has the form

$$
\frac{\partial u}{\partial t} = \sum_{j=1}^{s} D_j \frac{\partial u}{\partial x_j} + \tau \sum_{j>k} A_j \frac{\partial A_k}{\partial x_j} \frac{\partial u}{\partial x_k}
$$

$$
+ \frac{\tau}{2} \sum_{j=1}^{s} \left[\sum_{\alpha=1}^{2} (\lambda_\alpha^j)^2 B_\alpha^j - A_j^2 \right] \frac{\partial^2 u}{\partial x_j^2} + \frac{\tau}{2} \sum_{j>k} (A_j A_k - A_k A_j) \frac{\partial^2 u}{\partial x_j \partial x_k},
$$

where   (4.15)

$$
\sum_{\alpha=1}^{2} B_\alpha^j = I; \quad \sum_{\alpha=1}^{2} \lambda_\alpha^j B_\alpha^j = A_j; \quad j,k = 1,\dots,s.
$$

As the system of equations (4.15) is not totally parabolic it follows that every system of equations of the form

$$
\frac{\partial u}{\partial t} = D_j \frac{\partial u}{\partial x_j} + C_{2j} \frac{\partial^2 u}{\partial x_j^2},
$$ (4.16)

is not totally parabolic, where

$$
C_{2j} = \frac{\tau}{2} \left[\sum_{\alpha=1}^{2} (\lambda_\alpha^j)^2 B_\alpha^j - A_j^2 \right]; \quad j = 1,\dots,s.
$$

As the system of equations (4.16) is not totally parabolic it follows that the matrices

$$
B_1^j = (\lambda_1^j - \lambda_2^j)^{-1} (A_j - \lambda_2^j I),
$$
$$
B_2^j = (\lambda_1^j - \lambda_2^j)^{-1} (-A_j + \lambda_1^j I)
$$

are nonnegative and therefore according to Friedrichs' theorem [8] for the operator

$$
\Omega_j = \sum_{\alpha=1}^{2} B_\alpha^j T_j^{\tau \lambda_\alpha^j}
$$

we have the bound

$$
\|\Omega_j\| = 1 + O(\tau); \quad (j = 1,\dots,s),
$$

and then

$$
\|\Omega\| = \|\Omega_s \Omega_{s-1}\dots\Omega_1\| = 1 + O(\tau).
$$

Therefore, the simple scheme with splitting is stable under the assumptions. The theorem is proved.   □

In an analogous way one can prove the following theorems [72, 73, 103].

· ·

Theorem 4.8. Let us consider the majorant scheme with splitting [30, 34], i.e.

$$
\Omega_j = \kappa_j A_j^+ T_j + (I - \kappa_j |A_j|) E - \kappa_j A_j^- T_{-j},
$$

where

$$A_j = A_j^+ + A_j^-; \quad A_j^+ \geq 0; \quad A_j^- \leq 0;$$

$$|A_j| = A_j^+ - A_j^-; \quad \kappa_j := \frac{\tau}{h_j} = \text{const.}$$

If A_j are symmetric matrices, and if A_j^+, A_j^- are Lipshitz-continuous matrices and if the system of equations of the first differential approximation of the majorant scheme with splitting is not totally parabolic, then the scheme is stable.

. . . .

Theorem 4.9. If the matrix $\sum\limits_{j=1}^{s} \omega_j A_j(x)$ has only different eigenvalues $\mu_1(x, \omega), \ldots, \mu_m(x, \omega)$ for all real $\omega = (\omega_1, \ldots, \omega_s)$, $|\omega|^2 = \sum\limits_{j=1}^{s} \omega_j^2 = 1$ with $|\mu_j(x, \omega) - \mu_k(x, \omega)| > 0$ for $j \neq k$, then the strong parabolicity of the system of equations of the first differential approximation is a sufficient stability condition of the scheme with splitting (4.11), where

$$\Omega_j = \tfrac{1}{2}(T_j + T_{-j}) + \tfrac{1}{2}\kappa\, A_j(T_j - T_{-j}).$$

. . . .

4.4 The Case of Multi-dimensional Schemes

We will approximate the system of equations (1.1) by the following majorant scheme:

$$u^{n+1}(x) = \left(I - \sum_{j=1}^{s} \kappa_j |A_j|\right) u^n(x)$$

$$+ \sum_{j=1}^{s} (\kappa_j A_j^+ T_j - \kappa_j A_j^- T_{-j}) u^n(x). \tag{4.17}$$

The Π-form of the first differential approximation of the scheme (4.17) has the form

$$\frac{\partial u}{\partial t} = \sum_{j=1}^{s} D_j \frac{\partial u}{\partial x_j} - \frac{\tau}{2} \sum_{j \neq k} A_j A_k \frac{\partial^2 u}{\partial x_j \partial x_k}$$

$$+ \sum_{j=1}^{s} \frac{h_j}{2} |A_j|(I - \kappa_j |A_j|) \frac{\partial^2 u}{\partial x_j^2}, \tag{4.18}$$

and the Γ-form of the first differential approximation is given by

$$\frac{\partial^2 u}{\partial t^2} = \sum_{j,k=1}^{s} P_{jk} \frac{\partial^2 u}{\partial x_j \partial x_k} + N, \tag{4.19}$$

where

$$N = \frac{2}{\tau} \left(\sum_{j=1}^{s} A_j \frac{\partial u}{\partial x_j} - \frac{\partial u}{\partial t} \right);$$

$$P_{jj} = \frac{1}{\kappa_j} |A_j|; \qquad P_{jk} = 0; \qquad j \neq k.$$

Together with the systems of equations (1.1) and (4.19) we will consider the following one-dimensional system of equations:

$$\frac{1}{s} \frac{\partial u}{\partial t} = A_j \frac{\partial u}{\partial x_j}; \tag{4.20}$$

$$\frac{1}{s} \frac{\partial^2 u}{\partial t^2} = P_{jj} \frac{\partial^2 u}{\partial x_j^2}; \qquad j = 1, \ldots, s. \tag{4.21}$$

We designate by $[b_l, b_r]$ the region of dependence of a point (x, t) for the system of equations (1.1) and by $[b_l^*, b_r^*]$ the region of dependence of a point (x, t) for (4.19).

. . . .

Theorem 4.10. If the following conditions are satisfied

1) A_j are symmetric matrices,
2) A_j^+, A_j^- are Lipshitz-continuous matrices,
3) The system of equations (4.18) is not totally parabolic, and
4) $[b_l, b_r] \subset [b_l^*, b_r^*]$,

then the majorant scheme (4.17) is stable.

. . . .

Proof. Condition 4) means that the region of dependence of the system of equations (1.1) does not contain the region of dependence of the system of equations (4.19) and, consequently, the region of dependence $[b_{jl}, b_{jr}]$ of the system of equations (4.20) does not contain the region of dependence $[b_{jl}^*, b_{jr}^*]$ of the system of equations (4.21). As the characteristic curves of the system of equations (4.20, 21) the lines with slopes

$$\frac{dx}{dt} = s a^{(j)};$$

$$\frac{dx}{dt} = \pm \sqrt{s} \mu \left[\left(\frac{1}{\kappa_j} |A_j| \right)^{\frac{1}{2}} \right],$$

respectively, in the point (x_j, t) are considered, where $a^{(j)}$ are the eigenvalues of the matrix A_j; $\mu [(|A_j|/\kappa_j)^{1/2}]$ are the eigenvalues of the matrix $(|A_j|/\kappa_j)^{1/2}$. The inclusion $[b_{jl}, b_{jr}] \subset [b_{jl}^*, b_{jr}^*]$ is equivalent to the inequality

$$-\sqrt{\frac{s}{\kappa_j} |a^{(j)}|} \leqq s a^{(j)} \leqq \sqrt{\frac{s}{\kappa_j} |a^{(j)}|},$$

i.e.

$$\kappa_j|a^{(j)}| \le \frac{1}{s} \quad \text{and} \quad \frac{1}{s}I - \kappa_j|A_j| \ge 0, \quad (j = 1,\dots,s).$$

From the last inequality we get

$$I - \sum_{j=1}^{s} \kappa_j|A_j| \ge 0.$$

Thus the coefficients of the difference scheme (4.17) are symmetric, non-negative Lipshitz-continuous matrices and therefore, according to Friedrichs' theorem, the scheme is stable. $\qquad\square$

We consider for the system of equations (1.1) the difference scheme due to Friedrichs [172]

$$u^{n+1}(x) = \sum_{j=1}^{s} \frac{1}{2}\left[\left(\frac{1}{s}I + \kappa_j A_j\right)T_j + \left(\frac{1}{s}I - \kappa_j A_j\right)T_{-j}\right]u^n(x). \tag{4.22}$$

The Γ-form of the first differential approximation of the scheme (4.22) has the form

$$\frac{\partial^2 u}{\partial t^2} = \sum_{j=1}^{s} \frac{1}{s\kappa_j^2}\frac{\partial^2 u}{\partial x_j^2} - \frac{2}{\tau}\left(\frac{\partial u}{\partial t} - \sum_{j=1}^{s} A_j \frac{\partial u}{\partial x_j}\right). \tag{4.23}$$

The regions of dependence of a point (x, t) for the system of equations (1.1) and (4.23) will be denoted by $[b_l, b_r]$ and $[b_l^*, b_r^*]$, respectively.

The following theorem holds.

....

Theorem 4.11. If A_j are Lipshitz-continuous symmetric matrices and $[b_l, b_r] \subset [b_l^*, b_r^*]$, then the difference scheme is stable.

....

The proof of this statement is carried out in an analogous way to the proof of Theorem 4.10.

Let us approximate the system of equations (1.1) by a simple difference scheme [169]

$$u^{n+1}(x) = \sum_{\alpha=1}^{s+1} B_\alpha u^n(x + \tau \lambda_\alpha). \tag{4.24}$$

In this case the compatibility conditions are represented by a system of linear algebraic equations of dimension $s + 1$ of the $s + 1$ unknowns $B_1,\dots, B_{s+1}$. Consequently, the matrices B_α are linear combinations of the matrices I, $A_1,\dots, A_s$.

We will denote by Q'', Q', and Q the regions of dependence of a point (x, t) for the system of equations of the Γ-form of the first differential approximation of the difference scheme (4.24), of the difference scheme (4.24) itself, and of the system of equations (1.1), respectively.

The Γ-form and the Π-form of the first differential approximation of the difference scheme (4.24) have the form (4.19) and (1.3), where

$$D_j = A_j - \frac{\tau}{2}\left(\frac{\partial A_j}{\partial t} + \sum_{k=1}^{s} A_k \frac{\partial A_j}{\partial x_k}\right);$$

$$C_{jk} = \frac{\tau}{2}\left(\sum_{\alpha=1}^{s+1} \lambda_\alpha^j \lambda_\alpha^k B_\alpha - A_j A_k\right) = \frac{\tau}{2}(P_{jk} - A_j A_k);$$

$$P_{jk} := \sum_{\alpha=1}^{s+1} \lambda_\alpha^j \lambda_\alpha^k B_\alpha.$$

. . . .

Theorem 4.12. If the conditions hold

1) A_j are symmetric, Lipshitz-continuous matrices,
2) The system of equations (1.3) ($\alpha = 1, \ldots, s + 1$) are strongly parabolic,
3) $Q'' \subset Q'$,
4) The vectors λ_α do not lie in a $(s - 1)$-dimensional hyperplane,

then the simple difference scheme (4.24) is stable.

. . . .

Proof. To characterize the regions of dependence Q, Q', and Q'' we introduce their support functions. As a support function z of a closed point set Q the function $z(\xi) = \max_{x \in Q} \{x\xi\}$ is defined [173]. Then the inclusion $Q'' \subset Q'$ means that $z''(\xi) \leq z'(\xi)$. It is easy to prove that under the assumptions made, the following inequality holds

$$z(\xi) \leq z''(\xi),$$

i.e. the inequality

$$z(\xi) \leq z'(\xi) \tag{4.25}$$

is true.

If λ_α does not lie in a $(s - 1)$-dimensional hyperplane then–as shown in [173]–the inequality (4.25) can be satisfied only under the condition that the matrices B_α are nonnegative, but then the difference scheme is stable.

The theorem is proved.   □

4.5 Two-level Difference Schemes

For the equation (2.6) we consider the two-level scheme

$$\sum_{\alpha=-q_i^1}^{q_r^1} b_\alpha^1 u^{n+1}(x + \alpha h) = \sum_{\alpha=-q_i^0}^{q_r^0} b_\alpha^0 u^n(x + \alpha h),$$

where

$$\sum_{\alpha=-q_i^1}^{q_r^1} b_\alpha^1 = \sum_{\alpha=-q_i^0}^{q_r^0} b_\alpha^0;$$

q_i^j, q_r^j ($j = 0, 1$) are nonnegative integers. In this case [160] we get:

$$|\varrho|^2 = 1 - 4 z^d \frac{S_0(z)}{S_1(z)}, \tag{4.26}$$

with

$$S_0(z) := \sum_{j=0}^{r_0} \omega_j^0 z^j; \quad S_1(z) := \sum_{j=0}^{r_1} \omega_j^1 z^j;$$

$$\omega_0^0 = S_0(0) \neq 0; \quad \omega_0^1 = 1, \quad z = \sin^2(\xi/2);$$

$$r_1 = q_i^1 + q_r^1; \quad d + r_0 = m; \quad m = \max\{(r_1, q_i^0 + q_r^0)\}.$$

Indeed, we have:

$$|\varrho|^2 = \frac{\left|\sum_{\alpha=-q_i^0}^{q_r^0} b_\alpha^0 e^{i\alpha\xi}\right|^2}{\left|\sum_{\alpha=-q_i^1}^{q_r^1} b_\alpha^1 e^{i\alpha\xi}\right|^2} = \frac{\sum_{\alpha,\beta=-q_i^0}^{q_r^0} b_\alpha^0 b_\beta^0 e^{i(\alpha-\beta)\xi}}{\sum_{\alpha,\beta=-q_i^1}^{q_r^1} b_\alpha^1 b_\beta^1 e^{i(\alpha-\beta)\xi}} = \frac{\sum_{\alpha,\beta=-q_i^0}^{q_r^0} b_\alpha^0 b_\beta^0 \cos[(\alpha-\beta)\xi]}{\sum_{\alpha,\beta=-q_i^1}^{q_r^1} b_\alpha^1 b_\beta^1 \cos[(\alpha-\beta)\xi]}.$$

As

$$\cos(\gamma\xi) = \sum_{j=0}^{\gamma} \delta_j z^j; \quad \delta_0 = 1; \quad \delta_\gamma = (-1)^\gamma 2^{2\gamma-1},$$

we have

$$\sum_{\alpha,\beta=-q_i^0}^{q_r^0} b_\alpha^0 b_\beta^0 \cos[(\alpha-\beta)\xi] = \sum_{j=0}^{q_i^0+q_r^0} \bar\omega_j^0 z^j, \quad \bar\omega_0^0 = 1;$$

$$\bar\omega_{q_i^0+q_r^0}^0 = (-1)^{q_i^0+q_r^0} 2^{2(q_i^0+q_r^0)-1} (b_{-q_i^0}^0 b_{q_r^0}^0 + b_{q_r^0}^0 b_{-q_i^0}^0)$$
$$= (-1)^{q_i^0+q_r^0} 2^{2(q_i^0+q_r^0)} b_{-q_i^0}^0 b_{q_r^0}^0;$$

$$\sum_{\alpha,\beta=-q_i^1}^{q_r^1} b_\alpha^1 b_\beta^1 \cos[(\alpha-\beta)\xi] = \sum_{j=0}^{q_i^1+q_r^1} \omega_j^1 z^j = \sum_{j=0}^{r_1} \omega_j^1 z^j;$$

$$\omega_0^1 = 1; \quad r_1 = q_i^1 + q_r^1.$$

Consequently, we get

$$|\varrho|^2 - \frac{\sum_{j=0}^{q_i^0+q_r^0} \bar\omega_j^0 z^j}{\sum_{j=0}^{r_1} \omega_j^1 z^j} = 1 + \frac{\sum_{j=0}^{q_i^0+q_r^0} \bar\omega_j^0 z^j - \sum_{j=0}^{r_1} \omega_j^1 z^j}{\sum_{j=0}^{r_1} \omega_j^1 z^j} = 1 + \frac{\sum_{j=0}^{m} (\bar\omega_j^0 - \omega_j^1) z^j}{S_1(z)},$$

with

$$S_1(z) = \sum_{j=0}^{r_1} \omega_j^1 z^j; \quad m = \max\{(q_i^0 + q_r^0, r_1)\}.$$

If for $j < d$ $\bar\omega_j^0 = \omega_j^1$, and for $j = d$ $\bar\omega_j^0 \neq \omega_j^1$, then with

$$r_0 = m - d,$$
$$-4\omega_j^0 = \bar\omega_{j+d}^0 - \omega_{j+d}^1; \quad j = 1, 2, \ldots, r_0$$

we get

$$|\varrho|^2 = 1 + \frac{\sum\limits_{j=d}^{m} (\bar{\omega}_j^0 - \omega_j^1) z^j}{S_1(z)} = 1 - 4 z^d \frac{\sum\limits_{j=0}^{r_1} \omega_j^0 z^j}{S_1(z)} = 1 - 4 z^d \frac{S_0(z)}{S_1(z)}.$$

Thus we have shown that the relation (4.26) holds.

It is clear that an analogous formula can be found also in connection with the amplification matrix of the scheme. Then it is necessary to follow the order of matrix multiplication. When the coefficients of the scheme are matrices which commute the formula can be written down without any change.

For a single equation (2.6) (or, equivalently, when the coefficients b_α^0, b_α^1 are matrix functions of the matrix a) for the stability of the difference scheme it is necessary and sufficient that

$$|\varrho|^2 \leqq 1,$$

i.e.

$$S_0(0) > 0; \quad S_0(z) \geqq 0 \quad \text{for} \quad 0 < z \leqq 1 \quad (0 < \xi \leqq \pi). \tag{4.27}$$

As

$$|\varrho|^2 = \exp\left[2 \sum_{j=d}^{\infty} (-1)^j c_{2j} \frac{\tau}{h^{2j}} \xi^{2j}\right], \tag{4.28}$$

where c_{2j} are the coefficients of the differential representation, c_{2d} the first coefficient of the c_{2j} which is not zero, then by comparison of (4.26) and (4.28) we get

$$- 4 \frac{\xi^{2d}}{2^{2d}} S_0(0) = 2(-1)^d c_{2d} \frac{\tau}{h^{2d}} \xi^{2d},$$

i.e.

$$\frac{\tau}{h^{2d}} c_{2d} = (-1)^{d-1} 2^{1-2d} S_0(0).$$

From this we get

$$S_0(0) = (-1)^{d-1} 2^{2d-1} \frac{\tau}{h^{2d}} c_{2d}. \tag{4.29}$$

As already shown (see Sect. 3.3), the inequality

$$(-1)^{d-1} c_{2d} > 0 \tag{4.30}$$

is necessary for the stability and by virtue of (4.29) it is equivalent to the inequality

$$S_0(0) > 0. \tag{4.31}$$

It is not difficult to see that the following lemma holds.

Lemma. If the difference scheme which approximates equation (2.6) is such that its amplification factor is represented by (4.26) with $S_0(z) = \text{const}$, then the inequality (4.30) is a necessary and sufficient condition for the stability of the scheme.

In the case of a three-point difference scheme, e.g.,

$$\frac{\Delta_0 u^n(x)}{\tau} = a \frac{\Delta_1 + \Delta_{-1}}{2h} u^n(x) + \frac{\lambda h^2}{2\tau} \frac{\Delta_1 \Delta_{-1}}{h^2} u^n(x),$$

$S_0(z) = \text{const}$ for $\lambda^2 = \kappa^2 a^2$ ("corner"-scheme), and $\lambda = \kappa^2 a^2$ (Lax-Wendroff-scheme). Here the conditions $c_2 > 0$ and $c_4 < 0$, respectively, are necessary and sufficient conditions of the schemes above.

When $S_0(z)$ is a linear function of z:

$$S_0(z) = S_0(0) + \alpha z,$$

the necessary and sufficient condition has the form

$$S_0(z) > 0; \quad S_0(1) > 0,$$

i.e.

$$(-1)^{d-1} c_{2d} > 0;$$

$$(-1)^d c_{2d+2} + \frac{3+d}{12} h^2 (-1)^{d-1} c_{2d} - \tau \delta_d c_{2d}^2 > 0.$$

4.6 Remarks on Nonlinear Equations

The theorems formulated above on the connection of the stability of difference schemes and properties of their differential approximations are confined to the case of constant and variable coefficients. For nonlinear equations there are not such strong statements although at the moment the method of differential approximations has found a wide application in the stability analysis of difference schemes for nonlinear equations. Calculations have underlined that such a generalization is possible. Later a stability analysis will be given for difference schemes for nonlinear equations by the method of the differential approximation and the benefit of the given method investigating nonlinear effects will be shown. Besides that, calculations of problems of gas dynamics with difference schemes are carried out the stability of which was investigated only by the method of the differential approximation. In the paper of N. N. Anuchina, V. E. Petrenko, Yu. I. Shokin, N. N. Yanenko [95] this method was used for the analysis of difference schemes of PIC-type.

In the nonlinear case the method of differential approximation leads to bounds which depend on the gradients of the solution. Such bounds show the instability of a number of difference schemes which are encountered in calculations, but which can not be detected by a local Fourier analysis because the latter does not take into account the gradients [19–41, 47–54, 61–66, 114–160].

5. Dissipative Difference Schemes for Hyperbolic Equations

5.1 Different Definitions of Dissipativity

We consider the difference scheme

$$u^{n+1}(x) = \sum_{\alpha} B_{\alpha} u^n(x + \alpha h) \tag{5.1}$$

for the hyperbolic system of differential equations (1.2) with constant coefficients. Here B_{α} are constant $(m \times m)$-matrices.

Let the difference scheme (5.1) have the order of approximation p. Then

$$\sum_{\alpha} \alpha^j B_{\alpha} = \kappa^j A^j; \quad (j = 0, 1, \ldots, p),$$

and the Π-form of the differential representation of the scheme (5.1) has the form

$$\frac{\partial u}{\partial t} = A \frac{\partial u}{\partial x} + \sum_{j=p+1}^{\infty} C_j \frac{\partial^j u}{\partial x^j},$$

where C_j are matrices for which the algorithm to compute them was explained in I.2.

Definition. We will call the difference scheme (5.1) dissipative of order $2d$ in the sense of Kreiss [174]–where d is a nonnegative integer–if there exists such a positive number $\delta > 0$, that for all eigenvalues ϱ_j $(j = 1, \ldots, m)$ of the amplification matrix

$$G = \sum_{\alpha} B_{\alpha} e^{i\alpha\xi}, \quad \xi := kh$$

of the difference scheme the estimate holds:

$$|\varrho_j| \leqq 1 - \delta |\xi|^{2d}; \quad |\xi| \leqq \pi. \tag{5.2}$$

. . . .

Theorem. If the difference scheme (5.1) has the order of approximation $p = 2d - 1$ and if it is dissipative of order $2d$ in the sense of Kreiss, then the scheme is stable [174].

. . . .

In the publication [175] this result was generalized with respect to difference schemes, which have the order of approximation $p = 2d$ and the order of dissipation $2d + 2$.

Definition. We will call the difference scheme (5.1) dissipative in the sense of Roshdestvenskii-Yanenko-Richtmyer (R-Y-R) (see e.g. [8, 176]), if the eigenvalues of the amplification matrix satisfy the following conditions:

$$|\varrho_j| < 1 \quad \text{for} \quad \xi \neq 0; \quad (j = 1, \ldots, m). \tag{5.3}$$

Definition. The difference scheme (5.1) is called dissipative in the generalized sense of order $(2d, 2d + 2r)$–where d and r are positive integers–if there exists such a constant $\delta > 0$, that for all ξ, $|\xi| \leq \pi$, the following inequalities hold:

$$|\varrho_j| \leq 1 - \delta|\xi|^{2d}; \quad (j = 1, \ldots, l - 1, l + 1, \ldots, m);$$
$$|\varrho_l| \leq 1 - \delta|\xi|^{2d + 2r}. \tag{5.4}$$

5.2 Stability Theorem for Dissipative Schemes in the Generalized Sense

. . . .

Theorem 5.1. If the difference scheme (5.1) has the order of approximation $p = 2d - 1$ and is dissipative in the generalized sense of order $(2d, 2d + 2r)$, then it is stable.

. . . .

Proof. The proof is analogous to the proof of the theorem in the publication [174]. As the difference scheme (5.1) has the order of approximation $p = 2d - 1$, the amplification matrix has the form

$$G = I + \sum_{j=1}^{2d-1} \frac{1}{j!}(i\xi\kappa A)^j + O(|\xi|^{2d}) = I + G_1 + O(|\xi|^{2d}).$$

In accordance with out assumptions there exists a nonsingular matrix Q which transforms the matrix A in diagonal form such that

$$\max\{(|Q|, |Q^{-1}|)\} \leq N \tag{5.5}$$

and, consequently,

$$Q(I + G_1)Q^{-1} = \begin{pmatrix} v_1 & 0 & \cdots & 0 \\ 0 & v_2 & \cdots & 0 \\ \vdots & \vdots & \ddots & \vdots \\ 0 & 0 & \cdots & v_m \end{pmatrix}. \tag{5.6}$$

From (5.5) and (5.6) it follows that without loss of generality the matrix G_1 can be assumed to have diagonal form. Using the estimate of Gershgorin [177] for the eigenvalues of a matrix we can assume that the eigenvalues of the matrix G are such, that

$$|\varrho_j - \nu_j| \leq \text{const } |\xi|^{2d}.$$

Therefore,

$$
zG - I = \begin{pmatrix}
z\varrho_1 + zq_1 - 1 & zq_{12} & \cdots & zq_{1m} \\
zq_{21} & z\varrho_2 + zq_2 - 1 & \cdots & zq_{2m} \\
\vdots & \vdots & & \vdots \\
zq_{m1} & zq_{m2} & \cdots & z\varrho_m + zq_m - 1
\end{pmatrix},
$$

where

$$
\begin{aligned}
&|z| < 1, \quad |q_j| + |q_{j\beta}| \leq \text{const} \cdot |\xi|^{2d}, \\
&j \neq \beta; \quad (j, \beta = 1, \ldots, m).
\end{aligned}
\tag{5.7}
$$

Let $(zG - I) = \| a_{\alpha\beta} \|_1^m$ and $(zG - I)_{\alpha\beta}$ be a matrix, which is derived from matrix $zG - I$ by cancelling the row with number α and the line of number β. Then taking into account (5.7), we have

$$
|a_{\alpha\beta}| = \left| \frac{\det \{ (zG - I)_{\alpha\beta} \}}{\det \{ (zG - I) \}} \right| = \frac{|\det \{ (zG - I)_{\alpha\beta} \}|}{\prod\limits_{j=1}^{m} |z\varrho_j - 1|}.
\tag{5.8}
$$

We remove the denominator estimating those sums which can not contain multipliers of the form $(z\varrho_j - 1)$ by $\prod |z||q_{js}| \leq \text{const} \cdot |\xi|^{2d(m-1)}$ using the conditions (5.7). We put together terms containing $(z\varrho_j - 1)$ as a factor in order to form a separate sum which is estimated as before with coefficients $|z||q_{js}|$. Altogether we get

$$
|a_{\alpha\beta}| \leq \text{const} \cdot \frac{|\xi|^{2d(m-1)} + \sum\limits_{s=1}^{m-1} \sum\limits_{j_v} \left(\prod\limits_{v=1}^{s} |z\varrho_{j_v} - 1| \right) |\xi|^{2d(m-1-s)}}{\prod\limits_{j=1}^{m} |z\varrho_j - 1|}.
$$

According to the assumptions of the theorem we have

$$
\begin{aligned}
&|z\varrho_j - 1| \geq 1 - |z||\varrho_j| \geq \delta |\xi|^{2d}; \quad j \neq l; \\
&|z\varrho_l - 1| \geq 1 - |z||\varrho_l| \geq \delta |\xi|^{2d+2r}; \\
&|z\varrho_j - 1| \geq 1 - |z|, \quad (j = 1, \ldots, m).
\end{aligned}
$$

Then the quantity $|\xi|^{2d(m-1)} \left(\prod\limits_{j=1}^{m} |z\varrho_j - 1| \right)^{-1}$ is estimated by

$$
\begin{aligned}
|\xi|^{2d(m-1)} \left(\prod\limits_{j=1}^{m} |z\varrho_j - 1| \right)^{-1} &= |\xi|^{2d(m-1)} \left(|z\varrho_l - 1| \prod\limits_{\substack{j=1 \\ j \neq l}}^{m} |z\varrho_j - 1| \right)^{-1} \\
&\leq |\xi|^{2d(m-1)} [(1 - |z|) \delta^{m-1} |\xi|^{2d(m-1)}]^{-1} \\
&= [(1 - |z|) \delta^{m-1}]^{-1}.
\end{aligned}
$$

Estimating the rest of the sums two cases are to be distinguished:

Case 1: The factor with the index l is encountered also in the numerator. Then we cancel it with the same term in the denominator, and the same is done with the other common factors, one of the factors is estimated by $|z\varrho_j - 1| \geq 1 - |z|$ and the others by $|z\varrho_j - 1| \geq \delta |\xi|^{2d}$. Altogether we get

$$\frac{\prod\limits_{v=1}^{s} |\varrho_{j_v} - 1||\xi|^{2d(m-1-s)}}{\sum\limits_{j=1}^{m} |z\varrho_j - 1|} \leq \frac{|\xi|^{2d(m-1-s)}}{(1-|z|)(\delta|\xi|^{2d})^{m-1-s}}$$

$$\leq \delta^{1+s-m}(1-|z|)^{-1}. \tag{5.9}$$

Case 2: If the factor $(z\varrho_l - 1)$ is not contained in the numerator, we cancel all common factors in the numerator and denominator, and in the denominator the quantity $|z\varrho_l - 1|$ is estimated from below $|z\varrho_l - 1| \geq 1 - |z|$ and the rest of the factors is estimated by δ and ξ. Altogether we get again the estimate (5.9). Finally, we get

$$|a_{\alpha\beta}| \leq \text{const} \cdot \left[(1-|z|)^{-1}\left(\delta^{1-m} + \sum_{j=1}^{m-1} C_{m-1}^j \delta^{j+1-m}\right)\right]$$

$$\leq \text{const} \cdot \delta^{1-m}(1-|z|)^{-1}.$$

Consequently, this inequality is true also for the Euclidean norm of the matrix G.

By virtue of Kreiss' theorem on matrices [178] this inequality is equivalent to the inequality $|G^k| \leq M = \text{const}$, i.e. the difference scheme (5.1) is stable.

$\square$

An analogous statement can be proved in the case, when the difference scheme has the order of approximation $p = 2d$ and when it is dissipative in the generalized sense of order $(2d + 2, 2d + 2 + 2r)$. Also for the multidimensional case the statement can be proved.

5.3 Stability Theorem for Dissipative Schemes in the Sense of Roshdestvenskii-Yanenko-Richtmyer

For difference schemes which are dissipative in the sense of R-Y-R the connection between properties of the first differential approximation and the stability of the scheme can be established. The following theorem holds:

· ·

Theorem 5.2. If B_α are symmetric matrices, the difference scheme (5.1) has the order of approximation $p = 2d - 1$ and if it is dissipative in the sense of R-Y-R, and if the first differential approximation of the difference scheme is parabolic in the strict sense, then the difference scheme is stable.

· ·

Proof. The first differential approximation of the difference scheme (5.1) has in the case under consideration the form

$$\frac{\partial u}{\partial t} = A \frac{\partial u}{\partial x} + C_{2d} \frac{\partial^{2d} u}{\partial x^{2d}},$$

where

$$C_{2d} = \frac{h^{2d}}{\tau(2\,d)!} \tilde{C}_{2d} = \frac{h^{2d}}{\tau(2\,d)!} \left[\sum_\alpha \alpha^{2d} B_\alpha - \kappa^{2d} A^{2d} \right].$$

The parabolicity in the strict sense of the first differential approximation means that

$$(-1)^{d+1} C_{2d} > 0.$$

Besides that in the given case we get

$$G^* G = I + \frac{2(-1)^d \xi^{2d}}{(2\,d)!} \tilde{C}_{2d} + O(|\xi|^{2d+2}).$$

Then the difference scheme is stable based on the theorem due to Lax [179]. $\square$

An analogous statement holds also for difference schemes of even order of approximation.

5.4 Stability Theorem for a Partly Dissipative Scheme

We consider the case where the eigenvalues ϱ_j of the amplification matrix G satisfy the following conditions:

$$|\varrho_j| \leq 1 - \delta |\xi|^{2d}, \quad (j \neq l);$$
$$|\varrho_l| = 1, \quad (j = 1, \ldots, l-1, l+1, \ldots, m) \tag{5.10}$$

(partly dissipative difference scheme).

The following theorem holds:

· ·

Theorem 5.3. If the difference scheme (5.1) has the order of approximation $p = 2\,d - 1$ and the eigenvalues ϱ_j of the amplification matrix satisfy the conditions (5.10), then the difference scheme is stable.

· ·

The proof follows the line of the proof of Theorem (5.1), because all inequalities which have been used there also hold in the present case, especially

$$|z\varrho_j - 1| \geq 1 - |z||\varrho_j| \geq \delta |\xi|^{2d}; \quad (j \neq l);$$
$$|z\varrho_j - 1| \geq 1 - |z|; \quad (j = 1, \ldots, m).$$

A similar statement also holds for difference schemes of even order of approximation.

6. A Means for the Construction of Difference Schemes with Higher Order of Approximation

6.1 Convergence Theorem

In recent years difference schemes of higher order of approximation were used universally in practical calculations. We will describe a means for the construction of the above mentioned difference schemes using a sequence of differential relations of the original equations which consists in the simultaneous solution of the original differential equation and the equations for the derivatives of the solution by means of the same difference schemes which are used also for the original differential equations [59, 60]. Thus, such properties of the original difference scheme as conservation and homogeneity are preserved. An analogous way of treatment was proposed in [180].

We consider Cauchy's problem

$$\frac{\partial u}{\partial t} = Lu, \quad t > 0, \ x \in \mathbb{R}_1, \tag{6.1}$$

$$u(x, 0) = u_0(x), \tag{6.2}$$

where $L = L(\partial/\partial x)$ is a differential operator with constant coefficients; $u_0(x)$ is a sufficiently smooth function. We will assume that the problem (6.1, 2) is correct in a Banach-space $\mathbb{B}$.

The equation (6.1) is approximated by a difference scheme of the form

$$w^{n+1}(x) = S_h w^n(x) \tag{6.3}$$

of first order of approximation. Here $S_h = S_h(T_1)$ is the step operator of the scheme; $w^0(x) = u_0(x)$; $\tau/h^q = \text{const}$; q is the order of the operator L.

The differential representation of the difference scheme has the form

$$\frac{\partial u}{\partial t} = Lu + \tau R_1 u + \tau^2 R_2 u + \dots,$$

where $R_i = R_i(\partial/\partial x)$ is some differential operator.

Let the function $f''(x)$ approximate the term $R_1 u(n\tau, x)$ with an order of accuracy $O(\tau)$. Then the difference scheme

$$u^{n+1}(x) = S_h u^n(x) - \tau^2 f^n(x), \tag{6.4}$$

$$u^0(x) = u_0(x) \tag{6.5}$$

is approximating the equation (6.1) with a higher order of accuracy then the difference scheme (6.3). As L is an operator with constant coefficients, all derivatives $r_{(j)} := \partial^j u/\partial x^j$ satisfy the equation (6.1)

$$\frac{\partial r_{(j)}}{\partial t} = L r_{(j)}.$$

Therefore the function $f''(x)$ can be found with a difference scheme of type (6.3)

$$f^{n+1}(x) = S_h f^n(x), \tag{6.6}$$

$$f^0(x) = \tilde{A} u^0(x), \tag{6.7}$$

where $\tilde{A}$ is a difference approximation of the operator R_1 of order $O(\tau)$.

We denote by $u^n(x)$ and $f^n(x)$ the solutions of the difference schemes (6.4, 5) and (6.6, 7), respectively, and by $u(n\tau, x)$ the exact solution of (6.1, 2) in the moment $t = n\tau$.

The following theorem holds.

. . . .

Theorem 6.1.
1) The problem (6.1, 2) may be approximated by the difference scheme (6.3) with first order, which is stable in $\mathbb{B}$;
2) The step-operator of problem (6.1, 2) may commute with the operator S_h;
3) The derivatives of the solution up to some order may be smooth solutions of equation (6.1).

Then $\| u^n(x) - u(n\tau, x) \|_{\mathbb{B}} = O(\tau^2)$.

. . . .

Proof. Because of the second condition of the theorem we get

$$u^n(x) - u(n\tau, x) = n S_h^{n-1} [S_h - C(\tau) - \tau^2 \tilde{A}] u^0(x)$$

$$- \sum_{j=0}^{n-1} S_h^j \sum_{\alpha=0}^{n-j-2} S_h^\alpha [S_h - C(\tau)]^2 C[(n-j-\alpha-2)\tau] u^0(x),$$

where $C(\tau)$ is the step-operator of equation (6.1). Then

$$\| u^n(x) - u(n\tau, x) \| \leq n \| S_h^{n-1} \| \, \| [S_h - C(\tau) - \tau^2 \tilde{A}] u^0(x) \|$$

$$+ \sum_{j=0}^{n-1} \| S_h^j \| \sum_{\alpha=0}^{n-j-2} \| S_h^\alpha \| \, \| [S_h - C(\tau)]^2 C[(n-j-\alpha-2)\tau] u^0(x) \|.$$

The first sum on the right side of the last inequality is of order $n K_1 O(\tau^3)$ because of the choice of the operator $\tau^2 \tilde{A}$ and the first condition of the theorem.

According to the third condition of the theorem the function which is defined by the remainder of the Taylor-series which is gained by expanding the quantity $[S_h - C(\tau)]u(n\tau, x)$ in powers of τ and h belongs to the region where the operator $C(\tau)$ is defined and is a smooth function. As that function is of the form $\tau_2 R(\partial/\partial x)u(n\tau, x)$ the second sum is estimated by $n^2 K_2 O(\tau^4)$ $= O(\tau^2)$. $\qquad\qquad\qquad\qquad\qquad\qquad\qquad\qquad\qquad\qquad\qquad$ □

6.2 A Weakly Stable Difference Scheme

In the case when the difference scheme (6.4) can be written in the form

$$u^{n+1}(x) = \Phi_n u^n(x)$$

we get a usual difference scheme, which is weakly stable in the sense of Strang [181].

Example. Let us take for the equation

$$\frac{\partial u}{\partial t} = a \frac{\partial u}{\partial x}, \qquad a = \text{const} > 0 \tag{6.8}$$

as a first choice the Lax' scheme, i.e.

$$S_h = \frac{\kappa a + 1}{2} T_1 - \frac{\kappa a - 1}{2} T_{-1},$$

which is stable for $\kappa a \leq 1$.

The first differential approximation has the form:

$$\frac{\partial u}{\partial t} = a \frac{\partial u}{\partial x} + \frac{h^2}{2\tau}(1 - \kappa^2 a^2)\frac{\partial^2 u}{\partial x^2}.$$

We consider the following difference scheme of second order of approximation:

$$u^{n+1}(x) = S_h^{n+1} u^0(x) - \frac{n+1}{2}(1 - \kappa^2 a^2) S_h^{n+1} \Delta_1 \Delta_{-1} u^0(x), \tag{6.9}$$

which can be represented by

$$u^{n+1}(x) = S_h^{n+1} R_{n+1} u^0(x),$$

where

$$R_{n+1} := E - \frac{n+1}{2}(1 - \kappa^2 a^2) \Delta_1 \Delta_{-1}.$$

As

$$(R_n v, v) \geq (v, v) \qquad \text{for all } n = 0, 1, 2, \ldots,$$

the operators R_n can be inverted and the difference scheme (6.9) can be written in the following way:

$$u^{n+1}(x) = \Phi_n u^n(x),$$

where

$$\Phi_n(x) := S_h R_{n+1} R_n^{-1}.$$

The amplification factor is given by

$$\varrho_n = \varrho_{S_h} \varrho_{R_{n+1}} \varrho_{R_n}^{-1}.$$

Here ϱ_{S_h}, ϱ_{R_n} are the spectral representations of the operators S_h and R_n, respectively. Then

$$|\varrho_n|^2 = |\varrho_{S_h}|^2 \frac{|\varrho_{R_{n+1}}|^2}{|\varrho_{R_n}|^2}$$
$$= [1 - (1 - \kappa^2 a^2)\sin^2 \xi] \cdot \left[1 + \frac{2(1 - \kappa^2 a^2)\sin^2(\xi/2)}{1 + 2n(1 - \kappa^2 a^2)\sin^2(\xi/2)}\right]^2,$$

and the difference scheme (6.9) is weakly stable in $\mathbb{L}_2$.

6.3 Construction of a Third-order Difference Scheme

We will show how one can derive a difference scheme of third order of approximation for the equation (6.8) starting with the scheme of first order of approximation.

The second differential approximation of the difference scheme of first order of approximation has the form

$$\frac{\partial u}{\partial t} = a \frac{\partial u}{\partial x} + \tau c_2 \frac{\partial^2 u}{\partial x^2} + \tau^2 c_3 \frac{\partial^3 u}{\partial x^3},$$

where $c_2 = O(1)$, $c_3 = O(1)$.

For the construction of a difference scheme of third order of approximation it is necessary to know $f_2^n(x)$ and $f_3^n(x)$ which approximate $\partial^2 u/\partial x^2$ and $\partial^3 u/\partial x^3$ with an accuracy $O(\tau^2)$ and $O(\tau)$, respectively. They can be found with the difference scheme

$$f_2^{n+1}(x) = S_h f_2^n(x) - \tau^2 c_2 S_h^n f_2^0(x),$$

$$f_2^0(x) = \frac{\Delta_1 \Delta_{-1}}{h^2} u^0(x);$$

$$f_3^{n+1}(x) = S_h f_3^n(x),$$

$$f_3^0(x) = \frac{\Delta_1 \Delta_{-1}}{2h^3}(\Delta_1 + \Delta_{-1})u^0(x).$$

Then $u^n(x)$ can be found by the formula

$$u^{n+1}(x) = S_h u^n(x) - \tau^2 c_2 f_2^n(x) - \tau^3 c_3 f_3^n(x),$$

$$u^0(x) = u(x,0).$$

(6.10)

. . . .

Theorem 6.2. The equation (6.8) may be approximated by the difference scheme (6.10) with first order of approximation, which is stable in $\mathbb{B}$. If the conditions of Theorem 6.1 are satisfied then

$$\| u^n(x) - u(n\tau, x) \|_{\mathbb{B}} = O(\tau^3).$$

. . . .

Proof. The difference

$$u^n(x) - u(n\tau, x)$$

$$= \sum_{j=0}^{n-1} S_h^j \sum_{\alpha=0}^{n-j-2} S_h^\alpha \sum_{\beta=0}^{n-j-\alpha-3} (S_h - C)^3 C [(n-j-\alpha-\beta-3)\tau] u^0(x)$$

$$+ n S_h^{n-1} [S_h - C - \kappa^2 c_2 \Delta_1 \Delta_{-1} - \tfrac{1}{2}\kappa^3 c_3 \Delta_1 \Delta_{-1}(\Delta_1 + \Delta_{-1})] u^0(x)$$

$$- n(n-1) S_h^{n-2} (S_h - C)(S_h - C - \kappa^2 c_2 \Delta_1 \Delta_{-1}) u^0(x)$$

$$+ \frac{n(n-1)}{2} S_h^{n-2} (S_h - C - \kappa^2 c_2 \Delta_1 \Delta_{-1})^2 u^0(x).$$

is of order $O(\tau^3)$, when the corresponding derivatives of the solution belong to the region of the operator C and have the smoothness which is necessary. $\square$

6.4 Application to Nonlinear Equations

The basic idea of the method considered can be applied to the case of nonlinear equations. We consider the equation

$$\frac{\partial u}{\partial t} + u \frac{\partial u}{\partial x} = 0, \quad 0 < t \leqq t_0, \quad x \in \mathbb{R}_1,$$

$$u(x,0) = u_0(x),$$

(6.11)

where $u_0(x)$ is sufficiently smooth. The differential sequences of equation (6.11) do not contain equations of the same kind. The first, second, third, and fourth derivative of the solution with respect to x satisfy the following equations, respectively,

$$\frac{\partial r_{(1)}}{\partial t} + u \frac{\partial r_{(1)}}{\partial x} + r_{(1)}^2 = 0;$$

(6.12)

$$\frac{\partial r_{(2)}}{\partial t} + u \frac{\partial r_{(2)}}{\partial x} + 3 r_{(1)} r_{(2)} = 0; \quad \Big\} \tag{6.13}$$

$$\frac{\partial r_{(3)}}{\partial t} + u \frac{\partial r_{(3)}}{\partial x} + 4 r_{(1)} r_{(3)} + 3 r_{(2)}^2 = 0;$$

$$\frac{\partial r_{(4)}}{\partial t} + u \frac{\partial r_{(4)}}{\partial x} + 5 r_{(1)} r_{(4)} + 10 r_{(2)} r_{(3)} = 0.$$

We consider a difference scheme of first order of approximation

$$u^{n+1}(x) = S_h u^n(x), \tag{6.14}$$

where

$$S_h := E - \frac{\kappa}{2} [(u^n(x) - |u^n(x)|) T_1 + 2 |u^n(x)| E - (u^n(x) + |u^n(x)|) T_{-1}],$$

which is stable for $\kappa a \leq 1$. The first differential approximation of this scheme is

$$\frac{\partial u}{\partial t} + u \frac{\partial u}{\partial x} = \frac{h}{2} \left[|u| (1 - \kappa |u|) \frac{\partial^2 u}{\partial x^2} - 2 \kappa u \left(\frac{\partial u}{\partial x} \right)^2 \right].$$

Consequently, the difference scheme

$$u^{n+1}(x) = S_h u^n(x) + \tau g^{n+1}(x),$$

where $g^{n+1}(x)$ approximates the quantity

$$-f(x) = -\frac{h}{2} [|u| (1 - \kappa |u|) r_{(2)} - 2 \kappa u r_{(1)}^2],$$

has second order of approximation. To establish this it is sufficient to know the quantities $r_{(1)}$ and $r_{(2)}$ with an accuracy $O(\tau)$. The quantities $r_{(1)}$ and $r_{(2)}$ satisfy the equations (6.12, 13), respectively, and for their solution one can use the difference scheme (6.14).

6.5 Application of the Method to a Boundary Value Problem

We will show the application of the given method for the construction of schemes of higher order of approximation for boundary value problems

$$u(x, 0) = \phi(x), \quad 0 \leq x < +\infty.$$

We consider the boundary value problem for equation (6.8) and we approximate that equation by the difference scheme

$$u^{n+1}(x) = S_h u^n(x) - \tau c_2 r_{(2)}^{n+1}(x), \tag{6.15}$$

where

$$S_h := E + \frac{\kappa a}{2}(T_1 - T_{-1}) + \frac{\lambda}{2}\Delta_1 \Delta_{-1}, \tag{6.16}$$

$$\lambda^2 \leq \kappa^2 a^2 \leq \lambda.$$

The given scheme can be written in the form

$$R_n u^{n+1}(x) = S_h R_{n+1} u^n(x),$$

where

$$R_n := E - \frac{n}{2}(\lambda - \kappa^2 a^2)\Delta_1 \Delta_{-1}.$$

For its realization it is necessary to impose boundary conditions beyond the boundary:

$$u^n_{-1} = g^n_{-1}, \qquad u^n_{-2} = g^n_{-2}.$$

Here g^n_{-1}, g^n_{-2} satisfy the single estimate

$$\sup\{(|g^n_{-1}|, |g^n_{-2}|)\} \leq N_1 = \text{const.}$$

· ·

Theorem 6.3. If $\phi(x) \in \mathbb{C}_3(0, \infty)$ and $\phi(x)$ tends to zero for $x > N_2$, then there are constants N_3, N_4, and $\gamma > 0$ such that the solution of the difference problem can be written in the form

$$u^n_j = \bar{u}^n_j + \bar{\bar{u}}^n_j;$$

$$u^n_j = u^n(jh),$$

while the following estimate holds

$$\|\bar{u}^n_j - u(x, t)\|_0 \leq n N_3 h^2; \qquad \|f\|_0^2 = h \sum_{j=0}^{\infty} |f_j|^2,$$

$$|\bar{\bar{u}}^n_j| \leq N_4 (N_1 + \max_x \{|\phi(x)|\}) e^{-j\gamma}.$$

· ·

That this statement is true follows from the theorems which are proved in [182, 183] and from the fact, that for $\lambda^2 \leq \kappa^2 a^2 < \lambda$ the difference scheme (6.15, 16) is dissipative of order 4 (dissipative in the sense of Kreiss [174], see Chap. 5).

6.6 Stability Theorems for Dissipative Schemes

As a starting difference scheme one can choose a difference scheme of higher order of approximation and on its basis one can construct a difference scheme of still higher order of approximation.

....

Theorem 6.4. Let the difference scheme

$$u^{n+1}(x) = \sum_\alpha b_\alpha T_1^\alpha u^n(x) = S_h u^n(x)$$

be dissipative of order $2d$ and represent an approximation of the equation (6.8) with the order $2d - 1$, then the difference scheme

$$u^{n+1}(x) = S_h u^n(x) - \frac{1}{(2d)!} \bar{C}_{2d} S_h^{n+1} (\varDelta_1 \varDelta_{-1})^d u^0(x), \tag{6.17}$$

where

$$\bar{C}_{2d} = \sum_\alpha \alpha^{2d} b_\alpha - (\kappa a)^{2d},$$

is stable and has the order of approximation $2d$.

....

Proof. The difference scheme (6.17) according to the construction has the order of approximation $2d$. We will prove the stability. For this reason we write the scheme (6.17) in the following form:

$$u^{n+1}(x) = S_h^{n+1} \left[E + \frac{(-1)^{d+1}(n+1)}{(2d)!} \bar{C}_{2d}(-\varDelta_1 \varDelta_{-1})^d \right] u^0(x)$$

$$= S_h^{n+1} R_{n+1} u^0(x).$$

As a necessary stability condition of the operator S_h the following inequality must be satisfied:

$$(-1)^{d+1} \bar{C}_{2d} \geqq 0$$

(correctness of the first differential approximation, see I.3), and as the operator $(-\varDelta_1 \varDelta_{-1})^d$ is positive definit, the operator R_{n+1} has an inverse and the difference scheme (6.17) can be written in the form

$$u^{n+1}(x) = S_h R_{n+1} R_n^{-1} u^n(x).$$

We will show that the norms of the transition operator

$$C_{m+n, m} = S_h^n R_{m+n} R_m^{-1}$$

are bounded. Indeed, the spectral representation of the operator $C_{m+n, m}$ has the form

$$\varrho C_{m+n, m} = \varrho_{S_h}^n \left[1 + n \frac{4^d \dfrac{(-1)^{d+1}}{(2d)!} \bar{C}_{2d} \sin^{2d}\left(\dfrac{\xi}{2}\right)}{1 + m 4^d \dfrac{(-1)^{d+1}}{(2d)!} \bar{C}_{2d} \sin^{2d}\left(\dfrac{\xi}{2}\right)} \right].$$

As the operator S_h is dissipative, we get

$$|\varrho_{S_h}| \leqq 1 - \delta \sin^{2d} \frac{\xi}{2}, \qquad \delta = \text{const} > 0.$$

Therefore,

$$|\varrho_{C_{m+n,m}}| \leqq \left[1 - \delta \sin^{2d}\left(\frac{\xi}{2}\right)\right]^n \left[1 + n\beta \sin^{2d}\left(\frac{\xi}{2}\right)\right],$$

where

$$\beta = 4^d \frac{(-1)^{d+1}}{(2d)!} \bar{C}_{2d}.$$

We will show the uniformly boundedness

$$|\varrho_{C_{m+n,m}}| \leqq \text{const}.$$

For this purpose we will prove a lemma first.

Lemma. The set of numbers $\phi(\alpha, n) = n^k(1 - \delta\alpha)^n \alpha^k$ is uniformly bounded with respect to n and α, where $n = 0, 1, 2, \ldots$; $\alpha \in [0,1]$; $0 < \delta \leqq 1$; k is a natural number.

Indeed, for $\alpha = 0$, all $\phi(0, n) = 0$. If $\alpha = 1$, $\delta = 1$ then $\phi(\alpha, n) = 0$. For $\alpha = 1$, $\delta \neq 1$ the set of numbers $n^k(1 - \delta)^n$ can be represented in the form $n^k/(1 + q)^n$, where q is some positive number. As

$$\frac{n^k}{(1 + q)^n} \to 0 \quad \text{for} \quad n \to \infty,$$

for all n

$$\frac{n^k}{(1 + q)^n} < \text{const}.$$

For fixed n the extremum of the quantity $\phi(\alpha, n)$ can be reached only for α, for which

$$\frac{\partial \phi(\alpha, n)}{\partial \alpha} = 0.$$

This will be the case, if

$$\alpha^{k-1}[-n\delta\alpha + (1 - \delta\alpha)k] = 0, \quad \text{i.e.} \quad \alpha = 0 \quad \text{or} \quad \alpha = \frac{k}{\delta(n + k)}.$$

If $\alpha = 0$ then $\phi(0, n) = 0$, but for $\alpha = k/\delta(n+k)$

$$\phi\left(\frac{k}{\delta(n + k)}, n\right) = \left(1 - \frac{k}{n + k}\right)^n \frac{\left(\frac{n}{n + k}\right)^k k^k}{\delta^k} < \text{const} \quad \text{qed}.$$

By virtue of this lemma the uniform boundedness of $|\varrho_{C_{m+n,m}}|$ follows which ensures the stability of the difference scheme (6.17) in $\mathbb{L}_2$. The theorem is proved. □

Theorem 6.5. The difference scheme

$$u^{n+1}(x) = S_h u^n(x) - \frac{(-1)^{d+1} \mu}{(2d)!} (\Delta_1 \Delta_{-1})^d S_h^{n+1} u^0(x),$$

which approximates the equation (6.8), where

$$S_h = \Lambda_1^{-1} \Lambda_0;$$

$$\Lambda_j = \cdot \sum_{\alpha = -q_1}^{q_2} \beta_{\alpha j} T_1^\alpha, \quad (j = 0, 1);$$

$$\sum_{\alpha = -q_1}^{q_2} [\beta_{\alpha_1}(\alpha + \kappa a)^l - \beta_{\alpha_0} \alpha^l] = 0, \quad (l = 0, 1, \dots, 2d - 1; d = q_1 + q_2);$$

$$\sum_{\alpha = -q_1}^{q_2} [\beta_{\alpha_1}(\alpha + \kappa a)^{2d} - \beta_{\alpha_0} \alpha^{2d}] = (-1)^d \mu;$$

$$\mu = \gamma(a\kappa) \prod_{i = -d+1}^{d-1} (\kappa a - i),$$

has the order of approximation $2d$ and is stable for $\mu > 0$.

Proof. Indeed, the first differential approximation of the original difference scheme

$$u^{n+1}(x) = S_h u^n(x)$$

can be written in the form

$$\frac{\partial u}{\partial t} = a \frac{\partial u}{\partial x} + (-1)^{d+1} \frac{\mu h^{2d}}{(2d)! \tau} \cdot \frac{\partial^{2d} u}{\partial x^{2d}}.$$

Consequently, the given scheme has the order of approximation $2d - 1$ and is dissipative of order $2d$ for $\mu > 0$. Then our statement follows from Theorem 6.4. □

Remark 1. The demonstrated method for the construction of difference schemes of increased order of approximation can easily be extended to hyperbolic systems of equations with constant coefficients under the assumption that the coefficient-matrices of the original difference scheme commute.

Remark 2. In the case of the equations of gas dynamics if we take as the original difference scheme one which is conservative, which can be derived by the well-known integro-interpolational means [10], the property of conservation is preserved also for the scheme of increased order of approximation which is derived according to the method which is described above.

Part II

Investigation of the Artificial Viscosity
of Difference Schemes

7. *K*-property of Difference Schemes

7.1 Introduction

Solving problems of mathematical physics (especially problems of gas dynamics) by finite-difference methods introduces effects of artificial viscosity, i.e. a viscosity which is generated by the structure of the finite-difference schemes. Calculations have shown that properties of difference schemes are especially marked in regions where discontinuities occur.

Depending on the dissipative properties difference schemes can be divided into two classes: The first class contains schemes which admit the calculation of contact surfaces. In this case a family of standing harmonic waves exists upon which the artificial viscosity does not act and which, consequently, do not die out [8].

An "entropy layer" corresponds to those standing waves, i.e. very large differences of entropy and density compared with the exact values are encountered, while pressure and velocity of the difference solution come out very close to the exact values. An entropy layer remains in the vicinity of a contact surface and has the character of a strong non-monotonic deviation from the true density and entropy profiles. An example for such schemes is the von Neumann-Richtmyer-scheme [8].

The schemes belonging to the second class are those which do not admit contact surfaces. In this case discontinuities of density and entropy are smoothed out. For such schemes the entropy layers disappear but also the contact surfaces disappear (are smeared out) [8]. The Lax' scheme can serve as an example for this kind of difference scheme.

A large number of investigations have been devoted to the discussion of dissipative properties of difference schemes. For more detailed information about questions which arise in this context we refer to the publications [184–190] where a nearly complete collection of references can be found. We will on the other hand investigate the artificial viscosity of difference schemes using the concept of the differential approximation. Here and in the following we will define a contact surface as follows: A contact surface is a discontinuity surface through which the gas flow is zero, i.e. the gas does not stream through the surface.

The contact surface is moving with the gas and separates two regions with different density (temperature) but pressure and velocity of the flow on both the sides of the discontinuity are the same.

7.2 Definition of K-property

We consider a hyperbolic system of equations with constant coefficients

$$\frac{\partial w}{\partial t} = A \frac{\partial w}{\partial x},$$ (7.1)

where $w = w(x, t)$ is a vector-function with m components; A is a $(m \times m)$-matrix which has different eigenvalues $\mu_1, \ldots, \mu_m$. We denote by X_j and Y_j the left and right eigenvectors of the matrix A, respectively, which correspond to the eigenvalue μ_j

$$X_j A = \mu_j X_j; \quad A Y_j = \mu_j Y_j.$$

By r_j we denote the Riemann invariant which is transferred without change along the jth characteristic with the slope $dx/dt = \mu_j$ and which satisfies the equation

$$\frac{\partial r_j}{\partial t} = \mu_j \frac{\partial r_j}{\partial x}.$$ (7.2)

Therefore, if the invariant r_j has a discontinuity in the first moment this discontinuity will be transferred at later times without alteration along the jth characteristic.

In the general case the first differential approximation of a difference scheme with an order of approximation p has the form:

$$\frac{\partial w}{\partial t} = A \frac{\partial w}{\partial x} + C_{rr}^{(p)} \frac{\partial^r w}{\partial x^r} + C_{rr-1}^{(p)} \frac{\partial^{r-1} w}{\partial x^{r-1}} + \ldots + C_{r1}^{(p)} \frac{\partial w}{\partial x},$$ (7.3)

where

$$C_{r\alpha}^{(p)} = O(h^p); \quad \alpha = 1, \ldots, r; \quad r \in \mathbb{N}.$$

Definition 7.1. A difference scheme has the property $K_j^{(p,\,1)}$ or–equivalently– has the property K along the jth characteristic in the first differential approximation (belongs to the class $K_j^{(p,\,1)}$), if

$$X_j C_{r\beta}^{(p)} = 0; \quad \left(\beta = 2, 4, \ldots, 2\left[\frac{r}{2}\right]\right).$$

If $C_{r\beta}^{(p)} = 0$ we will consider the second differential approximation of the difference scheme and we will speak of the property $K_j^{(p,\,2)}$, respectively or–equivalently–of the property K of a difference scheme of pth order of approximation along the jth characteristic in the second differential approximation.

In an analogous way the property $K_j^{(p,\,\gamma)}$ (class $K_j^{(p,\,\gamma)}$) can be defined.

Definition 7.1'. A difference scheme belongs to the class $K_j^{(p,\,\gamma)}$, if in the differential approximation of the difference scheme of the order of approximation p all terms with derivatives of even order disappear up to the order of $(\gamma - 1)$ inclusively, but in the differential approximation of order γ there is at least one term with a derivative of even order.

In the following we will normally omit the upper indices in the symbol for the property $K_j^{(p,\,\gamma)}$ and we will write simply K_j because it is evident which differential approximation is considered.

Definition 7.2. A difference scheme is called dissipative of order p along the jth characteristic if for at least one of the quantities $X_j C_{r\beta}^{(p)} Y_j$ we have

$$X_j C_{r\beta}^{(p)} Y_j \neq 0; \quad \left(\beta = 2, 4, \ldots, 2\left[\frac{r}{2}\right]\right).$$

We will denote by G the amplification matrix of a difference scheme.

Definition 7.3. A difference scheme has the strong property K_j (the strong property K along the jth characteristic) if

$$X_j G = e^{i\kappa\mu_j\xi} X_j.$$

7.3 Simple Difference Schemes

We consider for the system of equations (1.1) a class of simple difference schemes (see Sect. 4.1) of first order of approximation

$$w^{n+1}(x) = \sum_{\alpha=1}^{2} B_\alpha w^n(x + \tau\lambda_\alpha). \tag{7.4}$$

The following theorem holds [37, 78].

. . . .

Theorem 7.1. The simple difference scheme (7.4) possesses the property K_j if and only if $\lambda_1 = \mu_j$, and λ_2 is an arbitrary number, (or $\lambda_2 = \mu_j$, and λ_1 arbitrary).

. . . .

Proof. In the case of a simple scheme we have

$$C_2 = \frac{\tau}{2}[-A^2 + (\lambda_1 + \lambda_2)A - \lambda_1\lambda_2 I].$$

Then

$$X_j C_2 = \frac{\tau}{2}[-\mu_j^2 + (\lambda_1 + \lambda_2)\mu_j - \lambda_1\lambda_2]X_j$$

$$= -\frac{\tau}{2}(\mu_j - \lambda_1)(\mu_j - \lambda_2)X_j.$$

It follows if $X_j C_2 = 0$, then $\lambda_1 = \mu_j$, and λ_2 arbitrary (or if $\lambda_2 = \mu_j$, and λ_1 arbitrary).

In the other case if $\lambda_1 = \mu_j$, and λ_2 arbitrary (or $\lambda_2 = \mu_j$, and λ_1 arbitrary), then $X_j C_2 = 0$ and the finite difference scheme (7.4) has the property K_j. Thus the proof is complete. $\square$

. . . .

Theorem 7.2. The simple difference scheme (7.4) possesses the strong property K_j if and only if $\lambda_1 = \mu_j$, λ_2 arbitrary (or if $\lambda_2 = \mu_j$, and λ_1 arbitrary).

. . . .

Proof. In the case under consideration

$$G = \frac{(A - \lambda_2 I) e^{i\kappa\lambda_1\xi} + (\lambda_1 I - A) e^{i\kappa\lambda_2\xi}}{\lambda_1 - \lambda_2}.$$

Then

$$X_j G = \frac{(\mu_j - \lambda_2) e^{i\kappa\lambda_1\xi} + (\lambda_1 - \mu_j) e^{i\kappa\lambda_2\xi}}{\lambda_1 - \lambda_2} X_j.$$

Let

$$X_j G = e^{i\kappa\mu_j\xi} X_j.$$

From this we get

$$\frac{(\mu_j - \lambda_2) e^{i\kappa\lambda_1\xi} + (\lambda_1 - \mu_j) e^{i\kappa\lambda_2\xi}}{\lambda_1 - \lambda_2} = e^{i\kappa\mu_j\xi}$$

and, consequently,

$$(\mu_j - \lambda_2) e^{i\kappa(\lambda_1 - \mu_j)\xi} + (\lambda_1 - \mu_j) e^{-i\kappa(\mu_j - \lambda_2)\xi} = \lambda_1 - \lambda_2.$$

Putting

$$\lambda_1 - \mu_j := \beta, \qquad \lambda_2 - \mu_j := \gamma,$$

we get

$$- \gamma e^{i\kappa\beta\xi} + \beta e^{i\kappa\gamma\xi} = \beta - \gamma$$

or

$$\beta(1 + i\kappa\gamma\xi - \tfrac{1}{2}\kappa^2 \gamma^2 \xi^2 + \ldots) - \gamma(1 + i\kappa\beta\xi - \tfrac{1}{2}\kappa^2 \beta^2 \xi^2 + \ldots) = \beta - \gamma.$$

Then we get

$$\beta\gamma(\beta - \gamma) = 0;$$
$$\beta\gamma(\beta^2 - \gamma^2) = 0;$$
$$\beta\gamma(\beta^3 - \gamma^3) = 0;$$
$$\vdots$$

i.e. $\beta = 0$ or $\gamma = 0$. Thus from the condition $X_j G = [\exp(i\kappa\mu_j \xi)] X_j$ it follows that $\lambda_1 = \mu_j$ or $\lambda_2 = \mu_j$.

On the other side let $\lambda_1 = \mu_j$, λ_2 arbitrary (the case when $\lambda_2 = \mu_j$, λ_1 arbitrary can be considered in a similar way). Then

$$X_j G = \frac{(\mu_j - \lambda_2) e^{i\kappa\lambda_1\xi} + (\lambda_1 - \mu_j) e^{i\kappa\lambda_2\xi}}{\lambda_1 - \lambda_2} X_j$$

$$= \frac{1}{\mu_j - \lambda_2} (\mu_j - \lambda_2) e^{i\kappa\mu_j\xi} X_j = e^{i\kappa\mu_1\xi} X_j.$$

Thus the theorem is proved. ☐

Corollary 1. The simple finite difference scheme (7.4) has the strong property K_j if and only if it has the property K_j.

Corollary 2. The Lax' scheme is a simple difference scheme with $\lambda_1 = -\lambda_2 = h/\tau$, but has not the property K_j (strong property K_j) except in the case when $|\kappa\mu_j| = 1$.

Corollary 3. If $|\kappa\mu_j| \neq 1$ then the Lax' scheme is dissipative of second order along the jth characteristics.

Remark. The property K_j means that in the first differential approximation of the finite difference scheme the jth invariant is preserved. The strong property K_j means that the jth invariant is preserved by the difference scheme.

In the following if we will focus our consideration to the eigenvalue μ_j (omitting the others), the properties: K_j, strong property K_j, and others which will be introduced later which are connected with this special eigenvalue will be just called property K, strong property K etc.

Corollary 4. In the case of one equation

$$\frac{\partial w}{\partial t} = a \frac{\partial w}{\partial x} \tag{7.5}$$

a simple difference scheme which has the property K (and, consequently, also the strong property K) has the form:

$$w^{n+1}(x) = w^n(x + a\tau). \tag{7.6}$$

We consider the system of equations of gas dynamics in Lagrangean coordinates

$$\frac{\partial w}{\partial t} = A \frac{\partial w}{\partial x}, \tag{7.7}$$

where

$$w := \begin{pmatrix} u \\ v \\ p \end{pmatrix}; \quad A := \begin{pmatrix} 0 & 0 & -1 \\ 1 & 0 & 0 \\ -a^2 & 0 & 0 \end{pmatrix}; \tag{7.8}$$

u is the velocity of the gas, p the pressure, v the specific volume; $a^2 = (\partial \varepsilon / \partial v + p)/(\partial \varepsilon / \partial p)$; ε is the specific internal energy.

For the system of equations (7.7) we have

$$\mu_1 = 0; \qquad \mu_{2,3} = \pm a.$$

The eigenvalue $\mu_1 = 0$ corresponds to the left eigenvector

$$X = (0, a^2, 1),$$

and the invariant $r_1 = s$ is the entropy. From this we get

$$\frac{\partial s}{\partial t} = 0,$$

If we assume that $a^2 = \text{const} > 0$ or, what is equivalent, if we consider the system of equations (7.7) in each point (x, t) fixing the coefficients, then the analysis of this system of equations and of its approximating difference scheme gives only a qualitative estimation of the nonlinear system of gas dynamics and, indeed, it means a linearization of the system of equations in the vicinity of a flow with constant parameters. Nevertheless such qualitative conclusions are confirmed by practical calculations.

If we proceed to the invariants of the first differential approximation of a simple difference scheme for the system of equations (7.7) we get:

$$\frac{\partial r_j}{\partial t} = \mu_j \frac{\partial r_j}{\partial t} + \frac{\tau}{2}[-\mu_j^2 + (\lambda_1 + \lambda_2)\mu_j - \lambda_1 \lambda_2]\frac{\partial^2 r_j}{\partial x^2}.$$

We consider the case $\mu_j = 0$. If the conditions of Theorem 7.1 are satisfied then the scheme possesses the property K, and the corresponding invariant (the entropy s) is not affected by the artificial viscosity and this is propagated without changing $\partial s/\partial t = 0$. Whereas by the Lax' scheme the entropy is "smeared out":

$$\frac{\partial s}{\partial t} = \frac{h^2}{2\tau} \frac{\partial^2 s}{\partial x^2}.$$

Thus the existence of an artificial viscosity in connection with a difference scheme which preserves a contact discontinuity is equivalent to the fact that the scheme possesses the property K.

7.4 Three-point Schemes

Let us consider now a three-point difference scheme

$$w^{n+1}(x) = \sum_{\alpha = -1}^{1} B_\alpha w^n(x + \alpha h). \tag{7.9}$$

Theorem 7.3. The difference scheme (7.9) possesses the property K_j if and only if $X_j B_0 = (1 - \kappa^2 \mu_j^2) X_j$.

Proof. For the scheme (7.9) we have

$$C_2 = \frac{h^2}{2\tau}(I - B_0 - \kappa^2 A^2),$$

and therefore

$$X_j C_2 = \frac{h^2}{2\tau}[(1 - \kappa^2 \mu_j^2) X_j - X_j B_0].$$

From this it is easy to show that the theorem holds.   □

Theorem 7.4. Necessary and sufficient conditions for the difference scheme (7.9) to have the strong property K_j are:

$$\mu_j = 0 \quad \text{and} \quad X_j B_0 = X_j I$$

or

$$|\kappa \mu_j| = 1 \quad \text{and} \quad X_j B_0 = 0.$$

Proof. In the case under consideration we have

$$G = B_0 + (I - B_0)\cos \xi + i\kappa A \sin \xi.$$

Then it follows that

$$X_j G = X_j B_0 + X_j (I - B_0)\cos \xi + i\kappa \mu_j X_j \sin \xi.$$

Let

$$X_j G = e^{i\kappa \mu_j \xi} X_j, \quad \text{i.e.} \quad X_j G = X_j \cos(\kappa \mu_j \xi) + i X_j \sin(\kappa \mu_j \xi)$$

and, consequently,

$$\begin{aligned} X_j B_0 + X_j (I - B_0)\cos \xi &= X_j \cos(\kappa \mu_j \xi); \\ \kappa \mu_j X_j \sin \xi &= X_j \sin(\kappa \mu_j \xi). \end{aligned} \tag{7.10}$$

From the second equation we get

$$\mu_j = 0 \quad \text{or} \quad |\kappa \mu_j| = 1.$$

If $\mu_j = 0$ from the first equation (7.10) we get

$$X_j (I - B_0) \sin^2\left(\frac{\xi}{2}\right) = 0, \quad \text{i.e.} \quad X_j B_0 = X_j I.$$

For $|\kappa \mu_j| = 1$ from the first equation (7.10) we get

$$X_j B_0 \sin^2\left(\frac{\xi}{2}\right) = 0, \quad \text{i.e.} \quad X_j B_0 = 0.$$

Thus the necessary conditions are proved.

Now we show the sufficiency of the conditions of the theorem. If $\mu_j = 0$, then $X_j B_0 = X_j I$ and

$$X_j G = X_j I = X_j e^{i\kappa\mu_j\xi}|_{\mu_j=0}.$$

For $|\kappa\mu_j| = 1$ and $X_j B_0 = 0$ we get

$$X_j G = X_j \cos\xi \pm i X_j \sin\xi = X_j e^{\pm i\xi} = X_j e^{i\kappa\mu_j\xi}|_{|\kappa\mu_j|=1}.$$

The theorem is proved. $\square$

Corollary 1. For $\mu_j = 0$ the three-point difference scheme possesses the strong property K_j if and only if it possesses the property K_j.

If $\mu_j \neq 0$ then the three-point difference scheme having the property K_j, possesses the strong property K_j for $|\kappa\mu_j| = 1$ and vice versa.

Corollary 2. A majorant scheme possesses the property K_j and the strong property K_j under the condition that $\mu_j = 0$.

Indeed, for a majorant scheme (see Sect. 4.2.) $B_0 = I - \kappa(A^+ - A^-)$. From the relation $X_j A = 0$ it follows that $X_j A^+ = X_j A^- = 0$. Then $X_j B_0 = X_j I$ and, consequently, the sufficient conditions of the Theorems 7.3 and 7.4 are satisfied.

From this it follows especially, a majorant scheme for the system of equations of gas dynamics in Lagrangean coordinates possesses the property K as the strong property K.

Now we consider the case of Eulerian coordinates. The system of equations of gas dynamics can be written in the form (7.7) where

$$w := \begin{pmatrix} u \\ \varrho \\ E \end{pmatrix}; \qquad A := \begin{pmatrix} \dfrac{u}{\varrho} p_\varepsilon - u & -\dfrac{1}{\varrho} p_\varrho & -\dfrac{1}{\varrho} p_\varepsilon \\[2mm] -\varrho & -u & 0 \\[2mm] \dfrac{u^2}{\varrho} p_\varepsilon - \dfrac{p}{\varrho} & -\dfrac{u}{\varrho} p_\varrho & -\dfrac{u}{\varrho} p_\varepsilon - u \end{pmatrix}; \qquad (7.11)$$

$$E := \varepsilon + \tfrac{1}{2} u^2.$$

The matrix A has the following eigenvalues

$$\mu_1 = -u; \qquad \mu_{2,3} = -u \pm c,$$

where

$$c^2 := p_\varrho + \frac{p p_\varepsilon}{\varrho^2}.$$

The eigenvalue $\mu_1 = -u$ corresponds to the left eigenvector of the matrix A

$$X_1 = (-u, -p/\varrho, 1),$$

and the invariant is the specific entropy s which is preserved along the characteristic with the slope $-u$:

$$\frac{\partial s}{\partial t} + u \frac{\partial s}{\partial x} = 0.$$

As long as in the general case $u \not\equiv 0$ a majorant scheme for the system of equations of gas dynamics in Eulerean coordinates does not possess the property K.

Corollary 3. In the case of a single equation (7.5) a three-point difference scheme which possesses the strong property K has the form (7.6).

7.5 Necessary and Sufficient Conditions for the Strong Property K

We consider now the following difference scheme:

$$w^{n+1}(x) = \sum_\alpha B_\alpha w^n(x + \alpha h). \tag{7.12}$$

For this scheme the amplification matrix G has the form:

$$G = \sum_\alpha B_\alpha e^{i\alpha\xi}.$$

We use the abbreviation

$$\bar{C}_l = \frac{h^l}{\tau l!} [\sum_\alpha \alpha^l B_\alpha - \kappa^l A^l]; \quad l = 2, 3, 4, \dots.$$

It is clear that by virtue of the compatibility conditions $\bar{C}_0 = \bar{C}_1 = 0$ and (see Sect. 2.3) the coefficients C_l of the differential representation are linear combinations of the coefficients $\bar{C}_2, \bar{C}_3, \dots, \bar{C}_l$. As

$$\sum_\alpha \alpha^l B_\alpha = \frac{\tau l!}{h^l} \bar{C}_l + \kappa^l A^l,$$

we get

$$G = \sum_l \frac{i^l}{l!} \frac{\tau l!}{h^l} \bar{C}_l \xi^l + \sum_l \frac{i^l}{l!} \kappa^l A^l \xi^l,$$

and, consequently,

$$G = e^{i\kappa A\xi} + \sum_l \frac{i^l \tau}{h^l} \bar{C}_l \xi^l.$$

Then

$$X_j G = X_j e^{i\kappa A\xi} + \sum_l \frac{i^l \tau}{h^l} X_j \bar{C}_l \xi^l = e^{i\kappa\mu_j\xi} + \frac{i^l \tau}{h^l} X_j \bar{C}_l \xi^l.$$

From the last equation together with the definition it follows that the following lemma is true:

Lemma. In order that the difference scheme (7.12) possesses the strong property K_j it is necessary and sufficient that the following conditions are satisfied:

$$X_j \bar{C}_l = 0; \quad l = 2, 3, \ldots \tag{7.13}$$

7.6 Predictor-Corrector Scheme

Let us approximate the system of equations (7.1) by a predictor-corrector scheme

$$w^*(x) = \sum_\alpha B_\alpha w^n(x + \tau^* \lambda_\alpha); \tag{7.14}$$

$$w^{n+1}(x) = w^n(x) + \kappa A \left[w^* \left(x + \frac{h}{2} \right) - w^* \left(x - \frac{h}{2} \right) \right], \tag{7.15}$$

where (7.14) is an arbitrary difference scheme, but (7.15) is a cross-scheme.

Eliminating the function w^* from the equations (7.14, 15) we get a scheme with integral steps:

$$w^{n+1}(x) = w^n(x) + \kappa A \sum_\alpha B_\alpha \left[w^n \left(x + \tau^* \lambda_\alpha + \frac{h}{2} \right) - w^n \left(x + \tau^* \lambda_\alpha - \frac{h}{2} \right) \right].$$

In this case (we will assume that $\tau^* \neq \tau/2$) we get

$$C_2 = \left(\tau^* - \frac{\tau}{2} \right) A^2;$$

$$G = I + \kappa A \sum_\alpha 2 i \sin \left(\frac{\xi}{2} \right) B_\alpha \exp \left(i \frac{\tau^*}{h} \lambda_\alpha \xi \right).$$

Then we have

$$X_j C_2 = \left(\tau^* - \frac{\tau}{2} \right) \mu_j^2 X_j;$$

$$X_j G = X_j + 2 \kappa \mu_j \sum_\alpha i \sin \left(\frac{\xi}{2} \right) \exp \left(i \frac{\tau^*}{h} \lambda_\alpha \xi \right) X_j B_\alpha.$$

From the last equation the proof of the following statement can be found [69, 73].

. . . .

Theorem 1.5.

1) The difference scheme (7.14, 15) possesses the property K_j if and only if $\mu_j = 0$.
2) For $\mu_j = 0$ the scheme (7.14, 15) possesses the strong property K_j.

. . . .

7.7 Implicit Difference Schemes

We consider for the system of equations (7.1) an implicit difference scheme

$$w^{n+1}(x) = w^n(x) + \sum_{\alpha=1}^{2} B_\alpha (\gamma T_0 + \delta E) w^n (x + \tau \lambda_\alpha), \tag{7.16}$$

where

$$\sum_{\alpha=1}^{2} B_\alpha = 0; \quad \sum_{\alpha=1}^{2} \lambda_\alpha B_\alpha = A; \quad \gamma + \delta = 1; \quad \gamma, \delta \geqq 0.$$

The matrix C_2 of the first differential approximation has the form:

$$C_2 = \frac{\tau}{2} [(\lambda_1 + \lambda_2) A + (2\gamma - 1) A^2],$$

and, consequently, $X_j C_2 = 0$ for $\mu_j = 0$ or $(\lambda_1 + \lambda_2) + (2\gamma - 1)\mu_j = 0$.
  The amplification matrix of the scheme (7.16) can be written in the form:

$$G = \left[I - \gamma \frac{e^{ik\tau\lambda_1} - e^{ik\tau\lambda_2}}{\lambda_1 - \lambda_2} A \right]^{-1} \left[I + \delta \frac{e^{ik\tau\lambda_1} - e^{ik\tau\lambda_2}}{\lambda_1 - \lambda_2} A \right].$$

If the eigenvalues μ_j of the matrix A satisfy the inequality

$$\left| \gamma \frac{e^{ik\tau\lambda_1} - e^{ik\tau\lambda_2}}{\lambda_1 - \lambda_2} \mu_j \right| \leq 1, \tag{7.17}$$

then the following decomposition holds [177]:

$$G = \sum_{l=0}^{\infty} d_l A^l, \quad d_l = d_l(\lambda_1, \lambda_2, \gamma, \delta, k).$$

Then

$$X_j G = \sum_{l=0}^{\infty} d_l \mu_j^l X_j$$

and, consequently,

$$X_j G = e^{i\kappa\mu_j\xi} X_j, \quad \text{if } \mu_j = 0$$

or

$$d_l = \frac{i^l \kappa^l}{l!} \xi^l \quad (l = 0, 1, 2, \ldots).$$

  Thus the following theorem is true:

. . . .

Theorem 1.6. The difference scheme (7.16) possesses the property K_j if $\mu_j = 0$ or $(\lambda_1 + \lambda_2) + (2\gamma - 1)\mu_j = 0$, and it has the strong property K_j if $\mu_j = 0$ or $d_l = \frac{i^l \kappa^l}{l!} \xi^l$ $(l = 0, 1, 2, \ldots)$ if the inequality (7.17) holds.

. . . .

Corollary. For the system of equations of gas dynamics in Lagrangean coordinates a difference scheme of form (7.16) possesses the property K.

We approximate the system of equations (7.1) by the following implicit scheme

$$w^{n+1}(x) = w^n(x) + \sum_{\alpha=-1}^{1} B_\alpha (\gamma T_0 + \delta E) T_1^\alpha w^n(x), \qquad (7.18)$$

where

$$\sum_{\alpha=-1}^{1} B_\alpha = 0; \qquad \sum_{\alpha=-1}^{1} \alpha B_\alpha = \kappa A; \qquad \gamma + \delta = 1; \qquad \gamma, \delta \geqq 0.$$

Theorem 7.7. The difference scheme (7.18) possesses the property K_j if and only if the equation

$$X_j B_0 = (\gamma - \delta) \kappa^2 \mu_j^2 X_j$$

is satisfied.

Indeed, in this case

$$C_2 = \frac{h^2}{2\tau} [-B_0 + \kappa^2 (2\gamma - 1) A^2]$$

and, consequently,

$$X_j C_2 = \frac{h^2}{2\tau} [-X_j B_0 + (\gamma - \delta) \kappa^2 \mu_j^2 X_j].$$

From the last equation it can be easily proved that the theorem holds.

In an analogous way the following theorem can be proved.

Theorem 7.8. The difference scheme

$$\sum_{\alpha=-1}^{1} B_\alpha^1 w^{n+1}(x + \alpha h) = \sum_{\alpha=-1}^{1} B_\alpha^0 w^n(x + \alpha h),$$

where

$$\sum_{\alpha=-1}^{1} B_\alpha^1 = \sum_{\alpha=-1}^{1} B_\alpha^0 = I; \qquad \sum_{\alpha=-1}^{1} \alpha (B_\alpha^0 - B_\alpha^1) = \kappa A,$$

possesses the property K_j if and only if

$$X_j (B_0^1 - B_0^0) = 2 X_j (B_1^1 - B_{-1}^1) \kappa A + \kappa^2 \mu_j^2 X_j.$$

7.8 Higher-order Difference Schemes

We consider the following difference scheme for the system of equations (7.1):

$$w^{n+1}(x) = \sum_{\alpha=-N}^{N} B_\alpha w^n(x + \alpha h), \qquad (7.19)$$

which has the order of approximation $2N$. The compatibility conditions have the form

$$\sum_{\alpha=-N}^{N} \alpha^\beta B_\alpha = \kappa^\beta A^\beta; \qquad \beta = 0, 1, \ldots, 2N.$$

Then

$$B_\alpha = \prod_{\substack{l=-N \\ l \neq \alpha}}^{N} \frac{\kappa A - lI}{\alpha - l}.$$

The second differential approximation has the form

$$\frac{\partial w}{\partial t} = A \frac{\partial w}{\partial x} + C_{2N+1} \frac{\partial^{2N+1} w}{\partial x^{2N+1}} + C_{2N+2} \frac{\partial^{2N+2} w}{\partial x^{2N+2}},$$

where

$$C_{2N+1} := \frac{h^{2N+1}(-1)^{N+1}}{\tau(2N+1)!} \kappa A (I - \kappa^2 A^2) \ldots (N^2 I - \kappa^2 A^2);$$

$$C_{2N+2} := -\frac{h^{2N+2}(-1)^{N+1}(2N+1)}{\tau(2N+2)!} \kappa^2 A^2 (I - \kappa^2 A^2) \ldots (N^2 I - \kappa^2 A^2).$$

In the present case one can speak of the property $K_j^{(2N,\,2)}$. As

$$X_j C_{2N+2} = -\frac{h^{2N+2}(-1)^{N+1}(2N+1)}{\tau(2N+2)!}$$
$$\cdot \kappa^2 \mu_j^2 (1 - \kappa^2 \mu_j^2) \ldots (N^2 - \kappa^2 \mu_j^2) X_j,$$

consequently, the scheme (7.19) possesses the property $K_j^{(2N,\,2)}$ if and only if

$$\kappa^2 \mu_j^2 = \gamma^2; \qquad (\gamma = 0, 1, 2, \ldots, N).$$

From this statement the two following corollaries can be derived:

Corollary 1. The Lax-Wendroff-scheme for the system of equations of gas dynamics in Lagrangean coordinates possesses the property $K^{(2,\,2)}$ (in the second differential approximation the dissipation does not act on the entropy).

Corollary 2. For the system of equations of gas dynamics in Eulerian coordinates the Lax-Wendroff-scheme possesses the property $K^{(2,\,2)}$ for $\kappa^2 u^2 = 1$.

7.9 Application to Gas Dynamics

The results gained in the previous sections can be used to construct difference schemes which possess the property K and which approximate the system of equations of gas dynamics.

Let us consider, for example, the system of equations of gas dynamics in Eulerian coordinates for a polytropic gas $[(\gamma - 1)\,\varepsilon\varrho = p]$:

$$\frac{\partial w}{\partial t} = A\,\frac{\partial w}{\partial x}. \tag{7.20}$$

Here

$$w := \begin{pmatrix} u \\ \varrho \\ p \end{pmatrix}; \qquad A := \begin{pmatrix} -u & 0 & -1/\varrho \\ -\varrho & -u & 0 \\ -\gamma p & 0 & -u \end{pmatrix}.$$

The left eigenvalue $X = (0, -c^2, 1)$ with $c^2 = \gamma p/\varrho$ corresponds to the left eigenvalue $-u$ of the matrix A.

We approximate the system of equations (7.20) by a three-point difference scheme (7.9) with $B_\alpha = B_\alpha(w)$.

The first differential approximation has the form:

$$\frac{\partial w}{\partial t} = D\,\frac{\partial w}{\partial x} + C_2\,\frac{\partial^2 w}{\partial x^2},$$

where

$$D = A - \frac{\tau}{2}\left[\frac{\partial A}{\partial t} + A\,\frac{\partial A}{\partial x}\right], \qquad C_2 = \frac{h^2}{2\tau}\left(I - B_0 - \kappa^2 A^2\right).$$

Let $B_0 = \|b_{ij}^0\|_1^3$ then

$$C_2 = \frac{h^2}{2\tau}\begin{pmatrix} 1 - b_{11}^0 - \kappa^2(u^2 + c^2) & -b_{12}^0 & -b_{13}^0 - 2\kappa^2 u/\varrho \\ -b_{21}^0 - 2\varrho u\kappa^2 & 1 - b_{22}^0 - \kappa^2 u^2 & -b_{23}^0 - \kappa^2 \\ -b_{31}^0 - 2\gamma p u\kappa^2 & -b_{32}^0 & 1 - b_{33}^0 - \kappa^2(u^2 + c^2) \end{pmatrix}.$$

From the results of this paragraph it follows that the condition $XC_2 = 0$ is equivalent to the equation $XB_0 = (1 - \kappa^2 u^2)X$, i.e.

$$\begin{aligned} b_{31}^0 &= c^2 b_{21}^0; \\ b_{32}^0 &= c^2(1 - b_{22}^0 - \kappa^2 u^2); \\ b_{33}^0 &= 1 - \kappa^2 u^2 + c^2 b_{23}^0. \end{aligned} \tag{7.21}$$

Thus, if for the three-point difference scheme (7.9) the matrix B_0 satisfies the conditions (7.21), then the scheme possesses the property K. We get a whole family of difference schemes with this property which depends on the parameters $b_{1j}^0, b_{2j}^0; j = 1, 2, 3$. The method of differential approximation permits us to select from those schemes the stable ones. For these it is necessary to prove the condition of non-total parabolicity of the system of equations of parabolic form of the first differential approximation and the condition on the connection of the region of dependence of the original system of equations and the system of equations of the hyperbolic form of the first differential approximation. These requirements lead to the inequalities

$$\beta_0 \geq 0, \quad \delta \geq 0,$$

where β_0, δ are eigenvalues of the matrix B_0 and C_2, respectively. In particular, we get

$$\kappa^2 u^2 \leq 1; \quad \kappa^2 (u^2 \pm c^2) \leq 1.$$

In addition to the conditions $\beta_0 \geq 0, \delta \geq 0$ we get the following inequalities

$$2 - b_{11}^0 - b_{33}^0 - 2\kappa^2 (u^2 + c^2)$$

$$\pm \left[(b_{11}^0 - b_{33}^0)^2 + 16\kappa^4 c^2 u^2 + b_{13}^0 b_{31}^0 + 2\kappa^2 \frac{u}{\varrho} b_{13}^0 + 2\kappa^2 \gamma pub_{13}^0 \right]^{1/2} \geq 0,$$

$$\kappa^2 u^2 \leq 1;$$

$$\pm \sqrt{(b_{11}^0 - b_{33}^0)^2 + 4 b_{13}^0 b_{31}^0} \leq b_{11}^0 + b_{33}^0.$$

7.10 Connection Between Partly Dissipative Difference Schemes and Those with the Strong Property K

Now we focus our interest on the connection between a partly dissipative difference scheme (see Sect. 5.4) and schemes which possess the strong property K.

If the difference scheme possesses the strong property K then, by definition, $X_j G = [\exp(i\kappa\mu_j \zeta)] X_j$, i.e. one of the eigenvalues of the matrix G has the modulus 1. If the other eigenvalues of the matrix G have a modulus not larger than $1 - \delta |\xi|^{2d}$, according to Theorem 5.3 the difference scheme is stable. The dissipative property of the difference scheme in the generalized sense means, on the one hand, that one of the invariants of the system of equations is calculated with higher accuracy than the others, but, on the other hand, that the difference scheme belongs to class (see Part I, Eq. (5.4))

$$K_l^{(2r + 2d - 1, 1)} \cap K_j^{(2d - 1, 1)}; \quad j = 1, 2, \dots, l - 1, l + 1, \dots, m.$$

7.11 The Property $P_j^{(p, 1)}$

Definition 7.4. A difference scheme of order of approximation p possesses the property $P_j^{(p, 1)}$ or–equivalently–possesses the transition property along the jth characteristic in the first differential approximation (belongs to the class

$P_j^{(p,\,1)})$ if

$$X_j C_{r1}^{(p)} Y_j \neq 0.$$

Then along the jth characteristic a transition with overtaking (property $P_{j+}^{(p,\,1)}$) takes place, if $(\mu_j < 0)$

$$X_j C_{r1}^{(p)} Y_j < 0,$$

and a transition with delay takes place (property $P_{j-}^{(p,\,1)}$), if

$$X_j C_{r1}^{(p)} Y_j > 0.$$

If in the differential approximations of the difference scheme up to the order $(\gamma - 1)$ inclusive, terms of the form $C_{r1}^{(p)} \partial w/\partial x$ are not encountered, but in the differential approximation of order γ such a term appears, then we will say that the difference scheme has the property $P_j^{(p,\,\gamma)}$ (class $P_j^{(p,\,\gamma)}$). In the following where it is clear, we will call any differential approximation of the difference scheme which has the property $P_j^{(p,\,\gamma)}$ simply a scheme with the property P_j.

7.12 The Property $\mathscr{D}_j^{(p,\,1)}$

Definition 7.5. A difference scheme possesses the property $\mathscr{D}_j^{(p,\,1)}$ or $\mathscr{D}$ (dispersion) along the jth characteristic in the first differential approximation (belongs to the class $\mathscr{D}_j^{(p,\,1)}$), if at least one of the quantities satisfies the inequalitiy

$$X_j C_{r\beta}^{(p)} Y_j \neq 0; \quad \left(\beta = 3, 5, \ldots, 2\left[\frac{r+1}{2}\right] - 1\right).$$

In an analogous way the property $\mathscr{D}_j^{(p,\,\gamma)}$ is defined, if in the differential approximations up to the order $(\gamma - 1)$ inclusive, all terms with odd derivatives vanish starting with the third, but in the differential approximation of order γ there is at least one term with odd order of the derivatives which is not smaller than three.

We remark that if the difference scheme belongs to class $K_j^{(p,\,1)}$ and does not belong to class

$$\mathscr{D}_j^{(p,\,1)} \cap P_j^{(p,\,1)},$$

then the corresponding invariant r_j in the first differential approximation will be propagated without change along the jth characteristic with the slope μ_j and, consequently, satisfies the equation (7.2) in the first differential approximation.

8. Investigation of Dissipation and Dispersion of Difference Schemes

8.1 Dissipation and Dispersion of Difference Schemes

In the preceding section we have defined the properties $K_j^{(p,\,\gamma)}$, $P_j^{(p,\,\gamma)}$ and $\mathscr{D}_j^{(p,\,\gamma)}$ of difference schemes which – as shown – mean, that in the differential approximation of order γ the invariant r_j is not affected by the artificial viscosity by virtue of the dissipation and is affected by the dispersion and by an additional transport, respectively. Thus we will confine our investigation to a more detailed discussion of the case of a single equation.

For Cauchy's problem

$$\frac{\partial u}{\partial t} + a\frac{\partial u}{\partial x} = 0, \quad -\infty < x < \infty, \quad t > 0, \quad a = \text{const} > 0, \tag{8.1}$$

$$u(x,0) = u_0(x),$$

we consider a two-level difference scheme

$$u^{n+1}(x) = \sum_\alpha b_\alpha u^n(x + \alpha h), \tag{8.2}$$

which approximates the differential equation (8.1) and, consequently, especially the following compatibility equations are satisfied:

$$\sum_\alpha b_\alpha = 1; \quad \sum_\alpha \alpha b_\alpha = -\kappa a.$$

The Π-form of the differential representation of the difference scheme (8.2) has the form:

$$\frac{\partial u}{\partial t} + a\frac{\partial u}{\partial x} = \sum_{l=2}^\infty c_l \frac{\partial^l u}{\partial x^l}. \tag{8.3}$$

The impulse or signal of arbitrary form on the basis of a representation in integral form or as a Fourier series can be represented by a superposition of harmonic waves $\exp[i(kx - \omega t)]$, where ω is the frequency, $k = 2\pi/\lambda$ the wave number (dual variable), λ is the wave length. The velocity with which the phase of the wave $kx - \omega t$ propagates in space is named phase velocity of the wave and is equal to $v = \omega/k$. The dependence of the phase velocity of the wave spreading on the wave length λ (or on the wave number) is called the dispersion of the wave.

If $dv/d\lambda > 0$, i.e. the waves with a larger wave length propagate with a higher phase velocity, the dispersion is called normal; if $dv/d\lambda < 0$, i.e. the larger the wave length the smaller is the phase velocity of the propagating waves, the dispersion is called anomalous.

For $dv/d\lambda = 0$ the dispersion vanishes. In the case of dispersion the harmonic components of a signal are superposed and as a consequence of it the profile of the signal is changed. In the solution of the differential equation (8.1) the Fourier component $\exp(ikx)$ with the wave length $2\pi/k = 2\pi h/\xi$ ($\xi = kh$) in the time τ changes by an amount

$$\Omega = e^{-i\kappa a\xi},$$

the modulus of which is one and the argument is given by

$$\Phi = \arg\{\Omega\} = -\kappa a\xi.$$

It is clear that here Ω is the Fourier transform of the step-operator for equation (8.1); Φ is the phase change in the solution of the equation (8.1) in the time τ.

The Fourier transform of the step-operator of the difference scheme (8.2) [or of the differential representation (8.3)] is equal to

$$\varrho = \sum_{\alpha} b_{\alpha} e^{i\alpha\xi} = e^{\tau s},$$

where

$$\tau s = -i\kappa a\xi + \tau \sum_{l=2}^{\infty} (ik)^l c_l.$$

In the solution of the differential representation (8.3) [or of the difference scheme (8.2)] the Fourier component of the solution with a wave length $2\pi/k$ during the time τ changes by an amount ϱ while the change in the phase of the solution is equal to

$$\Phi_h = \arg\{\varrho\} = \arctg\left(\frac{\text{Im}\{\varrho\}}{\text{Re}\{\varrho\}}\right) = \tau \,\text{Im}\,\{s\}$$

$$= -\kappa a\xi + \tau \sum_{l=1}^{\infty} (-1)^l k^{2l+1} c_{2l+1}.$$

The value of the damping factor in the solution of the scheme for the time τ is

$$|\varrho| = \exp(\tau)\,\text{Re}\,\{s\} = \exp(\tau) \sum_{l=1}^{\infty} (-1)^l k^{2l} c_{2l}.$$

Definition. As the phase error (dispersion) of the difference scheme (8.2) we denote the difference

$$\Delta\Phi_h := \Phi_h - \Phi.$$

It is clear that

$$\Delta \Phi_h = \tau \operatorname{Im}\{s\} = \tau \sum_{l=1}^{\infty} (-1)^l k^{2l+1} c_{2l+1}.$$

Definition. As the dissipation of the difference scheme (8.2) we denote the quantity:

$$\chi := |\Omega| - |\varrho|.$$

In the case considered we get

$$\chi = 1 - \exp(\tau \operatorname{Re}\{s\}) = 1 - \exp\left[\tau \sum_{l=1}^{\infty} (-1)^l k^{2l} c_{2l}\right].$$

Thus dissipative effects of a difference scheme can be qualitatively calculated by a quantity $|\varrho|$ which is smaller than 1. The convective errors are given by the phase error.

A solution of the differential representation (8.3) by an elementary wave-like solution has the form:

$$\exp\left[-c_e k^2 t + ik(x - a_e t)\right] = (1 - \chi)^n \exp\left[ik(x - at) + \frac{1}{\tau k}\Delta \Phi_h t\right],$$

where

$$c_e := -\sum_{j=1}^{\infty} (-1)^j k^{2j-2} c_{2j}$$

is the coefficient of the artificial viscosity;

$$a_e = a - \sum_{j=1}^{\infty} (-1)^j k^{2j} c_{2j+1}$$

is the effective velocity of the wave propagation.

As it was shown, if for some wave number k, $c_e < 0$, then the equation (8.3) is not correct and the numerical process is unstable. The condition $(-1)^{q+1} c_{2q} > 0$ (where c_{2q} is the first non-vanishing coefficient c_l with even index) is a necessary condition for stability.

If for all j, $(-1)^{j+1} c_{2j} \geq 0$, then a more critical stability condition will be $(-1)^{q+1} c_{2q} \geq 0$. If for some j $(-1)^{j+1} c_{2j} < 0$, then an additional investigation is necessary and the conditions $(-1)^{q+1} c_{2q} > 0$ are not sufficient for stability.

8.2 Dissipation and Dispersion of Differential Approximations

Let the jth differential approximation of the difference scheme (8.2) have the form:

$$\frac{\partial u}{\partial t} + a \frac{\partial u}{\partial x} = \sum_{l=2}^{j+1} c_l \frac{\partial^l u}{\partial x^l}. \tag{8.4}$$

Then in the solution of the equation (8.4) the harmonic wave with a wave length $2\pi/k$ during the time τ changes by the quantity

$$\varrho^{(j)} = \exp(\tau s^{(j)})$$

where

$$\tau s^{(j)} = -i\kappa a\xi + \tau \sum_{l=2}^{j+1} (ik)^l c_l.$$

The modulus of $\varrho^{(j)}$ is equal to

$$|\varrho^{(j)}| = \exp(\tau \operatorname{Re} \{s^{(j)}\})$$

and the argument is given by:

$$\tilde{\Phi}_h^{(j)} = -\kappa a\xi + \tau \operatorname{Im} \{s^{(j)}\}.$$

The phase error (dispersion) of the jth differential approximation (8.4) of the difference scheme (8.2) is denoted by

$$\Delta\tilde{\Phi}_h^{(j)} := \tilde{\Phi}_h^j - \Phi.$$

It is clear that

$$\Delta\tilde{\Phi}_h^{(j)} = \tau \operatorname{Im} \{s^{(j)}\}.$$

As the dissipation of the jth differential approximation of the difference scheme (8.2) we denote the quantity

$$\chi^{(j)} := |\Omega| - |\varrho^{(j)}|.$$

If the order of the approximation of the difference scheme (8.2) is equal to p then the first differential approximation has the form

$$\frac{\partial u}{\partial t} + a\frac{\partial u}{\partial x} = c_{p+1}\frac{\partial^{p+1} u}{\partial x^{p+1}}$$

and for small ξ we get

$$\chi = \chi^{(p-2[(p-1)/2])} + O(\xi^{2[p/2]+4});$$

$$\Delta\Phi_h = \Delta\Phi_h^{(p+1-2[p/2])} + O(\xi^{2[(p+1)/2]+3}),$$

where

$$\chi^{(p-2[(p-1)/2])} = (-1)^{[p/2]}\frac{\tau}{h^{2[(p+2)/2]}} c_{2[(p+1)/2]}^{(p)} \xi^{2[(p+2)/2]},$$

$$\Delta\Phi_h^{(p+1-2[p/2])} = (-1)^{[(p+1)/2]}\frac{\tau}{h^{2[(p+1)/2]+1}} c_{2[(p+1)/2]+1}^{(p)} \xi^{2[(p+1)/2]+1};$$

$c_l^{(p)}$ is the value of the coefficient c_l in the scheme of order of approximation p; $[q]$ means the integral part of the number q.

We consider, for example, for equation (8.1) a three-point difference scheme of the form (8.2): $\alpha = -1, 0, 1$. In this case the difference scheme can be represented in the following way:

$$\frac{\Delta_0 u^n(x)}{\tau} + a \frac{\Delta_1 + \Delta_{-1}}{2h} u^n(x) = \frac{vh^2}{2\tau} \frac{\Delta_1 \Delta_{-1}}{h^2} u^n(x), \qquad (8.5)$$

where

$$b_0 = 1 - v;$$
$$b_1 = \tfrac{1}{2}(v - \kappa a);$$
$$b_{-1} = \tfrac{1}{2}(v + \kappa a).$$

For $v = 1$ we have the Lax' scheme of first order of approximation, for $v = \kappa^2 a^2$ we get the maximum order of approximation of a difference scheme in the given form (Lax-Wendroff-scheme). Between these two extreme schemes we find an infinite numer of difference schemes of first order. Especially for $v = \kappa a$ we get the linear analogy of *Godunov's scheme* [191] (in the following we will call it simply the Godunov scheme).

The stability condition of the difference scheme (8.5) has the form

$$\kappa^2 a^2 \leqq v \leqq 1.$$

For small values of ξ we have

$$\chi = \begin{cases} \tfrac{1}{2}(v - \kappa^2 a^2)\,\xi^2 + O(\xi^4) & \text{for } v > \kappa^2 a^2; \\[2mm] \tfrac{1}{8}(1 - \kappa^2 a^2)\kappa^2 a^2\,\xi^4 + O(\xi^6) & \text{for } v = \kappa^2 a^2; \end{cases}$$

$$\Delta\Phi_h = \tfrac{1}{6}(1 + 2\kappa^2 a^2 - 3v)\kappa a \xi^3 + O(\xi^5).$$

We consider now the difference scheme (8.5). As $c_3 = O(\tau)c_2$, $c_4 = O(\tau)c_3$, in the scheme of first order of approximation the effects of dissipation prevail against the effects of dispersion but in the scheme of second order of approximation, on the contrary, the effects of dispersion prevail over the effects of dissipation. Moreover the dissipative effects of schemes of first order of approximation are more pronounced than for schemes of second order, but parasitic oscillations for second order schemes originate from the overweight of the dispersion over the dissipation. These consequences are well proved experimentally by calculations according to the scheme of type (8.5) both for the linear and for nonlinear equations.

For the differential equation (8.1), the differential representation (8.3) and for the jth differential approximation the phases are, respectively:

$$-kat + kx;$$

$$-kat + kx + t \sum_{l=1}^{\infty} (-1)^l k^{2l+1} c_{2l+1};$$

$$-kat + kx + t \sum_{l=1}^{[(j-1)/2]} (-1)^l k^{2l+1} c_{2l+1}.$$

Then the phase velocity for the kth harmonic in the solution of the equations (8.1, 3, 4) have the following form:

$$v = \frac{\omega}{k} = a;$$

$$v_h = \frac{\omega_h}{k} = a - \sum_{l=1}^{\infty} (-1)^l k^{2l} c_{2l+1};$$

$$v_h^{(j)} = a - \sum_{l=1}^{[(j-1)/2]} (-1)^l k^{2l} c_{2l+1},$$

and, consequently,

$$\frac{dv}{d\lambda} = -\frac{2\pi}{\lambda^2} \frac{dv}{dk} = 0, \quad \lambda = \frac{2\pi}{k};$$

$$\frac{dv_h}{d\lambda} = -\sum_{l=1}^{\infty} \frac{l}{\pi} (-1)^l k^{2l+1} c_{2l+1};$$

$$\frac{dv_h^{(j)}}{d\lambda} = \sum_{l=1}^{[(j-1)/2]} \frac{l}{\pi} (-1)^l k^{2l+1} c_{2l+1}.$$

Thus the dispersion in the solution of the differential equation (8.3) [or of the difference scheme (8.2)] is normal, if

$$\sigma_h = \sum_{l=1}^{\infty} (-1)^{l+1} \frac{l}{\pi} k^{2l+1} c_{2l+1} > 0,$$

and anomalous, if $\sigma_h < 0$. For $\sigma_h = 0$ the dispersion vanishes. Especially if $c_{2l+1} = 0$, for all $l = 1, 2, \ldots$, then the dispersion in the solution of the difference scheme vanishes.

In an analogous way the dispersion in the solution of the jth differential approximation is normal, if $\sigma_h^{(j)} > 0$, and anomalous, if $\sigma_h^{(j)} < 0$. Here

$$\sigma_h^{(j)} = \sum_{l=1}^{[(j-1)/2]} \frac{l}{\pi} (-1)^{l+1} k^{2l+1} c_{2l+1}.$$

For $\sigma_h^{(j)} = 0$ the dispersion in the solution of equation (8.4) vanishes. Especially, this will be the case if

$$c_3 = c_5 = \ldots = c_{2[(j-1)/2]+1} = 0.$$

For the difference scheme (8.2) of the order of approximation p its second differential approximation has the form

$$\frac{\partial u}{\partial t} + a \frac{\partial u}{\partial x} = c_{p+1}^{(p)} \frac{\partial^{p+1} u}{\partial x^{p+1}} + c_{p+2}^{(p)} \frac{\partial^{p+2} u}{\partial x^{p+2}},$$

the phase of its solution is equal to

$$- kat + kx + t(-1)^{[(p+1)/2]} k^{2[(p+1)/2]+1} c_{2[(p+1)/2]+1}^{(p)}.$$

Then

$$v_h^{(2)} = a + (-1)^{[(p+1)/2]} k^{2[(p+1)/2]} c_{2[(p+1)/2]+1}^{(p)},$$

and, consequently,

$$\frac{dv_h^{(2)}}{d\lambda} = \frac{[(p+1)/2]}{\pi}(-1)^{[(p+1)/2]} k^{2[(p+1)/2]+1} c_{2[(p+1)/2]+1}^{(p)}.$$

Thus in the solution of the second differential approximation of the difference scheme (8.2) of pth order of approximation the dispersion vanishes if and only if

$$c_{2[(p+1)/2]+1}^{(p)} - 0.$$

The effective velocity of propagation of waves, a_e, differs from the velocity of propagation of waves a in the solution of equation (8.1). The wave of the error which originates in some point, propagates with a velocity which depends on the wave number k. By virtue of dispersion the local error will influence everywhere the results in the region of calculations. In the numerical results of calculations of gasdynamic problems such dispersive errors appear in form of oscillations downstream of a shock wave and they are created by errors in the region of the shock wave. One can arrive at an essential decreasing of the amplitude of such oscillations if one introduces into the difference scheme additional viscous terms of the order $O(h)$ into a difference scheme of first order of approximation or–equivalently–by approximating instead of (8.1) the equation

$$\frac{\partial u}{\partial t} + a\frac{\partial u}{\partial x} = \mu\frac{\partial^2 u}{\partial x^2}, \quad \mu = \text{const} \cdot h.$$

The logarithmic decrement of the error wave during the time of propagation Δt is equal to $c_e k^2 \Delta t$. We denote by Δx the distance of propagation of the error wave from the place of its generation where the amplitude of the wave is reduced by the factor h. Then

$$h = \exp(-c_e k^2 \Delta x/a_e).$$

From this we get

$$\Delta x = |\ln h|\frac{|a_e|}{|c_e k^2|} \approx h|\ln h|\frac{(2q)!}{(kh)^{2q}}, \tag{8.6}$$

where $2q$ is the index of the first nonzero coefficient c_{2l}. The relation (8.6) underlines the importance of the product kh and the order of dissipativity $2q$.

We estimate the magnitude of the reduction in the amplitude of the error wave during that time, in which the wave propagates over a distance which has the length of its wave length λ. For a difference scheme of first order of approximation the damping is given by:

$$\exp(-c_e k^2 t) = \exp(-c_e k^2 \lambda/a_e)$$
$$\approx \exp(-c_2 k^2 \lambda/a_e) = \exp\left[-(2\pi)^2 \frac{h}{\lambda} \frac{c_2}{ha} \frac{a}{a_e}\right].$$

Thus all waves with a wave length $\lambda \le 10\,h$, $[10\,h \approx \pi^2\,h, c_2/ah = O(1)]$ will be suppressed and can not travel out. For $\lambda > 10\,h$ the waves are such that their length is comparable with a characteristic length and they can not be detected if their amplitude is not very large. Therefore in the solution of a difference scheme of first order of approximation the disturbances of propagating waves vanish.

In the case of a difference scheme of second order of approximation we get

$$\exp(-c_e k^2 t) \approx \exp\left[(2\pi)^4 \left(\frac{h}{\lambda}\right)^3 \frac{c_4}{ah^3} \frac{a}{a_e}\right].$$

For the Lax-Wendroff-scheme $c_4 = -\dfrac{h^4}{8\tau} \kappa^2 a^2 (1 - \kappa^2 a^2)$ and, consequently, the waves with $\lambda = 4\,h$ are not noticeably damped (the damping coefficient is of order $e^{-5/4}$), but waves with $\lambda = 2\,h$ are strongly damped (the damping coefficient is of order e^{-4}). Thus for shock wave calculations the errors of a wave with $\lambda = 2\,h$ have no noticeable influence, but errors of a wave with $\lambda = 4\,h$ appear clearly in form of oscillations which are damped the farther the distance to the shock front (compare also the discussion of Eq. (8.5).

8.3 Relative Dissipative Error and Dispersion

For the analysis of properties like dissipation and dispersion of difference schemes one can use the relative dissipative error

$$\varepsilon_\chi := \frac{|\varrho| - |\Omega|}{\Omega}$$

and the relative dispersive error

$$\varepsilon_{\Delta\Phi_h} := \frac{\Phi_h - \Phi}{\Phi} = \frac{\Delta\Phi_h}{\Phi}.$$

In the case considered we have $\varepsilon_\chi = -\chi$.

If $\varepsilon_{\Delta\Phi_h} > 0$ for a fixed ξ the corresponding harmonic wave will have a wave velocity which is large compared with the wave velocity of the exact solution; if $\varepsilon_{\Delta\Phi_h} < 0$ for a fixed ξ the wave velocity of the corresponding harmonic wave will be smaller than the wave velocity of the exact solution. For $\varepsilon_{\Delta\Phi_h} = 0$ the dispersion vanishes.

It can easily be shown that

$$\varepsilon_{\Delta\Phi_h} = -\frac{1}{a}\sum_{l=1}^{\infty}(-1)^l k^{2l} c_{2l+1}.$$

Let c_{2q+1} be the first nonzero coefficient with odd index. Then

$$\varepsilon_{\Delta\Phi_h} = -\frac{(-1)^q}{a}\frac{1}{h^{2q}}c_{2q+1}\,\xi^{2q} + O(\xi^{2q+1}), \tag{8.7}$$

and we will say that the difference scheme has the order of dispersion $2q$ and, consequently, has the property $D^{(p,\,\gamma)}$. For an explicit determination of p and γ we have to investigate a special given scheme.

Thus, if the difference scheme (8.2) is of pth order of approximation, then its second differential approximation has the form:

$$\frac{\partial u}{\partial t} + a\frac{\partial u}{\partial x} = c_{p+1}^{(p)}\frac{\partial^{p+1}u}{\partial x^{p+1}} + c_{p+2}^{(p)}\frac{\partial^{p+2}u}{\partial x^{p+2}}.$$

Consequently, for $p = 2m$ the difference scheme has the property $\mathscr{D}^{(2m,\,1)}$ (if $c_{2m+2}^{(p)} = 0$, then it has the property $K^{(2m,\,2)}$), for $p = 2m + 1$ the difference scheme does not possess the property $K^{(2m,\,1)}$, and has the property $\mathscr{D}^{(2m,\,2)}$, if $c_{2m+3}^{(p)} \neq 0$. As (see Sect. 8.2)

$$\frac{dv_h^{(2)}}{d\lambda} = \frac{[(p+1)/2]}{\pi}(-1)^{[(p+1)/2]}k^{2[(p+1)/2]+1}c_{2[(p+1)/2]+1}^{(p)},$$

for small ξ the dispersion is normal, if

$$(-1)^{[(p+1)/2]}c_{2[(p+1)/2]+1}^{(p)} > 0.$$

Although this representation holds only for small wave numbers, nevertheless it gives practical criteria to find out whether we get in the numerical solution an effect of acceleration or retardation. We consider, for example, the *Rusanov-scheme* [192] of third order of approximation:

$$\frac{\Delta_0 u^n(x)}{\tau} + a\frac{\Delta_1 + \Delta_{-1}}{2h}u^n(x) = \frac{a^2\tau}{2}\frac{\Delta_1\Delta_{-1}}{h^2}u^n(x) + \frac{ah^2}{3}(1 - \kappa^2 a^2)$$

$$\cdot\frac{\Delta_1 + \Delta_{-1}}{2h}\frac{\Delta_1\Delta_{-1}}{h^2}u^n(x)$$

$$-\frac{h^4}{24\tau}(\omega - 3\kappa^2 a^2)\left(\frac{\Delta_1\Delta_{-1}}{h^2}\right)^2 u^n(x).$$

The second differential approximation of this scheme has the form

$$\frac{\partial u}{\partial t} + a\frac{\partial u}{\partial x} = c_4^{(3)}\frac{\partial^4 u}{\partial x^4} + c_5^{(3)}\frac{\partial^5 u}{\partial x^5},$$

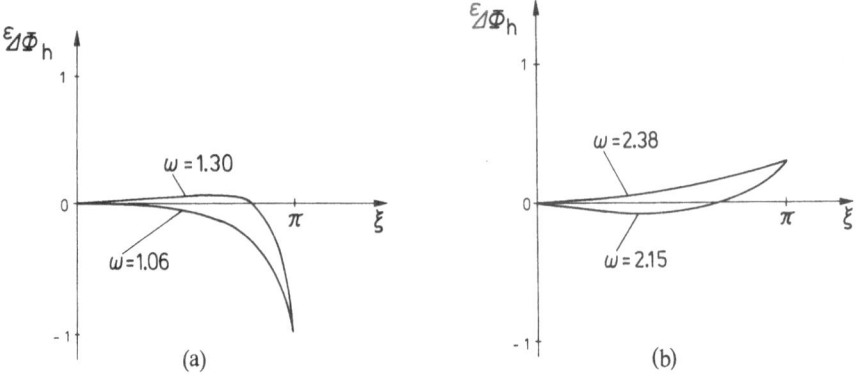

Fig. 8.1. (a) Relative dispersive error, $\kappa a = 0.3$; (b) Relative dispersive error, $\kappa a = 0.8$

where

$$c_4^{(3)} = -\frac{h^4}{24\tau}(\omega - 4\kappa^2 a^2 + 4\kappa^4 a^4);$$

$$c_5^{(3)} = \frac{ah^4}{120}[-5\omega + (1 + 4\kappa^2 a^2)(4 - \kappa^2 a^2)];$$

ω is a parameter of the scheme which is chosen empirically taking into account the stability condition.

The choice of the parameter ω influences the behaviour of the dispersion. In Fig. 8.1 plots of $\varepsilon_{\Delta\Phi_h}$ as a function of ξ are given for the parameter values $\kappa a = 0.3$ (see Fig. 8.1a), and $\kappa a = 0.8$ (see Fig. 8.1b). The calculations show that the minimum of $\varepsilon_{\Delta\Phi_h}$ is encountered for $c_5^{(3)} = 0$.

8.4 Geometrical Illustration of Dissipative and Dispersive Errors

We consider now some geometrical illustrations of dissipative and dispersive features of difference schemes using as an example the three-point difference schemes of the form (8.5), frequently encountered in practice. In this case we get

$$\varrho = 1 - \nu(1 - \cos\xi) + i\kappa a \sin\xi.$$

For fixed values of ν and κa the graph of ϱ has the form of an ellipse ($\xi \in [0, 2\pi]$)

$$\frac{[\text{Re}\{\varrho\} - (1 - \nu)]^2}{\nu^2} + \frac{[\text{Im}\{\varrho\}]^2}{\kappa^2 a^2} = 1,$$

with semiaxes ν and κa and the centre on the real axis in the point $1 - \nu$.

For all values of κa and ν the ellipse intersects the unit circle in the point $(1,0)$ which is in conformity with the compatibility condition. As already shown (see Sect. 8.1) the dissipative effects of a difference scheme can quantitatively be expressed by the quantity χ in which $|\varrho|$ is different from 1, and therefore the geometrical quantity of the dissipation is equal to the distance between corresponding points of the ellipse and the unit circle. In the point $(1,0)$ the radius of curvature is equal to $\kappa^2 a^2/v$. For the Lax-Wendroff-scheme the radius of curvature in the point $(1,0)$ is equal to 1 which underlines the high accuracy of the given scheme among the considered class of three-point formulas. Especially for small ξ the deviation of the ellipse from the unit circle is of order ξ^4. For the other three-point schemes this deviation is of order ξ^2 (see Sect. 8.2). The ellipse which corresponds to the Lax' scheme ($v = 1$) deviates very rapidly from the unit circle, and the radius of curvature of the ellipse in the point $(1,0)$ is equal to $\kappa^2 a^2$. This ellipse intersects for all values of κa the point $(-1,0)$.

All ellipses for fixed values of κa but different v are encountered in a region which is defined by the left half of the ellipse of the Lax' scheme and the right half of the ellipse of the Lax-Wendroff-scheme, and they connect straight lines which intersect the points $(0, \pm \kappa a)$ parallel to the real axis. The centres of the ellipses are situated on the real axis between the points 0 and $1 - \kappa^2 a^2$ (between the centres of the ellipses of the Lax' scheme and the Lax-Wendroff-scheme).

The ellipse which corresponds to the Lax' scheme extends along the real axis. All schemes for which $v > \kappa a$ show the same property.

In the case of the Lax-Wendroff-scheme the ellipse has its major axis along a straight line which is parallel to the imaginary axis. All schemes for which $v < \kappa a$ show the same property. The Godunov scheme with $v = \kappa a$ leads to a circle for all values of κa and the given scheme can be considered as a central scheme between the Lax' and the Lax-Wendroff-scheme. For $\kappa a = 1$ all ellipses coincide with the unit circle, and therefore in this case the dissipative and dispersive errors vanish and the schemes lead to an exact solution of the original problem.

8.5 Classification of Difference Schemes According to Dissipative Properties

Using the differential representations difference schemes for the hyperbolic equations can be classified according to their dissipative character:

1) If $\chi > 0$ for all $0 < |\xi| \leq \pi$, and if $\chi = 0$ for $\xi = 0$, then the difference scheme is strongly dissipative [dissipative in the sense of Rozhdestvenskii-Yanenko-Richtmyer (see Sect. 5.1)] of order $2d$, where C_{2d} is the first nonzero coefficient with even index in the differential representation,

2) If $\chi \geq 0$ for all $|\xi| \leq \pi$, then the difference scheme is dissipative,

3) If $\chi = 0$ for all $|\xi| \leq \pi$, then the difference scheme is called conservative.

In the case of two-level difference schemes for equation (8.1) we get

$$\chi = 4 z^d \frac{S_0(z)}{S_1(z)}, \quad 0 \le z \le 1,$$

$$z = \sin^2\left(\frac{\xi}{2}\right),$$

(see Sect. 4.4) and the strong dissipativity is equivalent to the fact that the inequality $S_0(z) > 0$ is satisfied.

8.6 Some Remarks on Using Finite Number of Terms of the Differential Approximation

We know already that keeping only a finite number of terms in the differential representation (we consider the corresponding differential approximation) leads already to a definite information about the difference scheme.

We will concentrate our interest on a special way of treatment which is devoted to the possibility of getting information taking into account a finite number of terms in the differential representation. We consider for the differential equation (8.1) the difference scheme (8.2), the differential representation of which has the form (8.3), whereas $c_j = O(h^{j-1})$. The representation of a continuous function on a net is bounded in the sense that it corresponds to a finite choice of wave numbers $2h \le \lambda \le O(1)$; the shortest wave length is $2h$ and the longest is of the order of the dimension of the region. For $h \to 0$ the ratio of the wave length λ to the step size h is bounded between 2 and $O(1/h)$: $2 \le \lambda/h \le O(1/h)$. Consequently, it is reasonable to consider the behaviour of harmonic waves for different limit processes if $h/\lambda \to 0$ in an arbitrary way. We put, for example, $h/\lambda = O(h^{1-m})$; $(m = \frac{1}{2}, \frac{2}{3}, \ldots,)$. If we assume that $\lambda h^{-m} = $ const and $\zeta = h/\lambda \to 0$, in the differential representation (8.3) we may omit terms of higher order in ζ assuming that all derivatives with respect to ζ are equal to 1. Such a treatment means that we omit wave numbers k which are larger than the considered above whereas small-scale effects are not taken into account. Under such assumptions the term $c_j \frac{\partial^j u}{\partial x^j}$ in equation (8.3) will be equivalent (up to a constant factor) to the expression

$$\frac{h^{j-1}}{\lambda^j} \bar{c}_j \cdot \frac{\partial^j u}{\partial x^j},$$

where $\bar{c}_j = O(1)$; $c_1 = \bar{c}_1 = a$ or–equivalently–to the expression

$$h^{j-jm-1} \bar{c}_j \frac{\partial^j u}{\partial x^j}.$$

Then for $m = 1/2$ in equation (8.3) one can cancel all terms containing c_k $(k \geq 3)$, for $m = 2/3$ all terms containing c_k $(k \geq 4)$ and for $m = 3/4$ all terms containing c_k $(k \geq 5)$ etc. i.e. one considers the first, second, third etc. differential approximation. In the limit $\lambda/h = \text{const}\,(m = 1)$ it is also necessary to consider all differential representations, i.e. the difference scheme. Thus not taking into account small-scale effects the behaviour of a difference solution in the limit can be investigated not using the difference scheme.

In an analogous way also in the case of a differential equation not of hyperbolic type the same considerations can be carried out. We introduce, for example, the difference scheme

$$\frac{u^{n+1}(x) - u^n(x)}{\tau} = \sigma \frac{u^n(x+h) - 2u^n(x) + u^n(x-h)}{h^2}.$$

for the parabolic equation:

$$\frac{\partial u}{\partial t} = \sigma \frac{\partial^2 u}{\partial x^2}.$$

The differential representation in this case has the form:

$$\frac{\partial u}{\partial t} = \sigma \frac{\partial^2 u}{\partial x^2} + h^2 \bar{c}_4 \frac{\partial^4 u}{\partial x^4} + h^4 \bar{c}_6 \frac{\partial^6 u}{\partial x^6} + \ldots,$$

$$\frac{\tau}{h^2} = \text{const}.$$

Then treating the problem in an analogous way as in the example before we will prove that for $m = 1/2$ all terms with coefficients $c_{2l}(l \geq 2)$ in the differential representation can be eliminated, for $m = 2/3$ all terms with $c_{2l}(l \geq 3)$ etc. i.e. one considers the original differential equation, the first, second etc. differential approximation.

8.7 Connection Between Dispersion, Dissipation and Errors of Difference Schemes

We consider the question of the connection between dispersion, dissipation and the error of difference schemes.

The error satisfies the relation

$$\int\limits_{-\infty}^{\infty} [u(x, t + \tau) - u^{n+1}(x)]^2 \, dx = \frac{2\pi}{h^2} \int\limits_{-\infty}^{\infty} |\hat{u}(k,t)|^2 \, |\Omega - \varrho|^2 \, d\xi,$$

where $\hat{u}(k, t)$ is the Fourier transform of $u(x, t)$. Thus, for defining the minimum of the error it is sufficient to define the minimum of the functional

$$\Psi := \int\limits_{-\infty}^{\infty} \phi\left(\frac{\xi}{h}\right) |\Omega - \varrho|^2 \, d\xi.$$

where the weight function should be equal to the modulus of the square of the Fourier transform of the exact solution in the moment t.

The following chain of equations shows the connection between the magnitude of the error and the dissipation and dispersion of a difference scheme:

$$|\Omega - \varrho| = |\Omega||1 - \varrho\Omega^{-1}| = |1 - |\varrho| \, e^{i\Delta\Phi_h}| = |1 - (1-\chi) \, e^{i\Delta\Phi_h}|.$$

Consequently, we get

$$\Psi = \int_{-\infty}^{\infty} \phi\left(\frac{\xi}{h}\right) |1 - (1-\chi) e^{i\Delta\Phi_h}|^2 \, d\xi.$$

It is clear that the smaller the functional Ψ the better the difference scheme approximates the solution of the differential problem. Now the question arises how to minimize the functional Ψ. In practise the exact solution is unknown, but a series of conclusions can be derived also with an a priori chosen weight-function [193].

The following theorem holds which is analogous to the theorem of the paper [193], but its proof is carried out in terms of the differential representations.

· ·

Theorem 8.1. For a stable difference scheme of form (8.2) a necessary condition for the minimum of the error is the maximal order of approximation for a given set of points which build up a difference star. This is the same as the minimum of the functional Ψ for $\phi(x) = \delta(x)$, where $\delta(x)$ is the δ-function.

· ·

Proof. We consider two difference schemes of the form (8.2) which may have the order of approximation p and q, respectively, $(p < q)$; the differential representations are written in the following way:

$$\frac{\partial u}{\partial t} + a \frac{\partial u}{\partial x} = \sum_{l=p+1}^{\infty} c_l^{(1)} \frac{\partial^l u}{\partial x^l};$$

$$\frac{\partial u}{\partial t} + a \frac{\partial u}{\partial x} = \sum_{l=q+1}^{\infty} c_l^{(2)} \frac{\partial^l u}{\partial x^l}.$$

Then

$$|\Omega - \varrho^{(1)}| = |1 - e^{\tau r^{(1)}}| = \left| \sum_{l=p+1}^{\infty} \beta_l^{(1)} \xi^l \right|;$$

$$|\Omega - \varrho^{(2)}| = |1 - e^{\tau r^{(2)}}| = \left| \sum_{l=q+1}^{\infty} \beta_l^{(2)} \xi^l \right|,$$

where

$$\tau r^{(1)} = \sum_{l=p+1}^{\infty} (-i\xi)^l \frac{\tau}{h^l} c_l^{(1)};$$

$$\tau r^{(2)} = \sum_{l=q+1}^{\infty} (-i\xi)^l \frac{\tau}{h^l} c_l^{(2)};$$

$$\Sigma_1 := \sum_{l=p+1}^{\infty} \beta_l^{(1)} \xi^l = -\tau r^{(1)} + \frac{\tau^2 [r^{(1)}]^2}{2} + \dots$$

$$= -\frac{\tau}{h^{p+1}} c_{p+1}^{(1)} (-i\xi)^{p+1} + O(\xi^{p+2});$$

$$\Sigma_2 := \sum_{l=q+1}^{\infty} \beta_l^{(2)} \xi^l = -\frac{\tau}{h^{q+1}} c_{q+1}^{(2)} (-i\xi)^{q+1} + O(\xi^{q+2}).$$

The series Σ_1 and Σ_2 are uniformly convergent in an arbitrary finite region $|\xi| \leq R$.

The δ-function can be defined by a sequence of the following functions:

$$\phi_n(k) = \frac{n}{h\sqrt{\pi}} e^{-n^2 k^2}.$$

Then

$$\Psi_2 - \Psi_1 = I_R + I_\infty,$$

where

$$I_R := \frac{n}{h\sqrt{\pi}} \int_{-R}^{R} e^{-n^2 k^2} (|\Sigma_2|^2 - |\Sigma_1|^2) d\xi;$$

$$I_\infty := \frac{n}{h\sqrt{\pi}} \left[\int_{-\infty}^{-R} e^{-n^2 k^2} (|\Sigma_2|^2 - |\Sigma_1|^2) d\xi + \int_{R}^{\infty} e^{-n^2 k^2} (|\Sigma_2|^2 - |\Sigma_1|^2) d\xi \right].$$

Further, following the paper [193] we will show that $\Psi_2 - \Psi_1 < 0$ if n is sufficiently large. This is the proof of the theorem. □

If h is so small that $\phi(\xi)$ is different from zero only in a small vicinity of $\xi = 0$, from the theorem follows that the difference scheme is optimal with the largest order of approximation. But in practise h is finite and it is desired to run the calculation for large values of h. Then $\phi(\xi)$ should not be zero in some finite vicinity of $\xi = 0$ and, consequently, it is necessary to minimize the functional Ψ with a weight-function which is different from the δ-function. For the minimization of the functional Ψ with some weight-function some estimates of the coefficients of the scheme are derived. These estimates determine the difference scheme.

We choose, for example, a three-point difference scheme of form (8.2) with $\alpha = -1, 0, 1$:

$$u^{n+1}(x) = \sum_{\alpha=-1}^{1} b_\alpha u^n(x + \alpha h).$$

The compatibility conditions lead to the following connection between the coefficients b_1, b_0, b_{-1}:

$$b_1 = \tfrac{1}{2}(1 - \kappa a - b_0);$$
$$b_{-1} = \tfrac{1}{2}(1 - \kappa a + b_0).$$

As a priori given weight-functions we consider the following functions:

1) $\phi(x) = \delta(x)$;
2) $\phi(x) = 1/x^2$;
3) $\phi(x) = \delta(x - \pi/h)$.

Then the minimization of the functional Ψ with the weight-functions 1)–3) leads to the following choice of the coefficient

1) $b_0 = 1 - \kappa^2 a^2$,   (Lax-Wendroff-scheme),
2) $b_0 = 1 - |\kappa a|$,   (majorant-scheme),
3) $b_0 = \cos^2(\kappa a \pi/2)$.

As expected on the basis of the theorem the minimization of the functional Ψ with a weight-function $\phi(x) = \delta(x)$ leads to the maximal order of approximation of the difference scheme for the given points of the difference star.

Using $\phi(x) = x^{-2}$ as a weight-function (this is the negative of the Fourier transform of a step-function) leads to schemes which give better results of the calculation for one time step with initial data given in the shape of a step-function.

In the case of a weight-function $\phi(x) = \delta(x - \pi/h)$ we get difference schemes which resolve well short waves.

9. Application of the Method of Differential Approximation to the Investigation of the Effects of Nonlinear Transformations

9.1 Introduction

Investigating difference schemes for nonlinear hyperbolic systems of equations it may happen that nonlinear transformations of the given system of equations lead to different weak solutions. Thus the same difference scheme which approximates two systems of nonlinear hyperbolic equations which are connected by a nonlinear transformation may lead to two different weak solutions [161]. For the difference scheme due to Lax and Lax-Wendroff it was shown how to introduce additional terms in the transformation of the scheme for the conservation of the weak solution. In the following it is shown that the choice of additional terms for the calculation of nonlinear transformations is connected with the differential approximation of difference schemes, expecially with the notation of the equivalence of difference schemes.

9.2 Equivalence of Difference Schemes

We consider in some region G of the variables x and t two systems of differential equations:

$$F_1\left(x, t, w, \frac{\partial w}{\partial t}, \frac{\partial w}{\partial x}\right) = 0; \tag{9.1}$$

$$F_2\left(x, t, w, \frac{\partial w}{\partial t}, \frac{\partial w}{\partial x}\right) = 0, \tag{9.2}$$

where $w = w(x, t)$ and F_j are m-dimensional vector functions ($j = 1, 2$). We assume that the systems of equations (9.1, 2) can be solved in $\mathbb{C}_1(G)$ for $\partial w/\partial t$. Let be

$$BF_1\left(x, t, w, \frac{\partial w}{\partial t}, \frac{\partial w}{\partial x}\right) = F_2\left(x, t, w, \frac{\partial w}{\partial t}, \frac{\partial w}{\partial x}\right),$$

where $B = B(x, t, w)$ is a matrix which has an inverse in $\mathbb{C}(G)$.

Definition. Two difference schemes

$$\Lambda_1(w, R_1, \tau) = 0; \qquad (9.3)$$
$$\Lambda_2(w, R_2, \tau) = 0, \qquad (9.4)$$

which approximate the systems of equations (9.1, 2), respectively, will be called equivalent if there exists a difference operator B_τ which has an inverse in $\mathbb{C}(G)$ that B_τ and B_τ^{-1} can be represented as follows:

$$B_\tau w = Bw + O(\tau);$$
$$B_\tau^{-1} w = B^{-1} w + O(\tau);$$

and

$$B_\tau \Lambda_1(w, R_1, \tau) = \Lambda_2(w, R_2, \tau).$$

Here τ and h are the steps of the difference scheme with respect to t and x, respectively; $t = n\tau$; $\tau/h = \kappa = \text{const}$; R_j is the fixed set of net points ($j = 1, 2$).

If the difference schemes (9.3, 4) approximate the system of equations (9.1, 2) with an order γ, then their first differential approximations have the form

$$\Phi_1 = F_1 + \tau^\gamma F_1^{(1)};$$
$$\Phi_2 = F_2 + \tau^\gamma F_2^{(1)}.$$

The following lemma holds.

Lemma. If the difference schemes (9.3, 4) are equivalent, then also their first differential approximations are equivalent, i.e.

$$B\Phi_1 = \Phi_2$$

or

$$BF_1^{(1)} = F_2^{(1)}. \qquad (9.5)$$

Thus we have derived a necessary criterion for equivalence of difference schemes. The control of this condition is formal and simple. To find the first differential approximation of a difference scheme does not lead to difficulties especially the described problem can be solved by a computer.

The notation of equivalence of difference schemes uses the important idea of Samarskii and Tikhonov on fully conservative difference schemes [191, 194, 195], for which the basic assumption is the demand on difference schemes to reflect those basic features of the medium which are described by the system of differential equations.

Definition. The difference schemes (9.3, 4) are called weakly equivalent, if their first differential approximation is equivalent.

Thus for the equivalence of difference schemes their weak equivalence is necessary. In the following examples will be given when the weak equivalence is sufficient for the equivalence of difference schemes.

9.3 The Fluid Equations Including Gravity

We consider the system of equations of an incompressible, nonviscous, fluid [75, 161] under gravitational forces

$$F_1 := \frac{\partial w}{\partial t} + \frac{\partial f}{\partial x} + q = 0, \tag{9.6}$$

where

$$w = \begin{pmatrix} m \\ v \end{pmatrix}; \quad f = \begin{pmatrix} um + \frac{gv^2}{2} \\ m \end{pmatrix}; \quad q = \begin{pmatrix} gvl_x \\ 0 \end{pmatrix}; \quad m = uv;$$

u is the velocity, g is the gravitational acceleration, v is the height of the level of the fluid above the lower boundary $l = l(x)$.

By the help of a nonlinear transformation which is defined by the matrix

$$B = \begin{pmatrix} \dfrac{1}{v} & -\dfrac{u}{v} \\ 0 & 1 \end{pmatrix}$$

the system of equations (9.6) is transformed into the form

$$F_2 := \frac{\partial \tilde{w}}{\partial t} + \frac{\partial \tilde{f}}{\partial x} + \tilde{q} = 0. \tag{9.6a}$$

where

$$\tilde{w} = \begin{pmatrix} u \\ v \end{pmatrix}; \quad \tilde{f} = \begin{pmatrix} \dfrac{u^2}{2} + gv \\ uv \end{pmatrix}; \quad \tilde{q} = \begin{pmatrix} gl_x \\ 0 \end{pmatrix}.$$

We approximate the systems of equations (9.6, 6a) by the following difference schemes, respectively:

$$\frac{\Delta_0 w^n(x)}{\tau} + \frac{\Delta_1 + \Delta_{-1}}{2h} f^n(x) + q^n(x)$$

$$= \frac{h^2}{2\tau} \frac{\Delta_1 \Delta_{-1}}{h^2} [\lambda w^n(x) + \mu f^n(x)]; \tag{9.7}$$

$$\frac{\Delta_0 \tilde{w}^n(x)}{\tau} + \frac{\Delta_1 + \Delta_{-1}}{2h} \tilde{f}^n(x) + \tilde{q}^n(x)$$

$$= \frac{h^2}{2\tau} \frac{\Delta_1 \Delta_{-1}}{h^2} [\lambda \tilde{w}^n(x) + \mu \tilde{f}^n(x)]. \tag{9.7a}$$

Here λ and μ are constants (especially for $\lambda = 1$, $\mu = 0$ we get the Lax' scheme, for $\lambda = 0$, $\mu = \sigma\kappa$ the Rusanov-scheme).

The first differential approximations of the schemes (9.7, 7a) have the forms

$$F_1 = \frac{h^2}{2\tau}(\lambda I + \mu A - \kappa^2 A^2)\frac{\partial^2 w}{\partial x^2} + \frac{h^2}{2\tau}[\mu A_x - \kappa^2 (A^2)_x]\frac{\partial w}{\partial x} - \frac{\tau}{2}(Aq)_x$$
$$- \frac{\tau}{2}\tilde{q}\cdot\frac{\partial uv}{\partial x};\tag{9.8}$$

$$F_2 = \frac{h^2}{2\tau}(\lambda I + \mu \tilde{A} - \kappa^2 \tilde{A}^2)\frac{\partial^2 \tilde{w}}{\partial x^2} + \frac{h^2}{2\tau}[\mu \tilde{A}_x - \kappa^2 (\tilde{A}^2)_x]\frac{\partial \tilde{w}}{\partial x} - \frac{\tau}{2}(\tilde{A}\tilde{q})_x.$$

$$A:=\frac{df}{dw}; \qquad \tilde{A}:=\frac{d\tilde{f}}{d\tilde{w}}.\tag{9.8a}$$

The system of equations (9.6a) leads to solutions which are analogous to the solutions of the systems of equations (9.6) for smooth flows, but if a discontinuity arises the system of equations (9.6a) in contrast to the system of equations (9.6) leads to unsteady solutions which have no physical meaning.

The question arises how to alter the difference scheme (9.7a) so that it will not lead to solutions which have no physical counterpart and that it will give the same solutions as the difference scheme (9.7).

The difference schemes (9.7, 7a) approximate the equations (9.8, 8a) with second order of approximation, respectively, and it is clear that with an accuracy $O(\tau^2)$ the solutions of the schemes will coincide, if the schemes are weakly equivalent.

The difference schemes (9.7, 7a) are not equivalent. Indeed, the condition (9.5) is not satisfied, as in this case

$$\tau^y F_1^{(1)} = \frac{h^2}{2\tau}(\lambda I + \mu A - \kappa^2 A^2)\frac{\partial^2 w}{\partial x^2} + \frac{h^2}{2\tau}[\mu A_x - \kappa^2 (A^2)_x]\frac{\partial w}{\partial x}$$
$$- \frac{\tau}{2}(Aq)_x - \frac{\tau}{2}\tilde{q}\frac{\partial uv}{\partial x};$$

$$\tau^y F_2^{(1)} = \frac{h^2}{2\tau}(\lambda I + \mu \tilde{A} - \kappa^2 \tilde{A}^2)\frac{\partial^2 \tilde{w}}{\partial x^2} + \frac{h^2}{2\tau}[\mu \tilde{A}_x - \kappa^2 (\tilde{A}^2)_x]\frac{\partial \tilde{w}}{\partial x} - \frac{\tau}{2}(\tilde{A}\tilde{q})_x;$$

$$B\tau^y F_1^{(1)} = \tau^y F_2^{(1)} + \tau R,$$

where

$$R:=\begin{pmatrix} r_1 \\ r_2 \end{pmatrix};$$

$$r_1:=\frac{h^2 \lambda}{\tau^2}\frac{u_x v_x}{v} + \frac{h^2 \mu}{\tau^2}\left(\frac{u}{v}u_x v_x + u_x^2 + gv_x^2\right) - \frac{(uv)_x}{v}(uu_x + gl_x + gv_x);$$

$$r_2:=0.$$

If we choose instead of the difference scheme (9.7a) the difference scheme

$$\frac{\Delta_0 \tilde{w}^n(x)}{\tau} + \frac{\Delta_1 + \Delta_{-1}}{2\tau}\tilde{f}^n(x) + \tilde{q}(x) = \frac{h^2}{2\tau}\frac{\Delta_1 \Delta_{-1}}{h^2}[\lambda \tilde{w}^n(x) + \mu \tilde{f}^n(x)] + \tau R,\tag{9.9}$$

then its first differential approximation looks like

$$F_2 = \frac{h^2}{2\tau}(\lambda I + \mu\tilde{A} - \kappa^2\tilde{A}^2)\frac{\partial^2\tilde{w}}{\partial x^2} + \frac{h^2}{2\tau}[\mu\tilde{A}_x - \kappa^2(\tilde{A}^2)_x]\frac{\partial\tilde{w}}{\partial x} - \frac{\tau}{2}(\tilde{A}\tilde{q})_x + \tau R,$$

and the defined scheme (9.9) will be weakly equivalent to the scheme (9.7).

Thus the additional term which is necessary to introduce into the difference scheme (9.7a) so that the scheme with an accuracy $O(\tau^2)$ leads to the same solutions as the scheme (9.7) has the form

$$\tau R = \begin{pmatrix} \tau r_1 \\ 0 \end{pmatrix}.$$

For $\lambda = 1$, $\mu = 0$ (Lax' scheme) we get an expression which was found in the contribution [161].

In an analogous way additional terms are found for difference schemes of higher order of approximation.

9.4 The Equations of Gas Dynamics

The system of equations of one-dimensional gas dynamics in Lagrangean coordinates is normally considered in one of three equivalent forms:

$$\left.\begin{aligned} \frac{\partial u}{\partial t} + \frac{\partial p}{\partial x} &= 0, \\[2mm] \frac{\partial v}{\partial t} - \frac{\partial u}{\partial x} &= 0, \\[2mm] \frac{\partial \varepsilon}{\partial t} + p\frac{\partial u}{\partial x} &= 0; \end{aligned}\right\} \tag{9.10a}$$

$$\left.\begin{aligned} \frac{\partial u}{\partial t} + \frac{\partial p}{\partial x} &= 0, \\[2mm] \frac{\partial v}{\partial t} - \frac{\partial u}{\partial x} &= 0, \\[2mm] \frac{\partial E}{\partial t} + \frac{\partial(up)}{\partial x} &= 0; \end{aligned}\right\} \tag{9.10b}$$

$$\left.\begin{aligned} \frac{\partial u}{\partial t} + \frac{\partial p}{\partial x} &= 0, \\[2mm] \frac{\partial v}{\partial t} - \frac{\partial u}{\partial x} &= 0, \\[2mm] \frac{\partial \varepsilon}{\partial t} + p\frac{\partial v}{\partial t} &= 0. \end{aligned}\right\} \tag{9.10c}$$

The systems of equations (9.10b, 10c) follow from the systems of equations (9.10a) by multiplying the latter from left by the matrices:

$$B_1 = \begin{pmatrix} 1 & 0 & 0 \\ 0 & 1 & 0 \\ u & 0 & 1 \end{pmatrix} \quad \text{and} \quad B_2 = \begin{pmatrix} 1 & 0 & 0 \\ 0 & 1 & 0 \\ 0 & p & 1 \end{pmatrix},$$

respectively.

We consider three families of difference schemes of first order of approximation (9.11a–c) which approximate the systems of equations (9.10a–c), the first two equations of them coincide:

$$\Delta_0 \phi_1^1 + \kappa \Delta_1 \phi_1^4 = 0, \tag{9.11a}$$

$$\Delta_0 \phi_1^2 - \kappa \Delta_1 \phi_2^1 = 0, \tag{9.11b}$$

but the third family has the form

$$\Delta_0 \phi_1^3 + \kappa \phi_2^4 \Delta_1 \phi_3^1 = 0;$$
$$\Delta_0 (\phi_2^3 + \tfrac{1}{2} \phi_4^1 \phi_5^1) + \kappa \Delta_1 (\phi_3^4 \phi_6^1) = 0, \tag{9.11c}$$
$$\Delta_0 \phi_3^3 + \kappa \phi_4^4 \Delta_0 \phi_2^2 = 0,$$

where

$$\phi_j^i = \sum_{\alpha \in \mathscr{I}_{ij}} a_\alpha T_0^{\alpha_0} T_1^{\alpha_1} u^i; \quad u^1 = u, \quad u^2 = v, \quad u^3 = \varepsilon, \quad u^4 = p;$$

$a_\alpha, \alpha_0, \alpha_1$ are real numbers; $\mathscr{I}_{ij}$ are some finite sets of double indices. In the case of two-layer difference schemes we get

$$A_{11}^0 = A_{21}^0 = A_{22}^0 = A_{31}^0 = 0,$$
$$A_{33}^0 = A_{32}^0 = A_{14}^0 = A_{15}^0 = 0,$$
$$A_{ij}^k = \sum_{\alpha \in \mathscr{I}_{ij}} \alpha_k a_\alpha; \quad k = 0, 1,$$

and in the first differential approximation of the schemes (9.11a–c) the first two equations coincide:

$$\frac{\partial u}{\partial t} + \frac{\partial p}{\partial x} + \tau \xi_{11} \frac{\partial}{\partial x} \left(a^2 \frac{\partial u}{\partial x} \right) + h \eta_{11} \frac{\partial^2 p}{\partial x^2} = 0;$$

$$\frac{\partial v}{\partial t} - \frac{\partial u}{\partial x} + \tau \xi_{21} \frac{\partial^2 p}{\partial x^2} - h \eta_{21} \frac{\partial^2 u}{\partial x^2} = 0;$$

and the third set of equations has the following form:

$$\frac{\partial \varepsilon}{\partial t} + p \frac{\partial u}{\partial x} + \tau \xi_{31} a^2 \frac{\partial^2 u}{\partial x^2} + \tau \xi_{32} p \frac{\partial^2 p}{\partial x^2} + h \eta_{31} p \frac{\partial^2 u}{\partial x^2} + h \eta_{32} \frac{\partial p}{\partial x} \frac{\partial u}{\partial x} = 0;$$

$$\frac{\partial E}{\partial t} + \frac{\partial (up)}{\partial x} + \tau \xi_{41} \frac{\partial (a^2 u u_x)}{\partial x} + \tau \xi_{42} \frac{\partial}{\partial x}\left(p\frac{\partial p}{\partial x}\right) + h\eta_{41}\frac{\partial}{\partial x}\left(u\frac{\partial p}{\partial x}\right)$$

$$+ h\eta_{42}\frac{\partial}{\partial x}\left(p\frac{\partial u}{\partial x}\right) = 0;$$

$$\frac{\partial \varepsilon}{\partial t} + p\frac{\partial v}{\partial t} + \tau \xi_{51}a^2\left(\frac{\partial u}{\partial x}\right)^2 + h\eta_{51}p\frac{\partial^2 u}{\partial x^2} + h\eta_{52}\frac{\partial p}{\partial x}\cdot\frac{\partial u}{\partial x} = 0,$$

where

$$\xi_{11} = \tfrac{1}{2} - A^0_{41}; \quad \xi_{21} = \tfrac{1}{2} - A^0_{12}; \quad \xi_{31} = \tfrac{1}{2} - A^0_{42}; \quad \xi_{32} = \tfrac{1}{2} - A^0_{13};$$

$$\xi_{41} = \tfrac{1}{2} - A^0_{43}; \quad \xi_{42} = \tfrac{1}{2} - A^0_{16}; \quad \xi_{51} = \tfrac{1}{2} - A^0_{44};$$

$$\eta_{11} = \tfrac{1}{2} - A^1_{11} + A^1_{41}; \quad \eta_{21} = \tfrac{1}{2} - A^1_{21} + A^1_{12}; \quad \eta_{31} = \tfrac{1}{2} - A^1_{31} + A^1_{33};$$

$$\eta_{32} = A^1_{42} - A^1_{31}; \quad \eta_{41} = \tfrac{1}{2}(1 - A^1_{14} + A^1_{15} + A^1_{43});$$

$$\eta_{42} = \tfrac{1}{2} - A^1_{32} - A^1_{33}; \quad \eta_{51} = A^1_{22} - A^1_{33}; \quad \eta_{52} = A^1_{44} - A^1_{33}.$$

The following theorem holds.

. . . .

Theorem 9.1.
1) A necessary condition for the equivalence of the difference scheme (9.11a) to the conservative difference scheme (9.11b) is, that

$$\xi_{11} = \xi_{31}; \quad \xi_{32} = 0; \quad \eta_{32} = \eta_{11} + \eta_{31},$$

a sufficient condition is, that

$$\phi^4_2 = \phi^4_1, \quad \tfrac{1}{2}(T_0 + E)\phi^1_1 = T_1\,\phi^1_3.$$

2) A necessary condition for the equivalence of the difference scheme (9.11a) to the difference scheme (9.11c) is

$$\xi_{21} = \xi_{32},$$

and a sufficient condition is

$$\phi^1_2 = \phi^1_3.$$

3) If the difference scheme (9.11a) is equivalent to the difference scheme (9.11c), then it has the property K.

4) If the difference scheme (9.11a) is equivalent to the difference scheme (9.11b) and has the property K then the artificial viscosity in the first differential approximation of the scheme (9.11a) joins additively with the pressure.

. . . .

In the last case the first differential approximation of the difference scheme (9.11a) has the form

$$\frac{\partial u}{\partial t} + \frac{\partial \bar{p}}{\partial x} = 0;$$

$$\frac{\partial v}{\partial t} - \frac{\partial u}{\partial x} = 0;$$

$$\frac{\partial \varepsilon}{\partial t} + \bar{p}\frac{\partial u}{\partial x} = 0,$$

where

$$\bar{p} = p + \tau \xi_{11} a^2 \frac{\partial u}{\partial x}.$$

From the formulated statements the following statements can be derived:

1) If the difference scheme (9.11a) is equivalent to the difference scheme (9.11b), then it is implicit.

2) A necessary and sufficient condition for the equivalence of Samarskii's and Popov's scheme [13, 196]

$$\Delta_0 u''(x) + \kappa \Delta_{-1} p^{(\sigma_1)}\left(x + \frac{h}{2}\right) = 0;$$

$$\Delta_0 v''\left(x + \frac{h}{2}\right) - \kappa \Delta_1 u^{(\sigma_2)}(x) = 0; \qquad (9.12)$$

$$\Delta_0 \varepsilon''\left(x + \frac{h}{2}\right) + \kappa p^{(\sigma_1)}\left(x + \frac{h}{2}\right)\Delta_1 u^{(\sigma_3)}(x) = 0$$

to the difference scheme (9.11b) is

$$\sigma_3 = 1/2.$$

3) A necessary and sufficient condition for the equivalence of the difference scheme (9.12) to the difference scheme (9.11c) is

$$\sigma_2 = \sigma_3.$$

4) The difference scheme (9.12) for $\sigma_3 = \sigma_2 = 1/2$ has the property K and the artificial viscosity in the first differential approximation is an additive term to the pressure p, and under the condition $\sigma_1 > 1/2$ this viscosity leads in the first differential approximation to a growth of the entropy.

Analogous statements are true also considering difference schemes for the systems of gasdynamic equations in Eulerian coordinates. The formulation of the statements represent criteria for checking the equivalence of difference schemes in terms of the first differential approximation.

10. Investigation of Monotonicity of Difference Schemes

10.1 Introduction

Applying difference schemes, it is very important to know in advance the character of the behaviour of a numerical solution in the vicinity of a discontinuity, especially whether or not oscillations will occur. In Godunov's publication [191] for linear equations the notation of a monotonic difference scheme is introduced. Constructing difference schemes for nonlinear equations normally demands the monotonic property of these schemes for linear equations. But calculations show that the monotonic property in the nonlinear case often is hurt due to causes which can not be discussed in the frame of linear theory. In Rusanov's publication [197] a method for estimating the monotonic property of difference schemes is presented based on the investigation of the limit profile which is produced by the scheme during the calculation of a solution with the character of a jump. Finally as a result this method leads to an estimate of the roots of some transcendental equation. The proposed method below for estimating the monotonic property uses the differential representation of a difference scheme.

10.2 Moving Shock with Constant Velocity

We consider for the equation

$$\frac{\partial u}{\partial t} + u\frac{\partial u}{\partial x} = 0$$

the problem of a steadily moving discontinuity with the velocity D

$$u(x, t) = \begin{cases} u_1, & x - Dt \leq 0, \\ u_2, & x - Dt > 0. \end{cases}$$

For the solution of the stated problem by virtue of a stable difference scheme we get a smeared-out profile with or without oscillations. The usual demand for monotonic property within the zone of the discontinuity according to a linear criterion does not always lead to a vanishing of the oscillations. As the solution of the original problem outside of the zone of the jump is constant the coefficients of the differential equation and of the difference scheme outside of the zone of the jump can be considered as constant and thus the problem of estimating the monotonicity of the nonlinear scheme for the nonlinear equation can be transferred to the estimate of the monotonic property of solutions of the corresponding scheme with constant coefficients for the equation

$$\frac{\partial u}{\partial t} + a \frac{\partial u}{\partial x} = 0. \tag{10.1}$$

Besides that as the solution of the original problem depends on the variable $x - Dt$ then also for the solution of the corresponding difference problem we will demand the dependence only on $x - Dt$.

We consider for the equation (10.1) the difference scheme

$$u^{n+1}(x) = S_h u^n(x) = \sum_\alpha b_\alpha u^n(x + \alpha h), \tag{10.2}$$

the differential representation of which has the form

$$\frac{\partial u}{\partial t} = \frac{1}{\tau} \ln S_h u = - a \frac{\partial u}{\partial x} + \sum_{l=2}^{\infty} c_l \frac{\partial^l u}{\partial x^l}. \tag{10.3}$$

We will look for a solution of equation (10.3) in the form of a function $v(\eta) = \tilde{u} + \omega(\eta)$, where $\eta = (x - Dt)/h$; $v(-\infty) = u_1$, $v(+\infty) = u_2$; under $\tilde{u}$ we understand either u_1 or u_2 depending on the sign of η. Then the function $\omega(\eta)$ satisfies the equation

$$- \kappa D \frac{d\omega}{d\eta} = \ln \sum_\alpha b_\alpha e^{\alpha(d/d\eta)} \omega, \tag{10.4}$$

where

$$\lim_{\eta \to \pm\infty} \omega(\eta) = 0.$$

Further we look for the solution of equation (10.4) in the form

$$\omega(\eta) = \mathrm{Re}\{c\,e^{v\eta}\},$$

where c is a constant, but v is a complex number. Then for v we get the equation

$$- \kappa D v = \ln \sum_\alpha b_\alpha e^{\alpha v},$$

or

$$e^{-\kappa D v} = \sum_\alpha b_\alpha e^{\alpha v}. \tag{10.5}$$

Equation (10.5) usually has an infinite number of roots whereas the real part of v defines how fast $\omega(\eta)$ approaches the limit value, but the imaginary part of v defines the character of the approaching (monotone or not). It is clear, that the asymptotic behaviour of $\omega(\eta)$ will be defined for $\eta \to +\infty$ by the root of equation (10.5) with maximal real part which lies in the left half-plane, but for $\eta \to -\infty$ by the root with minimal real part which lies in the right half-plane. The real character of the roots of equation (10.5) is the reason for the monotonicity.

The demand of monotonic property [real roots of equation (10.5)] bounds the choice of the coefficients b_α of the scheme and of the quantity κ and thus from the class of difference schemes of the form (10.5) in the considered sense monotonic schemes are selected.

If we set $z = e^v$ equation (10.5) leads to the equation

$$z^{-\kappa D} = \sum_\alpha b_\alpha z^\alpha. \tag{10.6}$$

In the case of three-point difference schemes ($\alpha = -1, 0, 1$) equation (10.6) has the form

$$\frac{1 - b_0}{2}(z - 2 + z^{-1}) + \tfrac{1}{2} \sum_{\alpha = -1}^{1} \alpha b_\alpha (z - z^{-1}) + 1 - z^{-\kappa D} = 0,$$

where $\sum_{\alpha = -1}^{1} \alpha b_\alpha$ is equal either to $-\kappa u_1$ or to $-\kappa u_2$ depending on the sign of η.
This equation is derived in the paper [197].

Numerical calculations underline the proposed access to an estimate of the feature of monotonicity of a difference scheme as the limitations which are imposed by this method lead to practically monotonic difference schemes.

11. Difference Schemes in an Arbitrary Curvilinear Coordinate System

11.1 Introduction

In this chapter a method is proposed for the construction of difference schemes on an arbitrary mesh starting from a difference scheme which is given on a uniform mesh in the same coordinate system. We study the question which of the characteristic properties of the original difference scheme are retained by the difference scheme on the arbitrary mesh.

The use of curvilinear coordinate systems for the solution of gasdynamic problems by means of finite difference methods admits the calculation of flows in regions of complex configuration taking into account the behavior of singularities of the flow. The application of curvilinear coordinate systems demands from the numerical analyst the solution of the problem of constructing difference schemes in such coordinate systems. The problem arises how to transform the difference schemes (when a coordinate transformation is given) which have good properties in a Cartesian coordinate system and especially how to transform difference schemes when the coordinate transformation is nonsingular.

Let $\mathbb{R}^n$ be the n-dimensional Euclidean space, $x, y \in \mathbb{R}^n$; let $\psi: \mathbb{R}^n \to \mathbb{R}^n$ $[\psi(x) = y]$ be a nonsingular mapping; $\mathbb{B}(\omega)$ and $\mathbb{B}(\tilde{\omega})$ are spaces of mesh functions which are defined on the meshes $\omega \in \mathbb{R}^n$ and $\tilde{\omega} \in \mathbb{R}^n$, respectively, with $\psi(\omega) = \tilde{\omega}$.

The restrictions of the mappings ψ and ψ^{-1} to the sets ω and $\tilde{\omega}$ induce mappings

$$\hat{\psi}: \mathbb{B}(\tilde{\omega}) \to \mathbb{B}(\omega) \quad \text{and} \quad \hat{\psi}^{-1}: \mathbb{B}(\omega) \to \mathbb{B}(\tilde{\omega})$$

defined by the equations

$$(\hat{\psi}\tilde{f})(x) = \tilde{f}(\psi(x)); \quad (\hat{\psi}^{-1}f)(x) = f(\psi^{-1}(y)).$$

Here $x \in \omega$, $y \in \tilde{\omega}$; $f \in \mathbb{B}(\omega)$, $\tilde{f} \in \mathbb{B}(\tilde{\omega})$. Therefore, if $\varLambda$ is an operator which maps $\mathbb{B}(\omega)$ into $\mathbb{B}(\omega)$, then the mapping

$$\psi: \omega \to \tilde{\omega}$$

defines a mapping of this operator into the operator $\tilde{\varLambda}$ which maps $\mathbb{B}(\tilde{\omega})$ into

$\mathbb{B}(\tilde{\omega})$ in the following way:

$$\tilde{\Lambda}\tilde{f} = \hat{\psi}^{-1}[\Lambda(\hat{\psi}\tilde{f})].$$

We note that in the present case the mesh in the new coordinate system depends on the old mesh $[\tilde{\omega} = \psi(\omega)]$.

11.2 Definition of a Mesh

Let $\mathbb{R}^{m+1}_+ = \{(t, x): t \geq 0, x \in \mathbb{R}^m\}$. We will call a discrete set of points $(t_n, x^n_j) \in \mathbb{R}^{m+1}_+$ a mesh, if there exists a nonsingular mapping

$$\psi: \mathbb{R}^{m+1}_+ \to \mathbb{R}^{m+1}_+$$

such that (t_n, x^n_j) is an inverse image (pre-image) (under the mapping ψ) of $(n\tau, j_1 h_1, \ldots, j_m h_m)$. Here $\tau, h = (h_1, \ldots, h_m)$ are steps on the respective axes of the space $\mathbb{R}^{m+1}_+$; $j = (j_1, \ldots, j_m)$; and $x^n_j \in \mathbb{R}^m$.

Thus we understand by a mesh the pre-image of a uniform mesh under a nonsingular mapping.

The essence of the proposed method consists in the following: Let $\mathbb{B}'$ be a Banach-space of functions defined on $\mathbb{R}^{m+1}_+$; let L be an operator $L: \mathbb{B}' \to \mathbb{B}'$. In a coordinate system we prescribe on a uniform mesh a difference scheme $\Lambda_{\tau h} u_{\tau h} = f_{\tau h}$ which approximates the equation $Lu = f$; $u, f, u_{\tau h}, f_{\tau h} \in \mathbb{B}'$. We denote by $\tilde{\omega}$ a uniform mesh in the coordinate system (t', x'), where τ is the step along the t'-axis; $h' = (h'_1, \ldots, h'_m)$, where h'_k is a step along the x'_k-axis, $k = 1, \ldots, m$. We assume that the coordinate systems (t, x) and (t', x') are related by a nonsingular transformation $t' = t, x' = x'(t, x)$ and that the parameters τ, h and τ', h' are related by the smooth transformation $\tau' = \tau, h' = h'(\tau, h)$ such that the Jacobian of the transformation does not vanish in the point $\tau = 0, h = 0$.

Let ω be the pre-image of the mesh $\tilde{\omega}$ under the mapping $(t, x) \to (t', x')$. On the mesh ω we construct a difference scheme

$$S_{\tau' h'} u' = f'; \quad u', f' \in \mathbb{B}(\omega),$$

$[\mathbb{B}(\omega)$ is the Banach-space of mesh functions] which approximates the problem $\Lambda_{\tau h} u_{\tau h} = f_{\tau h}$ with maximal order. Here by approximation we mean that the limit

$$\lim_{(\tau, h) \to 0} \|S_{\tau' h'} u_{\tau h} - f'\|_{\mathbb{B}(\omega)} = 0$$

exists and is equal to zero for solutions $u_{\tau h}$ of the problem $\Lambda_{\tau h} u_{\tau h} = f_{\tau h}$. From the difference scheme on the mesh ω we construct a difference scheme

$$\tilde{S}_{\tau' h'} \tilde{u} = \tilde{f}$$

on the mesh $\tilde{\omega}$. It is clear that the latter scheme is the transformed scheme $\Lambda_{\tau h} u_{\tau h} = f_{\tau h}$ and that it depends on the pattern used in the curvilinear coordinate system and on the functions $\tau' = \tau, h' = h'(\tau, h)$.

11.3 Closeness of Solutions of Difference Schemes on Different Meshes

In the following for simplicity we confine ourselves to the case $\tau' = \tau$, $h' = h$ and a nonsingular coordinate transformation $t' = t$, $x' = x'(t, x)$. We consider the difference scheme

$$\frac{u^{n+1}(x) - u^n(x)}{\tau} = \Lambda_{\tau h} u^n(x) \tag{11.1}$$

which approximates the differential equation

$$\frac{du}{dt} = Au, \quad 0 \leq t \leq T, \tag{11.2}$$

with the order $O(\tau' + |h|^k)$ for sufficiently smooth solutions $u(t, x) \in \mathbb{U}$. Here A is a linear operator $A: \mathbb{B} \to \mathbb{B}$; $\mathbb{U}$ is an everywhere dense subset of $\mathbb{B}$, and $\mathbb{U}$ contains infinitely often differentiable functions. By a solution of equation (5.2) we understand a one-parameter family of elements $u(t) \in \mathbb{B}$ which satisfy the condition:

$$\lim_{\Delta t \to 0} \left\| \frac{u(t + \Delta t) - u(t)}{\Delta t} - Au(t) \right\|_{\mathbb{B}} = 0, \quad t, t + \Delta t \in [0, T].$$

We consider also the difference scheme

$$\frac{v^{n+1}(x_j^{n+1}) - v^n(x_j^n)}{\tau} = \tilde{\Lambda}_{\tau h} v^n(x_j^n) \tag{11.3}$$

which approximates (11.1) with the order $O(\tau^{l_1} + |h|^{k_1})$ for the solutions $u^n(x) \in \mathbb{U}$, i.e. for all $u^n(x) \in \mathbb{U}$ the following inequality holds:

$$\left\| \frac{u^n(x_j^{n+1}) - u^n(x_j^n)}{\tau} - [\tilde{\Lambda}_{\tau h} u^n(x_j^n) - \Lambda_{\tau h} u^n(x_j^{n+1})] \right\|_{\mathbb{B}} \leq M(u^n)(\tau^{l_1} + |h|^{k_1}).$$

Here $\mathbb{B}(\omega)$ is the Banach-space of mesh functions which are defined at the nodes of the mesh $\omega = \{(n\tau, x_j^n)\}, n \geq 0, n\tau \leq T, x_j^n \in \mathbb{R}^m; j$ is a multi-index. In this case the difference scheme (11.3) approximates the equation (11.2) with the order $O(\tau^{l_2} + |h|^{k_2})$, where $l_2 = \min\{(l, l_1)\}, k_2 = \min\{(k, k_1)\}$.

. . . .

Theorem 11.1. Suppose that the difference scheme (11.1) approximates the equation (11.2) for a solution $u(t, x)$ of some standard class $\mathbb{U} \subset \mathbb{B}$ with the order $O(\tau^l + |h|^k)$ and suppose that the difference scheme (11.3) approximates the difference scheme (11.1) for the solution u from the class $\mathbb{U}$ with the order $O(\tau^{l_1} + |h|^{k_1})$, where $l_1 \geq l, k_1 \geq k$. Then the following statements hold:

(1) The difference scheme (11.3) approximates the equation (11.2) with the order $O(\tau^l + |h|^k)$,

(2) If the difference schemes (11.1) and (11.3) are stable, then

$$\|u^n - v^n\|_{\mathbf{B}(\omega)} \leq C(\tau^{l_1} + |h|^{k_1})$$

holds uniformly with respect to $t = n\tau \in [0, T]$, where $\{u^n\}$ is a solution of the difference scheme (11.1) with the initial condition $u^0(x) = u(x) \in \mathbf{U}$, and $\{v^n\}$ is a solution of the difference scheme (11.3) with the initial condition $v^0(x_j^0) = u(x_j^0)$,

(3) The dissipation and dispersion of the difference schemes (11.1) and (11.3) differ by the quantity $O(\tau^{l_1} + |h|^{k_1})$ for each harmonic.

The Π-forms of the differential representations of the difference schemes (11.1) and (11.3) are given by the following equations:

$$\frac{du}{dt} = \frac{1}{\tau} \ln(E + \tau \Lambda) u,$$

$$\frac{dv}{dt} = \frac{1}{\tau} \ln(E + \tau \tilde{\Lambda}) v.$$

. . . .

Let $\tau/h_j = \kappa_j = \text{const}$, $j = 1, \ldots, m$ (which holds in the case of hyperbolic equations) and suppose that the difference scheme (11.1) approximates the equation (11.2) with the order $O(\tau^l)$ and the difference scheme (11.3) approximates the equation (11.2) with the order $O(\tau^{l_1})$. Then

$$\tilde{\Lambda}_{\tau h} = \Lambda + Q,$$

where $Q v^n(x_j^n) = \tilde{\Lambda}_{\tau h} v^n(x_j^n) - \Lambda v^n(x_j^n)$. But for $Q u^n(x_j^n)$ we get the following expression:

$$Q u^n(x_j^n) = \frac{1}{\tau}(E + \tau \Lambda)[u^n(x_j^{n+1}) - u^n(x_j^n)] + \psi^n(u^n),$$

$$\psi^n(u^n) = \tilde{\Lambda}_{\tau h} u^n(x_j^n) - \Lambda u^n(x_j^{n+1}) - \frac{1}{\tau}[u^n(x_j^{n+1}) - u^n(x_j^n)].$$

Thus for a fixed mesh we have

$$Q u^n(x_j^n) = \psi^n(u^n)$$

and, consequently, for $u^n \in \mathbf{U}$

$$\|Q u^n\|_{\mathbf{B}(\omega)} \leq C(u^n) \tau^{l_1},$$

and then

$$\frac{dv}{dt} = \frac{1}{\tau} \ln(E + \tau \tilde{\Lambda}) v$$

$$= \sum_{j=1}^{\infty} \frac{(-1)^{j-1}}{j} \Lambda^j + \tau^{l_1} \sum_{j=1}^{\infty} \frac{(-1)^{j-1}}{j} \sum_{k=0}^{j-1} C_j^k \Lambda^k (\tau^{l_1})^{j-k-1} (Q')^{j-1},$$

where $Q' \quad \tau^{-l_1} Q$. Hence

$$\left[\frac{1}{\tau} \ln (E + \tau \tilde{A}) - \frac{1}{\tau} \ln (E + \tau A)\right] u = O(\tau^{l_1})$$

for sufficient smooth functions of the class $\mathbf{U}$.

Thus if the mesh $(n\tau, x_j^n)$ is fixed and if $\tau/h_j = k_j = \text{const}, j = 1, \ldots, m, k_1 \geq l_1 > \min \{(l, k)\}$, then the first differential approximations of the difference schemes (11.1) and (11.3) coincide. Then for difference schemes of odd order of approximation the correctness of the first differential approximation of the difference scheme (11.3) follows from the correctness of the first differential approximation of the difference scheme (11.1), which means that a necessary stability condition of the difference scheme (11.3) is satisfied. In the case of difference schemes of even order of approximation the same is true if $l_1 - 1 > \min \{(l, k)\}$. From the fact, that the first differential approximations of the difference schemes (11.1) and (11.3) coincide, a succession of properties of difference schemes follows which are defined by means of the first differential approximation (e.g. the so-called property K, property M, and the invariance of difference schemes).

Using the predictor-corrector method it is possible to obtain conservative difference schemes in curvilinear coordinate systems.

11.4 Example of Convective Equation

We consider in a Cartesian coordinate system the following difference scheme:

$$\frac{\Delta_0 u^n (x)}{\tau} = a \frac{\Delta_1 + \Delta_{-1}}{2h} u^n (x) + \frac{\lambda h^2}{2\tau} \frac{\Delta_1 \Delta_{-1}}{h^2} u^n (x), \tag{11.4}$$

which is an approximation of the differential equation

$$\frac{\partial u}{\partial t} = a \frac{\partial u}{\partial x}. \tag{11.5}$$

Let (t', x') be a new curvilinear coordinate system which is connected with the original coordinate system by the equations:

$$\begin{aligned} t' &= t, \\ x' &= x'(t, x). \end{aligned} \tag{11.6}$$

We assume that (11.6) represents a nonsingular transformation, i.e.

$$\det \left\| \begin{matrix} 1 & 0 \\ \dfrac{\partial x'}{\partial t} & \dfrac{\partial x'}{\partial x} \end{matrix} \right\| = \frac{\partial x'}{\partial x} \neq 0.$$

In the new coordinate system equation (11.5) has the form:

$$\frac{\partial v}{\partial t'} = \left(a\,\frac{\partial x'}{\partial x} - \frac{\partial x'}{\partial t}\right)\frac{\partial v}{\partial x'}. \tag{11.7}$$

In the (t', x') coordinate system we construct a difference scheme on a uniform mesh $(n\tau, jh)$

$$v_j^{n+1} = \sum_{\alpha=-1}^{1} b_{\alpha,j}^n\, v_{j+\alpha}^n, \tag{11.8}$$

which approximates equation (11.7).

Let $(n\tau, x_j^n)$ be the pre-image of the point $(n\tau, jh)$ under the transformation (11.6). On the mesh $(n\tau, x_j^n)$ in the (t, x) coordinate system we will find following difference scheme

$$u^{n+1}(x_j^{n+1}) = \sum_{\alpha=-1}^{1} b_{\alpha,j}^n\, u^n(x_{j+\alpha}^n), \tag{11.9}$$

which approximates the difference scheme (11.4) with maximal order. We assume that the function $u(t, x)$ is sufficiently smooth. Then using Taylor's expansion and the condition for approximation of the difference scheme we find that the coefficients $b_{\alpha,j}^n$ satisfy the following system of equations:

$$\sum_{\alpha=-1}^{1} b_{\alpha,j}^n = 1;$$

$$\sum_{\alpha=-1}^{1} b_{\alpha,j}^n (x_{j+\alpha}^n - x_j^n) = x_j^{n+1} - x_j^n + a\tau; \tag{11.10}$$

$$\sum_{\alpha=-1}^{1} b_{\alpha,j}^n (x_{j+\alpha}^n - x_j^n)^2 = (x_j^{n+1} - x_j^n)^2 + 2a\tau(x_j^{n+1} - x_j^n) + \lambda h^2.$$

Solving the system of equations (11.10) with respect to $b_{\alpha,j}^n$ ($\alpha = -1, 0, 1$) we get the following difference scheme which approximates the difference scheme (11.4) with second order and which, consequently, approximates the equation (11.5) on an arbitrary mesh $(n\tau, x_j^n)$ with first order:

$$\frac{u^{n+1}(x_j^{n+1}) - u^n(x_j^n)}{\tau}$$

$$= \left[a\,\frac{2x_j^{n+1} - x_j^n - x_{j-1}^n}{x_{j+1}^n - x_{j-1}^n} + \frac{\lambda h^2 + (x_j^{n+1} - x_j^n)(x_j^{n+1} - x_{j-1}^n)}{\tau(x_{j+1}^n - x_{j-1}^n)}\right]\frac{u^n(x_{j+1}^n) - u^n(x_j^n)}{x_{j+1}^n - x_j^n}$$

$$- \left[a\,\frac{2x_j^{n+1} - x_j^n - x_{j+1}^n}{x_{j+1}^n - x_{j-1}^n} + \frac{\lambda h^2 + (x_j^{n+1} - x_j^n)(x_j^{n+1} - x_{j+1}^n)}{\tau(x_{j+1}^n - x_{j-1}^n)}\right]\frac{u^n(x_j^n) - u^n(x_{j-1}^n)}{x_j^n - x_{j-1}^n}.$$
$$\tag{11.11}$$

From the facts discussed above it follows that the dissipation and dispersion of the difference schemes (11.4) and (11.11) coincide up to an order $O(\tau^3)$ for each harmonic.

The first differential approximation of the difference scheme (11.11) has the form:

$$\frac{\partial u}{\partial t} = a \frac{\partial u}{\partial x} + \frac{h^2}{2\tau}(\lambda - \kappa^2 a^2)\frac{\partial^2 u}{\partial x^2},$$

i.e. it coincides with the first differential approximation of the difference scheme (11.4). The difference scheme (11.11) can be written in the curvilinear (t', x') coordinate system in the following form:

$$\frac{v_j^{n+1} - v_j^n}{\tau} = \frac{1}{2}\left[a\left(\frac{\partial x'}{\partial x}\right)_j - \left(\frac{\partial x'}{\partial t}\right)_j\right]\left(\alpha_j \frac{v_{j+1}^n - v_j^n}{h} + \beta_j \frac{v_j^n - v_{j-1}^n}{h}\right)$$

$$+ \left(\frac{\partial x'}{\partial x}\right)_j\left\{\frac{\lambda h}{2\tau} + \kappa\left(\frac{\partial x}{\partial t'}\right)_j\left[a + \frac{1}{2}\left(\frac{\partial x}{\partial t'}\right)_j\right]\right\}$$

$$\times \left[\left(\frac{\partial x'}{\partial x}\right)_{j+1/2}\frac{v_{j+1}^n - v_j^n}{h} - \left(\frac{\partial x'}{\partial x}\right)_{j-1/2}\frac{v_j^n - v_{j-1}^n}{h}\right]. \qquad (11.12)$$

Here:

$$v_j^n := v(t', x')\big|_{\substack{t' = n\tau \\ x' = jh}}; \qquad \alpha_j := \left(\frac{\partial x'}{\partial x}\right)_{j+1/2}\left(\frac{\partial x}{\partial x'}\right)_{j-1/2};$$

$$\beta_j := \left(\frac{\partial x'}{\partial x}\right)_{j-1/2}\left(\frac{\partial x}{\partial x'}\right)_{j+1/2}; \qquad \left(\frac{\partial x'}{\partial x}\right)_j := \frac{2h}{x_{j+1}^n - x_j^n};$$

$$\left(\frac{\partial x'}{\partial x}\right)_{j-1/2} := \frac{h}{x_j^n - x_{j-1}^n}; \qquad \left(\frac{\partial x'}{\partial x}\right)_{j+1/2} := \frac{h}{x_{j+1}^n - x_j^n};$$

$$\left(\frac{\partial x}{\partial t'}\right)_j := \frac{x_j^{n+1} - x_j^n}{\tau}; \qquad \left(\frac{\partial x'}{\partial t}\right)_j := -\left(\frac{\partial x}{\partial t'}\right)_j\left(\frac{\partial x'}{\partial x}\right)_j.$$

The difference scheme (11.4) is monotone for $0 < \kappa a \leq \lambda \leq 1$. It is possible to prove that the difference scheme (11.12) is monotone for sufficiently small κ and $1 > \lambda\left(\frac{\partial x'}{\partial x}\right)_{j-1/2}\left(\frac{\partial x'}{\partial x}\right)_{j+1/2}$, and, consequently, that it is stable in the space $\mathbb{C}$.

Using the method described one can derive finite-difference relations on an arbitrary mesh which are analogous to those relations which were used in the Cartesian coordinate system. It is natural that the analogous formulas depend on the difference scheme used. For example, as an analogy of the difference relation

$$\frac{u_{j+1}^n - u_{j-1}^n}{2h}$$

we get in an arbitrary curvilinear coordinate system, if the difference scheme uses three points, the following formula:

$$\left(\frac{\partial x'}{\partial x}\right)_j \left\{ \frac{1}{2}\alpha_j \frac{v_{j+1}^n - v_j^n}{H} + \frac{1}{2}\beta_j \frac{v_j^n - v_{j-1}^n}{H} \right\},$$

where H is the step along the space variable x' in the new coordinate system. Using the analogies of difference schemes one can easily get difference schemes on an arbitrary mesh.

Thus in the present paragraph we have considered the investigation of dissipative and dispersive properties of difference schemes on the basis of the method of the differential approximation, which allows us to classify the difference schemes, to compare them and to select schemes with the necessary properties for the calculation of special problems. Such a classification for the difference schemes of gas dynamics is naturally a complicated problem, but one can solve it step by step. Firstly, one has to consider the analogies of the difference schemes of gas dynamics in the case of an equation for one time step with constant coefficients, secondly for a nonlinear equation for one time step and then one has to investigate – last not least – the difference schemes of gas dynamics. Such a splitting of the problem is natural and reflects the way in which a numerical analyst attacks the investigation of a difference scheme of gas dynamics, because it can help to be convinced of the usefulness of this procedure to choose this or that property for a difference scheme (and it is possible to exclude one of the properties) and to find a way to get better schemes for given practical problems.

Part III

Invariant Difference Schemes

The variety of difference schemes which are applied to problems of gas dynamics requires a classification of the schemes in different ways. It is well known that the equations of gas dynamics are invariant with respect to a group of point transformations in the space of independent and dependent variables. This invariance follows from the invariance of the laws of conservation, from which the equations of gas dynamics are derived. In addition each of the criteria which are applied to difference schemes of the equations of gas dynamics should have a physically based and therefore invariant character. Any difference scheme is realized on a special grid which by itself carries the noninvariance into the algorithm of the calculation. This noninvariance, for example, can appear in the calculation of singularities of the flow (shock waves, contact discontinuities, weak discontinuities) which travel in different directions with respect to the grid lines. It is well known that one of the main sources for the inaccuracy of the calculation stems from the lack of invariance of a difference scheme with respect to two types of transformation: Translation and rotation. In Harlow's scheme [64], for example, the process of selfoscillation is related to the noninvariance of the scheme with respect to a Galilei-transformation [95]. The transition to the difference scheme makes it difficult to perform an analysis of the group property of the scheme. Therefore it seems convenient to perform the group classification of difference schemes on the basis of their first differential approximation.

In the previous chapters the convenience of the use of the first differential approximation was demonstrated for the investigation of the stability and especially of the dissipative properties of difference schemes. The first differential approximation which plays an intermediate role between the original gasdynamic equations and the difference schemes approximating them, is a differential equation and the group theory can totally be applied to it [198]. Consequently, a classification of the difference schemes can be given on the basis of the concept of the first differential approximation using group theory.

Definition. We will say that a difference scheme admits a transformation group (or, equivalently, it is invariant with respect to a transformation group) if its first differential approximation admits this group [105–110].

According to this definition all difference schemes can be divided in two classes:

1) schemes which satisfy the group axioms,
2) schemes which do not satisfy the group axioms.

In the present chapter conditions are formulated under which the system of equations of parabolic form of the first differential approximation of difference schemes for the equations of gas dynamics admits the same transformations as the given system of equations of gas dynamics. The cases of one and two space variables are investigated (both in Lagrangean and in Eulerian coordinates). The conditions mentioned above allow the construction of classes of invariant difference schemes of arbitrary order of approximation. An investigation of the stability and the dissipative character of these schemes is performed, too.

Among the class of difference schemes mentioned above sub-classes of difference schemes are defined which satisfy the property K and the property of mass conservation in the first differential approximation. Results of numerical calculations are added which underline that it is correct to consider the invariance properties of difference schemes on the basis of their first differential approximation.

12. Some Basic Concepts of the Theory of Group Properties of Differential Equations

In this paragraph some basic concepts of the theory of group properties of differential equations and of the theory of Lie-groups of local transformations are cited which are necessary for the following discussion (for a more detailed information see the monograph [198]).

12.1 Infinitesimal Operator of G_r

We note by $\mathscr{E}(x, u)$ the Euclidean space of the independent variables x^i, $(i = 1, \ldots, m)$ and the dependent variables u^k, $(k = 1, \ldots, q)$ and by G_r the r-parameter Lie-group of point transformations of this space into itself, which are defined by the equations:

$$x'^i = f^i(x, u, a); \quad (i = 1, \ldots, m),$$
$$u'^k = \phi^k(x, u, a); \quad (k = 1, \ldots, q).$$

(12.1)

We will assume that the value zero of the parameter $a = (a^1, \ldots, a^r)$ of the transformation corresponds to the identical transformation. In any case we will concentrate on local Lie-groups but for the sake of brevity we will refer to them simply as Lie-groups.

A Lie-algebra of operators corresponds to the group G_r:

$$L = e^\alpha L_\alpha = \zeta^i \frac{\partial}{\partial x^i} + \eta^k \frac{\partial}{\partial u^k};$$
$$e^\alpha = \text{const}, \quad (\alpha = 1, \ldots, r),$$

(12.2)

with an operator basis

$$L_\alpha := \zeta^i_\alpha \frac{\partial}{\partial x^i} + \eta^k_\alpha \frac{\partial}{\partial u^k}, \quad (\alpha = 1, \ldots, r),$$

where $\zeta^i_\alpha(x, u)$ and $\eta^k_\alpha(x, u)$ are calculated by the formulas

$$\zeta^i_\alpha := \left. \frac{\partial f^i}{\partial a^\alpha} \right|_{a=0}; \quad \eta^k_\alpha := \left. \frac{\partial \phi^k}{\partial a^\alpha} \right|_{a=0},$$
$$(i = 1, \ldots, m; \ k = 1, \ldots, q; \ \alpha = 1, \ldots, r).$$

(12.3)

The operators L_α are also called infinitesimal operators of the group G_r. For the sake of brevity we will call the operators of this Lie-algebra simply operators of the group G_r. By the operators (12.2) it is possible to represent the transformations of the group G_r solving the Lie-equations under the initial conditions

$$
\frac{\partial x'^{\,i}}{\partial a^\alpha} = \zeta^i_\alpha(x',u'); \quad x'^{\,i}|_{a=0} = x^i;
$$

$$
\frac{\partial u'^{\,k}}{\partial a^\alpha} = \eta^k_\alpha(x',u'); \quad u'^{\,k}|_{a=0} = u^k. \tag{12.4}
$$

Therefore the group G_r can be given by its operators.

12.2 Invariant Subsets of G_r

Definition: The function $\mathscr{I}(x,u)$ will be called an invariant of the group G_r if the equation

$$
\mathscr{I}(x',u') = \mathscr{I}(x,u)
$$

holds.

. . . .

Theorem 12.1. For $\mathscr{I}(x,u)$ to be an invariant of the group G_r it is necessary and sufficient that the equation

$$
L\mathscr{I} = 0 \tag{12.5}
$$

holds for all operators of the group G_r.

. . . .

Definition: A subset $m \subset \mathscr{E}(x,u)$ is called an invariant subset of the group G_r if all transformations of the given group map any point of the subset m into a point of the same subset.

Assume that the subset m is regularly defined by the equations

$$
m: \ \Psi^\beta(x,u) = 0, \quad (\beta = 1, \ldots, d). \tag{12.6}
$$

. . . .

Theorem 12.2. A necessary and sufficient condition for the invariance of the subset m is, that the equations

$$
L\Psi^\beta|_m = 0, \quad (\beta = 1, \ldots, d), \tag{12.7}
$$

are satisfied for all operators of the group G_r.

. . . .

12.3 Necessary and Sufficient Conditions for the Invariance of the First-order Differential Equations

Let be given a system of first order differential equations:

$$\mathscr{F}^\gamma(x, u, p) = 0, \qquad (\gamma = 1, \ldots, q), \tag{12.8}$$

where

$$p_i^k = \frac{\partial u^k}{\partial x^i}.$$

The equations (12.8) define a subset in the space $\mathscr{E}(x, u, p)$. An extended group $\tilde{G}_r$ of transformations of the space $\mathscr{E}(x, u)$ corresponds to the group G_r of transformation of the space $\mathscr{E}(x, u, p)$. The operators of the group $\tilde{G}_r$ have the form:

$$\tilde{L} = L + \zeta_i^k(x, u, p)\frac{\partial}{\partial p_i^k}, \tag{12.9}$$

where

$$\zeta_i^k = \zeta_i^k(x, u, p) = D_i(\eta^k) - p_j^k D_i(\zeta^j);$$

$$D_i := \frac{\partial}{\partial x^i} + p_i^k \frac{\partial}{\partial u^k}. \tag{12.10}$$

The system of differential equations (12.8) admits a group G_r if the equation concerned define an invariant subset of the extended group $\tilde{G}_r$. This definition coincides with the usual definition of the invariance of the equations with respect to a transformation of the variables.

To find the group which is admitted by equations (12.8) one has to write down the necessary and sufficient condition for the invariance of the set $\mathscr{F}^\gamma = 0, (\gamma = 1, \ldots, q)$:

$$\tilde{L}\mathscr{F}^\gamma|_{\mathscr{F}^\gamma = 0} = 0, \qquad (\gamma = 1, \ldots, q). \tag{12.11}$$

By virtue of (12.10) the equations (12.11) are differential equations with respect to the functions $\zeta^i(x, u), \eta^k(x, u)$, and are called the defining equations of the group which is admitted by the system of equations (12.8). Finding the general solution of the defining equations we get a larger Lie-group of point transformations which the system of equations (12.8) admits. In the case of systems of equations of higher order it is necessary to perform an extension of the operator to the order which is necessary.

13. Groups Admitted by the System of the Equations of Gas Dynamics

13.1 Lie-Algebra for Two-dimensional Gas Dynamics

We consider the system of equations of gas dynamics in two dimensions:

$$\frac{\partial w}{\partial t} + \frac{\partial f}{\partial x} + \frac{\partial g}{\partial y} = 0, \tag{13.1}$$

where

$$w := \begin{pmatrix} \varrho u \\ \varrho v \\ \varrho \\ \varrho E \end{pmatrix}; \quad f := \begin{pmatrix} \varrho u^2 + p \\ \varrho u v \\ \varrho u \\ \varrho u E + u p \end{pmatrix}; \quad g := \begin{pmatrix} \varrho u v \\ \varrho v^2 + p \\ \varrho v \\ \varrho v E + v p \end{pmatrix};$$

p is the pressure; ϱ the density; u, v the components of the velocity vector parallel to the x- and y-direction of the coordinate system; $E := \varepsilon + (u^2 + v^2)/2$; ε is the specific internal energy. We will assume that the equation of state has the form:

$$p = p(\varepsilon, \varrho).$$

As the largest Lie-group of point transformations which are admitted by the equations (13.1) has the order 7 also the basis of the corresponding Lie-algebra G_7 consists of the following operators [198]:

$$L_1 := \frac{\partial}{\partial t}; \qquad\qquad L_2 := \frac{\partial}{\partial x};$$

$$L_3 := \frac{\partial}{\partial y}; \qquad\qquad L_4 := t \frac{\partial}{\partial x} + \frac{\partial}{\partial u};$$

$$L_5 := t \frac{\partial}{\partial y} + \frac{\partial}{\partial v}; \qquad L_6 := y \frac{\partial}{\partial x} - x \frac{\partial}{\partial y} + v \frac{\partial}{\partial u} - u \frac{\partial}{\partial v}; \tag{13.2}$$

$$L_7 := t \frac{\partial}{\partial t} + x \frac{\partial}{\partial x} + y \frac{\partial}{\partial y},$$

to which the following finite transformations correspond which preserve the system of equations (13.1)

1) Shifting along the t-axis;
2) Shifting along the x-axis;
3) Shifting along the y-axis;
4) Galilei-transformation along the x-axis;
5) Galilei-transformation along the y-axis;
6) Rotation;
7) Similarity transformation in the (t, x, y)-space (homothety).

13.2 One-dimensional Gas Dynamics

In the case of the system of equations of the one-dimensional gas dynamics

$$\frac{\partial w}{\partial t} + \frac{\partial f}{\partial x} = 0, \tag{13.3}$$

where

$$w := \begin{pmatrix} \varrho u \\ \varrho \\ \varrho E \end{pmatrix}; \quad f := \begin{pmatrix} p + \varrho u^2 \\ \varrho u \\ \varrho u E + up \end{pmatrix}; \quad E := \varepsilon + \tfrac{1}{2} u^2; \ p = p(\varepsilon, \varrho),$$

the largest Lie-group of point transformations has the order 4, and the basis of the corresponding Lie-algebra consists of the operators

$$L_1 := \frac{\partial}{\partial t},$$

$$L_2 := \frac{\partial}{\partial x},$$

$$L_3 := t \frac{\partial}{\partial x} + \frac{\partial}{\partial u}, \tag{13.4}$$

$$L_4 := t \frac{\partial}{\partial t} + x \frac{\partial}{\partial x}.$$

The following finite transformations correspond to those operators:

1) Shifting along t-axis,
2) Shifting along x-axis,
3) Galilei-transformation along the x-axis,
4) Similarity transformation in the (t, x)-space (homothety).

14. A Necessary and Sufficient Condition for Invariance of Difference Schemes on the Basis of the First Differential Approximation

Let us assume that the system of differential equations (12.8) is approximated by the following difference scheme of order l with respect to all variables on an equidistant grid

$$\Lambda^\gamma(x, u, h, Tu) = 0, \tag{14.1}$$

where

$$T := (T_1, \ldots, T_m),$$
$$h := (h^1, \ldots, h^m),$$
$$T_j u := u(x^1, \ldots, x^{j-1}, x^j + h^j, x^{j+1}, \ldots, x^m).$$

The first differential approximation of the difference scheme (14.1) is of the following form:

$$\mathscr{P}^\gamma(x, u, h, p^{(1)}, \ldots, p^{(\delta)}) = \mathscr{F}^\gamma(x, u, p) + (h^i)^l R_i^\gamma(x, u, p^{(1)}, \ldots, p^{(l_{\gamma i})}). \tag{14.2}$$

Here

$$p_{\beta_1 \ldots \beta_m}^{(\beta)k} := \frac{\partial^\beta u^k}{\partial^{\beta_1} x^1 \ldots \partial^{\beta_m} x^m},$$

$$(k = 1, \ldots, q; \quad \beta = 1, \ldots, \delta; \quad \beta := \beta_1 + \ldots + \beta_m, \quad \beta_j \geq 0),$$

$$p_i^k = p_{0 \ldots 010 \ldots 0}^{(1)k} = \frac{\partial u^k}{\partial x^i}; \quad \delta = \max_{\gamma, i} \{l_{\gamma i}\}.$$

We assume that the system of equations (14.1) admits a group G_r of transformations of the space $\mathscr{E}(x, u)$ with operators

$$L_\alpha = \zeta_\alpha^i \frac{\partial}{\partial x^i} + \eta_\alpha^k \frac{\partial}{\partial u^k}$$

forming a basis (see Sect. 12).

To the group G_r there corresponds another group G_r' of transformations of the space $\mathscr{E}(x, u, h)$ which differs from the group G_r by the transformation for h^i which follows from the transformation for x and y:

$$h'^i = \Psi^i(x, u, h, a).$$

Then the basis operators L'_α of the group G'_r have the form

$$L'_\alpha = L_\alpha + \Theta^i_\alpha \frac{\partial}{\partial h^i}$$

with

$$\Theta^i_\alpha := \frac{\partial \Psi^i}{\partial a^\alpha}\bigg|_{a=0}.$$

If, for example, in the group G_r a similarity transformation exists with an infinitesimal operator $x^i \dfrac{\partial}{\partial x^i}$ then in the group G'_r the infinitesimal operator $x^i \dfrac{\partial}{\partial x^i} + h^i \dfrac{\partial}{\partial h^i}$ corresponds to the operator $x^i \dfrac{\partial}{\partial x^i}$. Indeed, a transformation by the operator $x^i \dfrac{\partial}{\partial x^i}$ means a transformation of x according to the law $x'^i = ax^i$.

The points x'_1 and x'_2 correspond to the points x_1 and x_2 which are coupled by the relation $x_2 = x_1 + h$. Then

$$h' = x'_2 - x'_1 = ah,$$

i.e.

$$\Psi^i(x, u, h, a) = ah^i$$

and, consequently,

$$\Theta^i = \frac{\partial \Psi^i}{\partial a} = h^i.$$

Further, a group G_r of transformations of the space $\mathscr{E}(x, u)$ corresponds to the group $\tilde{G}'^{(\delta)}_r$ of transformations of the space $\mathscr{E}(x, u, h, p^{(1)}, \ldots, p^{(\delta)})$. The basis operators of the group $\tilde{G}'^{(\delta)}_r$ have the form

$$\tilde{L}'^{(\delta)}_\alpha = L'_\alpha + \sigma^{(\beta)k}_{\beta_1 \ldots \beta_m} \frac{\partial}{\partial p^{(\beta)k}_{\beta_1 \ldots \beta_m}}$$

and are obtained by an extension of the operator $\tilde{L}'_\alpha (\delta + 1)$ times.

The equations (14.2) lead to some manifolds in the space $\mathscr{E}(x, u, h, p^{(1)}, \ldots, p^{(\delta)})$. A necessary and sufficient condition for invariance of the system of equations (14.2) has the following form:

$$\tilde{L}'^{(\delta)}_\alpha \mathscr{P}^\gamma|_{\mathscr{P}^\gamma = 0} = 0; \quad (\alpha = 1, \ldots, r; \ \gamma = 1, \ldots, q). \tag{14.3}$$

These equations can be written in the form:

$$\tilde{L}'^{(\delta)}_\alpha \mathscr{P}^\gamma = \tilde{L}'^{(\delta)}_\alpha [\mathscr{F}^\gamma + (h^i)^l R^\gamma_i] = \tilde{L}_\alpha \mathscr{F}^\gamma + \tilde{L}'^{(\delta)}_\alpha (h^i)^l R^\gamma_i.$$

Thus the following theorem holds, [74].

. . . .

Theorem. A necessary and sufficient condition for the invariance of the first differential approximation (14.2) [or–equivalently–of the difference scheme (14.1)] is that the following relations are satisfied:

$$[\tilde{L}_\alpha \mathscr{F}^\gamma + \tilde{L}'^{(\delta)}_\alpha (h^i)^l R^\gamma_i]|_{\mathscr{P}^\gamma = 0} = 0, \quad (\alpha = 1, \ldots, r; \ \gamma = 1, \ldots, q). \tag{14.4}$$

. . . .

15. Conditions for the Invariance of Difference Schemes for the One-dimensional Equations of Gas Dynamics

15.1 The Class of Two-level Difference Schemes for the Eulerian Equations of Gas Dynamics

Assume that we approximate the system of one-dimensional gas dynamics in Eulerian coordinates (13.3) by the following first order difference scheme

$$\frac{\Delta_0 w^n(x)}{\tau} + \frac{\Delta_1 + \Delta_{-1}}{2h} [\alpha f^{n+1}(x) + (1 - \alpha) f^n(x)]$$

$$= \left[\Omega\left(x + \frac{h}{2}\right) \frac{\Delta_1}{h} - \Omega\left(x - \frac{h}{2}\right) \frac{\Delta_{-1}}{h} \right] [\beta w^{n+1}(x) + (1 - \beta) w^n(x)]. \quad (15.1)$$

Here $\Omega = \|\Omega_{ij}\|_1^3$ is for the moment an unknown matrix. The elements of it may depend on t, x, w and the derivatives with respect to x of the function w: w_x, w_{xx}. It may have the order $O(1)$; $0 \leq \alpha \leq 1, 0 \leq \beta \leq 1$.

The first differential approximation of the difference scheme (15.1) can be written in the form

$$\frac{\partial w}{\partial t} + \frac{\partial f}{\partial x} = \frac{\partial}{\partial w}\left(C \frac{\partial w}{\partial x}\right), \quad (15.2)$$

where

$$C := h\left[\Omega - \frac{\kappa}{2}(1 - 2\alpha) A^2\right]$$

$$= \bar{\Omega} - \frac{\tau}{2}(1 - 2\alpha) A^2$$

$$= \|\mu_{ij}\|_1^3;$$

$$\bar{\Omega} = \|\bar{\Omega}_{ij}\|_1^3 := h\Omega; \quad \kappa := \frac{\tau}{h} = \text{const};$$

$$A := \frac{df}{dw} = \left\| \begin{array}{ccc} u(2 - z) & -u^2 + \Theta + rz & z \\ 1 & 0 & 0 \\ E + m - u^2 z & -u(E + m - \Theta - rz) & u(1 + z) \end{array} \right\|;$$

$$r := u^2 - E; \quad \Theta := \frac{\partial p}{\partial \varrho}; \quad m := \frac{p}{\varrho}; \quad z := \frac{1}{\varrho} \cdot \frac{\partial p}{\partial \varepsilon};$$

$$\mu_{11} := \bar{\Omega}_{11} - d[3u^2(1-z) + \Theta + mz];$$

$$\mu_{12} := \bar{\Omega}_{12} + du(2u^2 - 2\Theta - 3rz + mz);$$

$$\mu_{13} := \bar{\Omega}_{13} - 3duz;$$

$$\mu_{21} := \bar{\Omega}_{21} - du(2-z);$$

$$\mu_{22} := \bar{\Omega}_{22} + d(u^2 - \Theta - rz);$$

$$\mu_{23} := \bar{\Omega}_{23} - dz;$$

$$\mu_{31} := \Omega_{31} - du(2E + 2m - 2u^2z + \Theta - Ez);$$

$$\mu_{32} := \bar{\Omega}_{32} + d[-u^2(\Theta - 2E - 2m + 2rz) - rEz - E\Theta - m\Theta + Emz];$$

$$\mu_{33} := \bar{\Omega}_{33} - d[u^2(1 + 2z) + z(E + m)];$$

$$d := \frac{\tau}{2}(1 - 2\alpha).$$

The system of equations (15.2) can be written in the following form

$$\mathscr{P}^1 := u_t + uu_x + \frac{1}{\varrho}p_x - \frac{1}{\varrho}N_{1_x} + \frac{u}{\varrho}N_{2_x} = 0;$$

$$\mathscr{P}^2 := \varrho_t + u\varrho_x + \varrho u_x - N_{2_x} = 0;$$

$$\mathscr{P}^3 := p_t + up_x + a^2u_x + ulN_{1_x} + bN_{2_x} - lN_{3_x} = 0,$$

where

$$N_j := v_{j1}u_x + v_{j2}\varrho_x + v_{j3}p_x;$$

$$v_{j1} := \varrho(\mu_{j1} + u\mu_{j3});$$

$$v_{j2} := u\mu_{j1} + \mu_{j2} + (E + \varrho\varepsilon_\varrho)\mu_{j3};$$

$$v_{j3} := \varrho\varepsilon_p\mu_{j3};$$

$$N - \begin{pmatrix} N_1 \\ N_2 \\ N_3 \end{pmatrix} = Cw_x = \bar{C}\bar{w}_x;$$

$$\bar{w} := \begin{pmatrix} u \\ \varrho \\ p \end{pmatrix}; \quad \bar{C} := \|v_{ij}\|_1^3;$$

$$a^2 := l(p - \varrho^2\varepsilon_\varrho);$$

$$b := l(\varepsilon + \varrho\varepsilon_\varrho - \tfrac{1}{2}u^2);$$

$$l := \frac{1}{\varrho\varepsilon_p}.$$

15.2 Condition for Invariance for the Difference Scheme (15.1)

We will derive conditions under which the difference scheme (15.1) admits an operator space with the basis (13.4). For this purpose we will use the theorem from Sect. 14. We need further the second extensions of the operators (13.4) in the space $\mathscr{E}\,(x,t,\bar{w},\tau,h,\bar{w}_t,\bar{w}_x,\bar{w}_{tt},\bar{w}_{xx},\bar{w}_{tx},\bar{w}_{xt})$ and the first extension in the space $\mathscr{E}\,(x,t,\bar{w},\bar{w}_t,\bar{w}_x)$:

$$\tilde{L}_1 := \frac{\partial}{\partial t};$$

$$\tilde{L}_2 := \frac{\partial}{\partial x};$$

$$\tilde{L}_3 := t\frac{\partial}{\partial x} + \frac{\partial}{\partial u} - u_x\frac{\partial}{\partial u_t} - \varrho_x\frac{\partial}{\partial \varrho_t} - p_x\frac{\partial}{\partial p_t};$$

$$\tilde{L}_4 := t\frac{\partial}{\partial t} + x\frac{\partial}{\partial x} - u_t\frac{\partial}{\partial u_t} - u_x\frac{\partial}{\partial u_x} - \varrho_t\frac{\partial}{\partial \varrho_t}$$
$$- \varrho_x\frac{\partial}{\partial \varrho_x} - p_t\frac{\partial}{\partial p_t} - p_x\frac{\partial}{\partial p_x};$$

$$\tilde{L}_1^{\prime(2)} := \frac{\partial}{\partial t};$$

$$\tilde{L}_2^{\prime(2)} := \frac{\partial}{\partial x};$$

$$\tilde{L}_3^{\prime(2)} := \tilde{L}_3 - (u_{tx} + u_{xt})\frac{\partial}{\partial u_{tt}} - u_{xx}\left(\frac{\partial}{\partial u_{xt}} + \frac{\partial}{\partial u_{tx}}\right)$$
$$- (\varrho_{tx} + \varrho_{xt})\frac{\partial}{\partial \varrho_{tt}} - \varrho_{xx}\left(\frac{\partial}{\partial \varrho_{xt}} + \frac{\partial}{\partial \varrho_{tx}}\right)$$
$$- (p_{tx} + p_{xt})\frac{\partial}{\partial p_{tt}} - p_{xx}\left(\frac{\partial}{\partial p_{tx}} + \frac{\partial}{\partial p_{xt}}\right);$$

$$\tilde{L}_4^{\prime(2)} = \tilde{L}_4 + \tau\frac{\partial}{\partial \tau} + h\frac{\partial}{\partial h} - 2u_{tx}\frac{\partial}{\partial u_{tx}} - 2u_{xt}\frac{\partial}{\partial u_{xt}}$$
$$- 2u_{tt}\frac{\partial}{\partial u_{tt}} - 2u_{xx}\frac{\partial}{\partial u_{xx}} - 2\varrho_{tt}\frac{\partial}{\partial \varrho_{tt}}$$
$$- 2\varrho_{tx}\frac{\partial}{\partial \varrho_{tx}} - 2\varrho_{xt}\frac{\partial}{\partial \varrho_{xt}} - 2\varrho_{xx}\frac{\partial}{\partial \varrho_{xx}}$$
$$- 2p_{tt}\frac{\partial}{\partial p_{tt}} - 2p_{tx}\frac{\partial}{\partial p_{tx}} - 2p_{xt}\frac{\partial}{\partial p_{xt}}$$
$$- 2p_{xx}\frac{\partial}{\partial p_{xx}}.$$

(15.3)

The following lemma holds:

Lemma 15.1. The difference scheme (15.1) is invariant if and only if the following relations are satisfied:

$$\frac{\partial}{\partial u} N_{1x} = N_{2x}; \quad \frac{\partial}{\partial u} N_{2x} = 0; \quad \frac{\partial}{\partial u} N_{3x} = N_{1x};$$

$$\frac{\partial}{\partial t} N_{jx} = 0; \quad \frac{\partial}{\partial x} N_{jx} = 0; \quad \bar{L}_4 N_{jx} = 2 N_{jx}; \quad j = 1, 2, 3; \tag{15.4}$$

where

$$\bar{L}_4 := u_x \frac{\partial}{\partial u_x} + \varrho_x \frac{\partial}{\partial \varrho_x} + p_x \frac{\partial}{\partial p_x}$$

$$+ 2 u_{xx} \frac{\partial}{\partial u_{xx}} + 2 \varrho_{xx} \frac{\partial}{\partial \varrho_{xx}} + 2 p_{xx} \frac{\partial}{\partial p_{xx}}. \tag{15.5}$$

Indeed, for the invariance of the difference scheme (15.1) it is necessary and sufficient that (see the theorem in Sect. 14).

$$[\tilde{L}_\alpha \mathscr{F}^\gamma + \tilde{L}_\alpha'^{(2)} hR^\gamma]|_{\mathscr{P}^\gamma = 0} = 0; \quad (\alpha = 1, 2, 3, 4; \; \gamma = 1, 2, 3); \tag{15.6}$$

where

$$hR^1 := -\frac{1}{\varrho}(N_{1x} - u N_{2x});$$

$$hR^2 := - N_{2x};$$

$$hR^3 := u l N_{1x} + b N_{2x} - l N_{3x}.$$

We consider the relations (15.6) for each α separately. As a preliminary remark we mention the fact that according to the construction of the first differential approximation (15.2) the matrix C and, consequently, the elements of the vector N do not depend on the derivatives $\bar{w}_t, \bar{w}_{tx}, \bar{w}_{xt}, \bar{w}_{tt}$. This remark remains valid also in the other paragraphs of the present chapter.

Thus for $\alpha = 1$ we get:

$$\tilde{L}_1'^{(2)} \mathscr{P}^1 = \frac{1}{\varrho}\left(-\frac{\partial N_{1x}}{\partial t} + u \frac{\partial N_{2x}}{\partial t}\right) = 0;$$

$$\tilde{L}_1'^{(2)} \mathscr{P}^2 = -\frac{\partial}{\partial t} N_{2x} = 0;$$

$$\tilde{L}_1'^{(2)} \mathscr{P}^3 = u l \frac{\partial N_{1x}}{\partial t} + b \frac{\partial N_{2x}}{\partial t} - l \frac{\partial N_{3x}}{\partial t} = 0.$$

From this it follows that

$$\frac{\partial N_{jx}}{\partial t} = 0; \quad j = 1, 2, 3. \tag{15.4a}$$

Similarly, we get for $\alpha = 2$:

$$\frac{\partial N_{jx}}{\partial x} = 0; \quad j = 1, 2, 3. \tag{15.4b}$$

For $\alpha = 3$ we get:

$$\tilde{L}_3^{\prime(2)} \mathscr{P}^1 = -\frac{1}{\varrho}\frac{\partial N_{1x}}{\partial u} + \frac{1}{\varrho} N_{2x} + \frac{u}{\varrho}\frac{\partial N_{2x}}{\partial u} = 0;$$

$$\tilde{L}_3^{\prime(2)} \mathscr{P}^2 = -\frac{\partial N_{2x}}{\partial u} = 0;$$

$$\tilde{L}_3^{\prime(2)} \mathscr{P}^3 = lN_{1x} - ulN_{2x} + ul\frac{\partial N_{1x}}{\partial u} + b\frac{\partial N_{2x}}{\partial u} - l\frac{\partial N_{3x}}{\partial u} = 0,$$

i.e.

$$\frac{\partial N_{1x}}{\partial u} = N_{2x}; \quad \frac{\partial N_{2x}}{\partial u} = 0; \quad \frac{\partial N_{3x}}{\partial u} = N_{1x}. \tag{15.4c}$$

In the case $\alpha = 4$ taking into account the relations (15.4a, b) and also $h\dfrac{\partial}{\partial h} N_{jx} = N_{jx}$ we find the relations

$$\tilde{L}_4^{\prime(2)} \mathscr{P}^1 = -\frac{2}{\varrho} N_{1x} + \frac{2u}{\varrho} N_{2x} + \frac{1}{\varrho}\bar{L}_4 N_{1x} - \frac{u}{\varrho}\bar{L}_4 N_{2x} = 0;$$

$$\tilde{L}_4^{\prime(2)} \mathscr{P}^2 = -2N_{2x} + \bar{L}_4 N_{2x} = 0;$$

$$\tilde{L}_4^{\prime(2)} \mathscr{P}^3 = 2ulN_{1x} + 2bN_{2x} - 2lN_{3x} - ul\bar{L}_4 N_{1x} - b\bar{L}_4 N_{2x} + l\bar{L}_4 N_{3x}$$
$$= 0.$$

Consequently,

$$\bar{L}_4 N_{jx} = 2N_{jx}; \quad j = 1, 2, 3. \tag{15.4d}$$

Thus, the lemma is proved. □

Corollary. A necessary and sufficient condition for the invariance of the difference scheme (15.1) has the form

$$\frac{\partial N_{2x}}{\partial t} = \frac{\partial N_{2x}}{\partial x} = \frac{\partial N_{2x}}{\partial u} = 0;$$

$$\bar{L}_4 N_{2x} = 2N_{2x};$$

$$N_{1x} = uN_{2x} + R_1;$$

$$N_{3x} = \tfrac{1}{2}u^2 N_{2x} + uR_1 + R_2; \tag{15.7}$$

$$\bar{L}_4 R_k = 2R_k;$$

$$\frac{\partial R_k}{\partial t} = \frac{\partial R_k}{\partial x} = \frac{\partial R_k}{\partial u} = 0; \quad k = 1, 2.$$

It can be proved easily that the above mentioned relations follow from conditions (15.4). We remark that the conditions on R_k mean especially that

$$R_k = h g_k (\varrho, p, \bar{w}_x, \bar{w}_{xx}).$$

If the quantities v_{2j} $(j = 1, 2, 3)$ do not depend on x, t, u, i.e.

$$\frac{\partial v_{2j}}{\partial t} = \frac{\partial v_{2j}}{\partial x} = \frac{\partial v_{2j}}{\partial u} = 0,$$

then

$$\frac{\partial N_{2x}}{\partial t} = \frac{\partial N_{2x}}{\partial x} = \frac{\partial N_{2x}}{\partial u} = 0. \tag{15.7a}$$

As

$$v_{21} := \varrho (\mu_{21} + u \mu_{23});$$

$$v_{23} := \varrho \varepsilon_p \mu_{23};$$

$$v_{22} := u \mu_{21} + \mu_{22} + \mu_{23} (E + \varrho \varepsilon_\varrho);$$

$$\mu_{21} := \bar{\Omega}_{21} - du (2 - z);$$

$$\mu_{23} := \bar{\Omega}_{23} - dz;$$

$$\mu_{22} := \bar{\Omega}_{22} + d (u^2 - \Theta - rz),$$

then putting

$$\bar{\Omega}_{21} := \omega_{21} + u (2d - \omega_{23});$$

$$\bar{\Omega}_{22} := \omega_{22} - u \omega_{21} + \tfrac{1}{2} u^2 \omega_{23} - du^2; \tag{15.8}$$

$$\bar{\Omega}_{23} := \omega_{23},$$

with

$$\frac{\partial \omega_{2j}}{\partial t} = \frac{\partial \omega_{2j}}{\partial x} = \frac{\partial \omega_{2j}}{\partial u} = 0,$$

we get

$$v_{21} = \varrho \omega_{21};$$

$$v_{22} = \omega_{22} + (\varepsilon + \varrho \varepsilon_\varrho)(\omega_{23} - dz) - d (\Theta - \varepsilon z);$$

$$v_{23} = \varrho \varepsilon_p (\omega_{23} - dz).$$

Thus, if the elements $\bar{\Omega}_{2j} (j = 1, 2, 3)$ satisfy the conditions (15.8) then (15.7a) is fulfilled.

From the facts given above the following theorem holds.

Theorem 15.1. If

$$
\left.
\begin{aligned}
&\bar{\Omega}_{21} = \omega_{21} + u(2d - \omega_{23}); \\
&\bar{\Omega}_{22} = \omega_{22} - u\omega_{21} + \tfrac{1}{2}u^2\omega_{23} - du^2; \\
&\bar{\Omega}_{23} = \omega_{23}; \\
&\frac{\partial \omega_{2j}}{\partial t} = \frac{\partial \omega_{2j}}{\partial x} = \frac{\partial \omega_{2j}}{\partial u} = 0; \quad j = 1, 2, 3;
\end{aligned}
\right\}
\tag{15.9}
$$

$$
\left.
\begin{aligned}
&N_{1x} = u N_{2x} + R_1; \\
&N_{3x} = \tfrac{1}{2}u^2 N_{2x} + u R_1 + R_2; \\
&\bar{L}_4 R_k = 2 R_k; \\
&\frac{\partial R_k}{\partial t} = \frac{\partial R_k}{\partial x} = \frac{\partial R_k}{\partial u} = 0; \\
&\bar{L}_4 N_{2x} = 2 N_{2x},
\end{aligned}
\right\}
\tag{15.10}
$$

then the difference scheme (15.1) is invariant.

In the following we will use the definition:

Definition. A difference scheme belongs to class $\mathscr{I}$ if it is invariant, and we will write $\Lambda_h \in \mathscr{I}$, where Λ_h is the step-operator of the scheme.

If the difference scheme is invariant only with respect to three transformations: Shift along the x- and t-axis and Galilei-transformation, then we will say that it belongs to the class $\Gamma\,(\Lambda_h \in \Gamma)$.

15.3 Property M of a Difference Scheme

Definition. We will say that the difference scheme (15.1) possesses the property M, i.e. it belongs to the class $M(\Lambda_h \in M)$ if in the first differential approximation the law of conservation of mass is satisfied, i.e.

$$
N_2 = 0.
\tag{15.11}
$$

The following theorem holds.

Theorem 15.2. If in the difference scheme (15.1) the matrix Ω is chosen such that

$$
\begin{aligned}
&\bar{\Omega}_{21} = du(2 - z); \\
&\bar{\Omega}_{22} = -d(u^2 - \Theta - rz); \\
&\bar{\Omega}_{23} = dz;
\end{aligned}
\tag{15.12}
$$

and besides that

$$N_{1x} = hg_1(\varrho, p, \bar{w}_x, \bar{w}_{xx});$$
$$N_{3x} = uN_{1x} + hg_2(\varrho, p, \bar{w}_x, \bar{w}_{xx});$$
$$\bar{L}_4 g_k = 2 g_k, \quad k = 1, 2,$$

then the difference scheme (15.1) possesses the property M and belongs to class $\mathscr{I}$.

· ·

Proof. Indeed, from the conditions (15.12) it follows that $\mu_{2j} = 0$ ($j = 1, 2, 3$) and therefore $v_{2j} = 0$. Consequently, we have $N_2 = 0$, i.e. the scheme possesses the property M. That the scheme belongs to class $\mathscr{I}$ follows from the fact that under the conditions of the Theorem 15.2 the conditions of Theorem 15.1 are satisfied with $R_k = hg_k(\varrho, p, \bar{w}_x, \bar{w}_{xx})$. $\qquad\qquad\square$

If we put

$$\bar{\Omega}_{11} := \omega_{11} - u\omega_{13} + 3\,du^2(1-z);$$
$$\bar{\Omega}_{12} := \omega_{12} - u\omega_{11} + \tfrac{1}{2}u^2\omega_{13} - du(2u^2 - \Theta - 3rz + 2mz); \qquad (15.13)$$
$$\bar{\Omega}_{13} := \omega_{13} + 3\,duz,$$

with

$$\frac{\partial \omega_{1j}}{\partial t} = \frac{\partial \omega_{1j}}{\partial x} = \frac{\partial \omega_{1j}}{\partial u} = 0; \quad j = 1, 2, 3; \qquad (15.14)$$

we get

$$v_{11} = \varrho\omega_{11} + d\varrho(\Theta + mz);$$
$$v_{12} = \omega_{12} + (\varepsilon + \varrho\varepsilon_\varrho)\omega_{13};$$
$$v_{13} = \varrho\varepsilon_p\omega_{13};$$

i.e.

$$\frac{\partial v_{1j}}{\partial t} = \frac{\partial v_{1j}}{\partial x} = \frac{\partial v_{1j}}{\partial u} = 0.$$

Thus the conditions

$$N_{1x} = hg_1(\varrho, p, \bar{w}_x, \bar{w}_{xx}),$$
$$\bar{L}_4 g_1 = 2 g_1;$$

in Theorem 15.2 can be replaced by the conditions (15.13, 14) and

$$\bar{L}_4 N_{1x} = 2 N_{1x}.$$

In this case we get

$$N_1 = \varrho[\omega_{11} + d(\Theta + mz)]u_x + [\omega_{12} + (\varepsilon + \varrho\varepsilon_\varrho)\omega_{13}]\varrho_x + \varrho\varepsilon_p\omega_{13}\,p_x.$$
$$(15.15)$$

Corollary. If in the difference scheme (15.1) the elements of the matrix Ω satisfy the conditions (15.12–14) and besides that if

$$N_{3x} = (u N_1)_x; \quad \bar{L}_4 N_{lx} = 2 N_{lx}; \quad \bar{L}_4 N_1 = N_1;$$

then the difference scheme possesses the property M, belongs to class $\mathscr{I}$ and the artificial viscosity is an additive term for p in the system of equations of the first differential approximation; under the condition $u_x N_1 > 0$ that artificial viscosity leads to an increase of the entropy in the first differential approximation.

That the statements above are true follows, indeed, from the fact that under these conditions the conditions of Theorem 15.2 are satisfied. In this case we get

$$hg_2 = u_x N_1,$$

and, consequently,

$$\bar{L}_4 hg_2 = \bar{L}_4 u_x N_1 = 2 u_x N_1.$$

The system of equations of the first differential approximation has then the following form

$$(\varrho u)_t + (\tilde{p} + \varrho u^2)_x = 0;$$

$$\varrho_t + (\varrho u)_x = 0; \qquad (15.16)$$

$$(\varrho E)_t + (\varrho u E + u\tilde{p})_x = 0,$$

where

$$\tilde{p} := p + N_1.$$

Thus the artificial viscosity enters the first differential approximation additively, analogous to the physical viscosity. The equation for the entropy s is derived from the original system of equations of gas dynamics by a formal multiplication of the system by such a vector $X_0 = (-u, u^2 - E - m, 1)$ that $X_0 A = u X_0$, and by using the second law of thermodynamics. After multiplication of the system of equations (15.16) from left by the vector X_0 we get

$$s_t + u s_x = \frac{1}{\varrho T} u_x N_1 > 0.$$

We remark that if the condition $N_{3x} = (u N_1)_x$ is not satisfied the artificial viscosity not at all enters the pressure p analogous to the physical viscosity and then the following relation holds:

$$T(s_t + u s_x) = \frac{h}{\varrho} g_2 (\varrho, p, \bar{w}_x, \bar{w}_{xx}).$$

15.4 Property K

In Part II the property K was defined and also the class K of difference schemes was defined which possess this property.

The following theorem holds.

· ·

Theorem 15.3. If in the difference scheme the elements of the matrix Ω are chosen such that the conditions (15.12–14) are satisfied and if, in addition,

$$\left.\begin{array}{l} \bar{L}_4 N_{1x} = 2 N_{1x}; \\ \bar{L}_4 N_{3x} = 2 N_{3x}; \end{array}\right\} \tag{15.17}$$

$$\left.\begin{array}{l} \bar{\Omega}_{31} = u\omega_{11} - u^2\,\omega_{13} + du(2\,E + 2\,m - 2\,u^2\,z - Ez - mz); \\ \bar{\Omega}_{32} = u\omega_{12} - u^2\,\omega_{11} + \tfrac{1}{2}u^3\,\omega_{13} + du^2\,(2\,rz - 2\,E - 2\,m - mz) \\ \qquad + d\,(rEz + E\Theta + m\Theta - Emz); \\ \bar{\Omega}_{33} = u\omega_{13} + d\,[u^2\,(1 + 2\,z) + z\,(E + m)], \end{array}\right\} \tag{15.18}$$

then the difference scheme (15.1) possesses the properties K, M and belongs to class $\mathscr{I}$.

· ·

Proof. The conditions (15.12)–as shown in Theorem 15.2–mean that the scheme possesses the property M. Because of the relations (15.13, 14) N_1 has the form (15.15) and, consequently,

$$\frac{\partial}{\partial u} N_{1x} = 0.$$

From the conditions (15.17) we get

$$\mu_{3j} = u\mu_{1j}; \quad j = 1,2,3; \tag{15.19}$$

and from this

$$v_{31}\,u_x + v_{32}\,\varrho_x + v_{33}\,p_x = uv_{11}\,u_x + uv_{12}\,\varrho_x + uv_{13}\,p_x,$$

i.e.

$$N_3 = uN_1. \tag{15.20}$$

From the equations (15.18) it follows that $X_0\,C = 0$ where $X_0\,A = u\,X_0$ and the difference scheme possesses the property K (see Part II, Sect. 7.2). The invariance of the scheme follows from the fact that the sufficient conditions (15.7) are satisfied. Thus the theorem is proved. ∎

Remark 1. The relations (15.15) and (15.19) mean (see Sect. 10.3) that $\bar{L}_4 N_1 = N_1$.

Remark 2. From the conditions of Theorem 15.3 can be derived that the artificial viscosity affects additively the pressure in the system of equations of the first differential approximation and if $u_x N_1 > 0$ this viscosity leads to an increase of the entropy in the first differential approximation.

Remark 3. That the difference scheme (15.1) belongs to the class K means that

$$N_3 = uN_1 - (r - m)N_2, \tag{15.21}$$

i.e.

$$\mu_{3j} = u\mu_{1j} - (r - m)\mu_{2j}; \quad j = 1, 2, 3. \tag{15.22}$$

15.5 Weak Solutions of Difference Scheme (15.1), $\alpha = \beta$

In the paragraphs above we considered how to construct difference schemes with defined group properties for the system of equations (13.3). We will assume that the initial conditions prescribed on the hyperplane $t = 0$ have the form

$$w(x, 0) = \phi(x). \tag{15.23}$$

A weak solution of the system of equations (13.3) is given by the integral relation [199]

$$\iint (\psi_t w - \psi_x f) \, dx \, dt + \int \psi(x, 0) \phi(x) \, dx = 0,$$

where ψ is an arbitrarily chosen smooth and finite vector-function.

We consider the difference scheme (15.1) for $\alpha = \beta$ and $\Omega = \Omega(w^n)$

$$\frac{\Delta_0 w^n(x)}{\tau} = \alpha \frac{g^{n+1}\left(x + \dfrac{h}{2}\right) - g^{n+1}\left(x - \dfrac{h}{2}\right)}{h}$$

$$+ \delta \frac{g^n\left(x + \dfrac{h}{2}\right) - g^n\left(x - \dfrac{h}{2}\right)}{h}, \tag{15.24}$$

where

$$g^n\left(x + \frac{h}{2}\right) := -\frac{f(w^n(x + h)) + f(w^n(x))}{2}$$

$$+ \Omega\left(w^n\left(x + \frac{h}{2}\right)\right)[w^n(x + h) - w^n(x)]; \quad \delta := 1 - \alpha.$$

Thus the function g can be thought of as a function of two vector variables

$$g(v_1, v_2) = -\frac{f(v_1) + f(v_2)}{2} + \Omega(v_1 - v_2),$$

where

$$g(w, w) = -f(w).$$

The function $g(v_1, v_2)$ is a continuous function of the variables v_1, v_2.

For the difference scheme (15.24) we consider Cauchy's problem with the initial value distribution ϕ:

$$w^0(x) = \phi(x). \tag{15.25}$$

The following theorem holds.

. . . .

Theorem 15.4. For $h, \tau \to 0$ let the function $v(x,t)$–a solution of the problem (15.24, 25)–converge uniformly nearly everywhere with respect to x and t to a function $w(x,t)$. Then $w(x,t)$ is a weak solution of the problem (13.3), (15.23).

. . . .

Proof. The proof is given in accordance with the publication [199]. We multiply equation (15.24) which is satisfied by the function $v(x,t)$ by an arbitrary, everywhere smooth (generally it is sufficient that it belongs to $\mathbb{C}_0^1$) finite vector-function $\psi(x,t)$. We integrate with respect to x and sum with respect to t. Using the summation formula and transforming the terms which contain g by changing the variable of integration x in $x + h/2$ and $x - h/2$, respectively, we get

$$\sum \int \frac{\psi(x, t-\tau) - \psi(x,t)}{\tau} v(x,t)\, dx\, \Delta t - \int \psi(x,0)\, v(x,0)\, dx$$

$$= -\alpha \sum \int \frac{\psi\left(x + \frac{h}{2}, t\right) - \psi\left(x - \frac{h}{2}, t\right)}{h} g\, dx\, \Delta t$$

$$- \delta \sum \int \frac{\psi\left(x + \frac{h}{2}, t-\tau\right) - \psi\left(x - \frac{h}{2}, t-\tau\right)}{h} g\, dx\, \Delta t, \quad \Delta t := \tau,$$

where g means $g(v_1, v_2)$; v_1, v_2 are values of the vector $v(x,t)$ in grid points.

If the function $v(x,t)$ converges nearly everywhere uniformly to the function $w(x,t)$ (in the components) then $g(v_1, v_2)$ converges towards the function $g(w,w) = f(w)$. Consequently, as a limit of the relation (15.26) we get the integral relation

$$\iint (\psi_t w - \psi_x f)\, dx\, dt + \int \psi(x,0)\, \phi(x)\, dx = 0. \qquad \square$$

15.6 One-dimensional System of the Equations of Gas Dynamics in Lagrangean Coordinates

We consider the one-dimensional system of equations of gas dynamics in Lagrangean coordinates:

$$\frac{\partial u}{\partial t} + \frac{\partial p}{\partial q} = 0;$$

$$\frac{\partial v}{\partial t} - \frac{\partial u}{\partial q} = 0; \qquad (15.27)$$

$$\frac{\partial E}{\partial t} + \frac{\partial (up)}{\partial q} = 0;$$

where u is the velocity of the gas, p the pressure, v the specific volume; $E :=$
$\varepsilon + \frac{1}{2} u^2$; ε the specific internal energy; q the Lagrangean coordinate

$$q := \int_{x(0,\,t)}^{x(q,\,t)} \varrho \, dx;$$

ϱ is the density of the gas in the point x at the time t, x the Eulerian coordinate.

Let us approximate the system of equations (15.27) by the following difference scheme:

$$\frac{\Delta_0 w^n (q)}{\tau} + \frac{\Delta_1 + \Delta_{-1}}{2h} [\alpha f^{n+1}(q) + (1 - \alpha) f^n(q)]$$

$$= \left[\Omega \left(q + \frac{h}{2} \right) \frac{\Delta_1}{h} - \Omega \left(q - \frac{h}{2} \right) \frac{\Delta_{-1}}{h} \right] [\beta w^{n+1}(q) + (1 - \beta) w^n(q)] \quad (15.28)$$

with

$$w := \begin{pmatrix} u \\ v \\ E \end{pmatrix}; \quad f := \begin{pmatrix} p \\ -u \\ up \end{pmatrix}; \quad f^n := f(w^n(q));$$

$\Omega = \| \Omega_{ij} \|_1^3$ is a matrix the elements of which may depend on t, q, w and the derivatives of the function w.

The first differential approximation of the difference scheme (15.28) has the form

$$\frac{\partial w}{\partial t} + \frac{\partial f}{\partial q} = \frac{\partial}{\partial q} \left(C \frac{\partial w}{\partial q} \right), \quad (15.29)$$

where

$$C := h\Omega - \frac{\tau}{2} (1 - 2\alpha) A^2 = \bar{\Omega} - dA^2 = \| \mu_{ij} \|_1^3;$$

$$\bar{\Omega} := h\Omega;$$

$$A := \frac{df}{dw};$$

$$d := \frac{\tau}{2} (1 - 2\alpha).$$

The elements μ_{ij} of the matrix C are expressed by the following terms:

$$\mu_{11} := \bar{\Omega}_{11} - da^2; \quad \mu_{12} := \bar{\Omega}_{12}; \quad \mu_{13} := \bar{\Omega}_{13};$$
$$\mu_{21} := \bar{\Omega}_{21} - du p_\varepsilon; \quad \mu_{22} := \bar{\Omega}_{22} + dp_v; \quad \mu_{23} := \bar{\Omega}_{23} + dp_\varepsilon;$$
$$\mu_{31} := \bar{\Omega}_{31} + du p_v; \quad \mu_{32} := \bar{\Omega}_{32} - dp p_v; \quad \mu_{33} := \bar{\Omega}_{33} - dp p_\varepsilon.$$

with $a^2 := pp_\varepsilon - p_v$.

We will find the conditions under which the difference scheme (15.28) admits the same transformation group as the original system of equations of gas dynamics (15.27). We remark for the moment that a transformation group of the system of equations of gas dynamics in Eulerian coordinates (13.3) corresponds to the transformation group of the system of equations of gas dynamics in Lagrangean coordinates (15.27), and vice versa.

The system of equations (15.29) can be written in the following way:

$$\mathscr{P}_1 := u_t + p_q - N_{1q} = 0;$$
$$\mathscr{P}_2 := v_t - u_q - N_{2q} = 0;$$
$$\mathscr{P}_3 := p_t + a^2 u_q + \frac{\varepsilon_v}{\varepsilon_p} N_{2q} + \frac{u}{\varepsilon_p} N_{1q} - \frac{1}{\varepsilon_p} N_{3q} = 0$$

(15.30)

with

$$N = \begin{pmatrix} N_1 \\ N_2 \\ N_3 \end{pmatrix} = C w_x;$$

$$N_k := v_{k1} u_q + v_{k2} v_q + v_{k3} p_q;$$
$$v_{k1} := \mu_{k1} + u \mu_{k3};$$
$$v_{k2} := \mu_{k2} + \varepsilon_v \mu_{k3};$$
$$v_{k3} := \varepsilon_p \mu_{k3}; \quad k = 1, 2, 3.$$

The following system of equations (15.30) corresponds to the system of equations in Eulerian coordinates

$$\tilde{\mathscr{P}}^1 := u_t + u u_x + \frac{1}{\varrho} p_x - \frac{1}{\varrho} \tilde{N}_{1x} = 0;$$
$$\tilde{\mathscr{P}}^2 := \varrho_t + u \varrho_x + \varrho u_x + \varrho \tilde{N}_{2x} = 0;$$

(15.31)

$$\tilde{\mathscr{P}}^3 := p_t + u p_x + \tilde{a}^2 u_x + \frac{u}{\varrho \varepsilon_p} \tilde{N}_{3x} - \frac{1}{\varrho \varepsilon_p} \tilde{N}_{3x} + \frac{\varepsilon_v}{\varrho \varepsilon_p} \tilde{N}_{2x} = 0;$$

with

$$N_k := \frac{1}{\varrho} \tilde{v}_{k1} u_x - \frac{1}{\varrho^2} \tilde{v}_{k2} \varrho_x + \frac{1}{\varrho} \tilde{v}_{k3} p_x; \quad k = 1, 2, 3.$$

The quantities $\tilde{v}_{kj}$ can be derived from the quantities v_{kj} replacing q by x using the formula

$$q := \int_{x(0)}^{x(t)} \varrho \, dx, \quad \tilde{a}^2 := \frac{p - \varrho^2 \varepsilon_\varrho}{\varrho \varepsilon_p}.$$

From this it is clear that in the case when we find necessary and sufficient conditions for the invariance of the equations (15.31) with respect to transformations which are admitted by the system of equations of gas dynamics in Eulerian coordinates (13.3), we can replace in these conditions the coordinate x by q and we will get necessary and sufficient conditions for invariance of the system of equations (15.30).

The conditions for invariance of the system of equations of the form (15.31) were found in 15.2 and therefore we can derive the required conditions performing the corresponding replacements.

The following theorem holds.

. . . .

Theorem 15.5. For the invariance of the difference scheme (15.28) it is necessary and sufficient that

$$
\left.
\begin{aligned}
&\frac{\partial N_{iq}}{\partial t} = \frac{\partial N_{iq}}{\partial q} = \frac{\partial N_{iq}}{\partial u} = 0; \quad i = 1, 2; \\
&N_{3q} = uN_{1q} + R; \\
&\frac{\partial R}{\partial t} = \frac{\partial R}{\partial q} = \frac{\partial R}{\partial u} = 0;
\end{aligned}
\right\}
\tag{15.32}
$$

$$
\bar{L}_4 \tilde{N}_{jx} = 2 \tilde{N}_{jx}; \quad j = 1, 2, 3.
\tag{15.33}
$$

. . . .

It is a consequence of the corollary to the Lemma 15.1 that this theorem holds.

. . . .

Theorem 15.6. If the elements of the matrix Ω satisfy the conditions (15.32), (15.33) and, if further

$$
\left.
\begin{aligned}
&\frac{\partial \bar{\Omega}_{1j}}{\partial t} = \frac{\partial \bar{\Omega}_{1j}}{\partial q} = \frac{\partial \bar{\Omega}_{1j}}{\partial u} = 0, \quad j = 1, 2, 3; \\
&\bar{\Omega}_{21} = \omega_{21} + du\, p_\varepsilon; \\
&\frac{\partial \omega_{21}}{\partial t} = \frac{\partial \omega_{21}}{\partial q} = \frac{\partial \omega_{21}}{\partial u} = 0; \\
&\frac{\partial \bar{\Omega}_{2k+1}}{\partial t} = \frac{\partial \bar{\Omega}_{2k+1}}{\partial q} = \frac{\partial \bar{\Omega}_{2k+1}}{\partial u} = 0; \quad k = 1, 2;
\end{aligned}
\right\}
\tag{15.34}
$$

then the difference scheme (15.28) is invariant.

. . . .

That this theorem is true follows from the Theorems 15.1 and 15.5.

Remark 1. The difference scheme (15.28) possesses the property M if

$$
\begin{aligned}
&\bar{\Omega}_{21} = du\, p_\varepsilon; \\
&\bar{\Omega}_{22} = -dp_v; \\
&\bar{\Omega}_{33} = dp_\varepsilon.
\end{aligned}
\tag{15.35}
$$

Indeed, in this case $\mu_{2j} = 0$; $(j = 1, 2, 3)$. Then $v_{2j} = 0$ and, consequently, $N_2 = 0$.

Remark 2. If the conditions (15.35) are satisfied and if the difference scheme (15.28) possesses the property K, then $N_3 = uN_1$.

Indeed, we consider the product $X_0 C$ where the vector $X_0 = (-u, p, 1)$ is such that $X_0 A = 0$. We get

$$X_0 C = X_0 \bar{\Omega} = (-u\mu_{11} + \mu_{31}, -u\mu_{12} + \mu_{32}, -u\mu_{13} + \mu_{33}) = 0,$$

i.e. $\mu_{3j} = u\mu_{1j}$; $j = 1, 2, 3$. Then $v_{3j} = uv_{1j}$ and, consequently, $N_3 = uN_1$. The latter means especially that the artificial viscosity enters the system of equations of the first differential approximation in an analogous way to that of the physical viscosity.

15.7 Polytropic Gas

In the case of a polytropic gas the equation of state is given by the relation

$$p = (\gamma - 1)\varepsilon\varrho, \tag{15.36}$$

where γ is the adiabatic exponent.

Besides the transformation with infinitesimal operators (13.4) the system of equations (13.3) admits for a polytropic gas two additional transformations with infinitesimal operators [198]:

$$L_5 := x\frac{\partial}{\partial x} + u\frac{\partial}{\partial u} - 2\varrho\frac{\partial}{\partial \varrho};$$

$$\tag{15.37}$$

$$L_6 := \varrho\frac{\partial}{\partial \varrho} + p\frac{\partial}{\partial p}.$$

We will find the conditions under which the difference scheme (15.1) together with the transformations (13.4) admits transformations (15.37). We need the operators

$$\tilde{L}_5^{\prime(2)} = L_5 + h\frac{\partial}{\partial h} + u_t\frac{\partial}{\partial u_t} - 2\varrho_t\frac{\partial}{\partial \varrho_t} - 3\varrho_x\frac{\partial}{\partial \varrho_x}$$

$$- p_x\frac{\partial}{\partial p_x} + u_{tt}\frac{\partial}{\partial u_{tt}} + u_{xt}\frac{\partial}{\partial u_{xt}} - 2\varrho_{tt}\frac{\partial}{\partial \varrho_{tt}} - 3\varrho_{tx}\frac{\partial}{\partial \varrho_{tx}}$$

$$- 2\varrho_{xt}\frac{\partial}{\partial \varrho_{xt}} - 3\varrho_{xx}\frac{\partial}{\partial \varrho_{xx}} - p_{tx}\frac{\partial}{\partial p_{tx}} - p_{xx}\frac{\partial}{\partial p_{xx}};$$

$$\tag{15.38}$$

$$\tilde{L}_6^{\prime(2)} = L_6 + \varrho_t\frac{\partial}{\partial \varrho_t} + \varrho_x\frac{\partial}{\partial \varrho_x} + p_t\frac{\partial}{\partial p_t} + p_x\frac{\partial}{\partial p_x}$$

$$+ \varrho_{tx}\frac{\partial}{\partial \varrho_{tx}} + \varrho_{tt}\frac{\partial}{\partial \varrho_{tt}} + \varrho_{xt}\frac{\partial}{\partial \varrho_{xt}} + \varrho_{xx}\frac{\partial}{\partial \varrho_{xx}} + p_{tt}\frac{\partial}{\partial p_{tt}}$$

$$+ p_{tx}\frac{\partial}{\partial p_{tx}} + p_{xt}\frac{\partial}{\partial p_{xt}} + p_{xx}\frac{\partial}{\partial p_{xx}}.$$

The first differential approximation of the difference scheme (15.1) has under these circumstances the following form (see (15.3)):

$$\mathscr{P}^1 := u_t + u u_x + \frac{1}{\varrho} p_x - \frac{1}{\varrho} N_{1x} + \frac{u}{\varrho} N_{2x} = 0;$$

$$\mathscr{P}^2 := \varrho_t + u\varrho_x + \varrho u_x - N_{2x} = 0;$$

$$\mathscr{P}^3 := p_t + u p_x + \gamma p u_x + u(\gamma - 1) N_{1x}$$
$$- \frac{\gamma - 1}{2} u^2 N_{2x} - (\gamma - 1) N_{3x} = 0.$$

$$(15.39)$$

Here the symbols are the same as in Sect. 15.1. In connection with the given explicit form of the equation of state (15.36) for the set of quantities in Sect. 15.1 one can get the following explicit expressions:

$$\Theta = m = (\gamma - 1)\varepsilon;$$

$$z = \gamma - 1; \quad a^2 = \gamma p;$$

$$l = \gamma - 1; \quad b = -\frac{\gamma - 1}{2} u^2.$$

Lemma 15.2. Let the difference scheme (15.1) admit an operator space with basis (13.4). For the invariance of the difference scheme (15.1) with respect to transformations with the infinitesimal operators (15.37) it is necessary and sufficient that

$$\bar{L}_5 N_{1x} = N_{1x} + u N_{2x} + h \frac{\partial N_{1x}}{\partial h};$$

$$\bar{L}_5 N_{2x} = 2 N_{2x} + h \frac{\partial N_{2x}}{\partial h};$$

$$\bar{L}_5 N_{3x} = u N_{1x} + h \frac{\partial N_{3x}}{\partial h};$$

$$\bar{L}_6 N_{jx} = N_{jx}; \quad j = 1, 2, 3,$$

$$(15.40)$$

with

$$\bar{L}_5 := 2\varrho \frac{\partial}{\partial \varrho} + 3\varrho_x \frac{\partial}{\partial \varrho_x} + p_x \frac{\partial}{\partial p_x} + 3\varrho_{xx} \frac{\partial}{\partial \varrho_{xx}} + p_{xx} \frac{\partial}{\partial p_{xx}},$$

$$\bar{L}_6 := \varrho \frac{\partial}{\partial \varrho} + p \frac{\partial}{\partial p} + \varrho_{xx} \frac{\partial}{\partial \varrho_{xx}} + p_{xx} \frac{\partial}{\partial p_{xx}}.$$

$$(15.41)$$

Proof. From the theorem (see Chap. 14) it follows that the necessary and sufficient conditions for invariance of the difference scheme (15.1) with respect to transformations with operators (15.38) can be written as follows:

$$\tilde{L}_k^{(2)} \mathscr{P}^j|_{\mathscr{P}^j = 0} = 0; \quad (j = 1, 2, 3; \ k = 5, 6).$$

From this we get:

$$\tilde{L}_5^{\prime(2)} \mathscr{P}^1|_{\mathscr{P}^j=0} = uN_{2x} - N_{1x} + uh\frac{\partial N_{2x}}{\partial h} - h\frac{\partial N_{1x}}{\partial h} + \bar{L}_5 N_{1x} - u\bar{L}_5 N_{2x} = 0;$$

$$\tilde{L}_5^{\prime(2)} \mathscr{P}^2|_{\mathscr{P}^j=0} = -2N_{2x} - h\frac{\partial N_{2x}}{\partial h} + \bar{L}_5 N_{2x} = 0;$$

$$\tilde{L}_5^{\prime(2)} \mathscr{P}^3|_{\mathscr{P}^j=0} = -u(\gamma-1)\bar{L}_5 N_{1x} + \frac{\gamma-1}{2}u^2\bar{L}_5 N_{2x} + (\gamma-1)\bar{L}_5 N_{3x}$$

$$+ u(\gamma-1)h\frac{\partial N_{1x}}{\partial h} - \frac{\gamma-1}{2}u^2h\frac{\partial N_{2x}}{\partial h} - (\gamma-1)h\frac{\partial N_{3x}}{\partial h} = 0;$$

$$\tilde{L}_6^{\prime(2)} \mathscr{P}^1|_{\mathscr{P}^j=0} = \frac{1}{\varrho}N_{1x} - \frac{u}{\varrho}N_{2x} - \frac{1}{\varrho}\bar{L}_6 N_{1x} + \frac{u}{\varrho}\bar{L}_6 N_{2x} = 0;$$

$$\tilde{L}_6^{\prime(2)} \mathscr{P}^2|_{\mathscr{P}^j=0} = N_{2x} - \bar{L}_6 N_{2x} = 0;$$

$$\tilde{L}_6^{\prime(2)} \mathscr{P}^3|_{\mathscr{P}^j=0} = (\gamma-1)N_{3x} - u(\gamma-1)N_{1x} + \frac{\gamma-1}{2}u^2 N_{2x}$$

$$+ u(\gamma-1)\bar{L}_6 N_{1x} - \frac{\gamma-1}{2}u^2\bar{L}_6 N_{2x} - (\gamma-1)\bar{L}_6 N_{3x} = 0,$$

and, consequently,

$$\bar{L}_6 N_{jx} = N_{jx}; \quad j = 1,2,3;$$

$$\bar{L}_5 N_{2x} = 2N_{2x} + h\frac{\partial N_{2x}}{\partial h};$$

$$\bar{L}_5 N_{1x} = N_{1x} + uN_{2x} + h\frac{\partial N_{1x}}{\partial h};$$

$$\bar{L}_5 N_{3x} = uN_{1x} + h\frac{\partial N_{3x}}{\partial h}.$$

Thus the lemma is proved.   □

Combining the sufficient conditions of invariance found in Theorems 15.1–15.3 together with the conditions (15.40) we get classes of difference schemes which possess the properties M, K and which admit the operator space with basis (13.4), (15.37).

Remark 1. For $\gamma = 3$ the system of equations (13.3) admits still a further transformation with the infinitesimal operator

$$L_7 = t^2\frac{\partial}{\partial t} + tx\frac{\partial}{\partial x} + (x - ut)\frac{\partial}{\partial u} - t\varrho\frac{\partial}{\partial\varrho} - 3tp\frac{\partial}{\partial p}, \tag{15.42}$$

which represents the operator of a projection in the space $\mathscr{E}(x,t,u,\varrho,p)$. The conditions for invariance of difference schemes with respect to this transformation can be found on the basis of the theorem in Chap. 14.

Remark 2. In the foregoing paragraphs the necessary and sufficient conditions for invariance of difference schemes of first order of approximation were found for which the first differential approximation has the form (15.2) (or, equivalently, (15.3)), and with their help classes of invariant difference schemes were constructed. The means of construction of an invariant scheme is the following:

It is necessary to choose the differential equation (15.2) such that the matrix $\Omega = C + \dfrac{\tau}{2} A^2$ satisfies, for example, the conditions of Theorem 15.1. Then the difference schemes with such a first differential approximation are invariant.

Remark 3. The conditions for invariance of difference schemes of higher order of approximation can be found by the same means as pointed out in the present paragraph. It is natural then that the extended operators of higher order must be used.

16. Investigation of Properties of the Artificial Viscosity of Invariant Difference Schemes for the One-dimensional Equations of Gas Dynamics

16.1 Γ-matrices in Eulerian Coordinates

Let us consider the difference scheme (15.1) the Π-form of the first differential approximation of which has the form (15.2). The investigations of such a system of equations which were carried out by a number of authors (see e.g. [8]) have shown that for the correctness of the system of equations a set of restrictions on the matrix C is necessary and, consequently, also on the matrix Ω; especially the matrix C was chosen such that the following conditions are satisfied [8]:

1) Cauchy's problem for the system of equations (15.2) should be correct;

2) For arbitrary piecewise continuous and piecewise smooth initial distributions the solutions of the system of equations (15.2) should be smooth for $t > 0$;

3) For $\tau, h \to 0$ the solution of the system of equations (15.2) should converge with respect to some norm to a stable generalized solution of the system of equations (13.3).

The question, which limitations on the matrix C are sufficient to satisfy the conditions 1)–3) can not be answered exactly now. But on the basis of some special results for the simplest systems of equations of the form (15.2) one is able to restrict the class of matrices C. Especially in accordance with the paper [200] we will postulate that the real parts of the eigenvalues of the matrices C and the diagonal elements of the matrix ZCZ^{-1} are nonnegative where the matrix Z

$$Z = \begin{pmatrix} -uz & (u^2 - E - m)z & z \\ c - uz & -uc + z(u^2 - F) + \Theta & z \\ -c - uz & uc + z(u^2 - E) + \Theta & z \end{pmatrix},$$

$$\det\{Z\} = 2cz(mz + uc + zu^2 + \Theta); \quad c^2 := \Theta + mz$$

is defined such that

$$ZAZ^{-1} = \tilde{A} = \begin{pmatrix} u & 0 & 0 \\ 0 & u+c & 0 \\ 0 & 0 & u-c \end{pmatrix};$$

c is the speed of sound; z, Θ, m are defined in Sect. 15.1.

The class Γ of invariant difference schemes which was introduced in the previous paragraph can completely by described by choosing the matrix

$$\Omega = \frac{\tau}{2} A^2 + C = \frac{\tau}{2} A^2 + \bar{C} H,$$

where

$$\bar{C} := CH^{-1}$$

$$H^{-1} = \begin{pmatrix} \varrho & u & 0 \\ 0 & 1 & 0 \\ \varrho u & E - \Theta/z & 1/z \end{pmatrix};$$

$$H = \begin{pmatrix} 1/\varrho & -u/\varrho & 0 \\ 0 & 1 & 0 \\ -uz & z(u^2 - E) + \Theta & z \end{pmatrix};$$

$$\det \{H\} = z/\varrho.$$

Indeed, from the corollary to the Lemma 15.1 it follows that the difference scheme (15.1) belongs to class Γ if and only if the matrix $\bar{C}$ has the form

$$\bar{C} = \begin{pmatrix} uv_{21} + r_1 & uv_{22} + r_2 & uv_{23} + r_3 \\ v_{21} & v_{22} & v_{23} \\ \frac{u^2}{2} v_{21} + ur_1 + k_1 & \frac{u^2}{2} v_{22} + ur_2 + k_2 & \frac{u^2}{2} v_{23} + ur_3 + k_3 \end{pmatrix}, \quad (16.1)$$

where

$$\frac{\partial v_{2i}}{\partial t} = \frac{\partial v_{2i}}{\partial x} = \frac{\partial v_{2i}}{\partial u} = 0;$$

$$\frac{\partial r_i}{\partial t} = \frac{\partial r_i}{\partial x} = \frac{\partial r_i}{\partial u} = 0;$$

$$\frac{\partial k_i}{\partial t} = \frac{\partial k_i}{\partial x} = \frac{\partial k_i}{\partial u} = 0; \quad i = 1, 2, 3.$$

Definition. In the following we will call matrices of the form (16.1) Γ-matrices.

As an example of a Γ-matrix the matrix H^{-1} can serve for which

$$v_{21} = v_{23} = 0;$$

$$v_{22} = 1;$$

$$r_1 = \varrho; \quad r_2 = r_3 = 0;$$

$$k_1 = 0; \quad k_2 = \varepsilon - \Theta/z; \quad k_3 = z^{-1}.$$

From this it follows that the difference scheme (15.1) with the matrix

$$\Omega = \tau I + \frac{\tau}{2} A^2$$

belongs to class Γ.

The following lemma holds.

Lemma 16.1. If the matrices C_1 and C_2 are Γ-matrices then also their sum represents a Γ-matrix.

That this statement is true follows from the definition of a Γ-matrix.

Lemma 16.2.
1) In the Γ-matrix (16.1) let

$$k_i = (\varepsilon + m) v_{2i}; \quad i = 1, 2, 3,$$

then the difference scheme (15.1) possesses the property K.
2) If in a Γ-matrix $v_{2i} = 0$, then the difference scheme (15.1) possesses the property M.

Indeed, in the present case (see Sects. 15.3, 4) $X_0 \bar{C} = 0$ and therefore

$$X_0 C = X_0 \bar{C} H^{-1} = 0,$$

i.e. the difference scheme (15.1) possesses the property K.

That the difference scheme (15.1) possesses the property M for $v_{2i} = 0$ follows from the definition of this property (see Sect. 15.3).

Corollary. Let the Γ-matrix $\bar{C}$ have the form

$$\begin{pmatrix} r_1 & r_2 & r_3 \\ 0 & 0 & 0 \\ ur_1 & ur_2 & ur_3 \end{pmatrix}$$

then the difference scheme (15.1) possesses the properties K and M.

16.2 Property $\bar{K}$

Definition. We will say that the difference scheme (15.1) possesses the property $\bar{K}$, if $X_0 N_x = 0$.

In the first differential approximation of the difference scheme which possesses the property $\bar{K}$ the equation for the entropy has the form

$$s_t + us_x = 0.$$

The following lemma holds.

Lemma 16.3. Let the difference scheme (15.1) possess the property K and

$$N_2 = \phi u_x; \quad N_1 = u N_2 + R; \quad R = -(\varepsilon + m)_x \phi,$$

where ϕ is an arbitrary function; then the difference scheme possesses the property $\bar{K}$. If besides that

$$\frac{\partial \phi}{\partial t} = \frac{\partial \phi}{\partial x} = \frac{\partial \phi}{\partial u} = 0, \tag{16.2}$$

then the difference scheme belongs to class Γ.

Proof. We assume that the difference scheme possesses the property K. Then

$$-uN_1 + (u^2 - E - m)N_2 + N_3 = 0.$$

If we differentiate this equation with respect to x we get

$$-uN_{1x} + (u^2 - E - m)N_{2x} + N_{3x} - u_x N_1 + (uu_x - \varepsilon_x - m_x)N_2$$
$$= X_0 N_x - u_x(N_1 - uN_2) - (\varepsilon + m)_x N_2 = 0.$$

From the conditions of the lemma it follows that

$$X_0 N_x = 0,$$

i.e. the difference scheme possesses the property $\bar{K}$.

That the difference scheme belongs to class Γ under the assumptions for the function ϕ follows from the results of this section and Sect. 15.4. □

Corollary. The difference scheme (15.1) which possesses the property M can not possess the property $\bar{K}$ and vice versa.

From the proved lemma it follows that a matrix Ω for which the difference scheme (15.1) possesses the properties K, $\bar{K}$ and Γ is uniquely determined with an accuracy up to the function ϕ which satisfies the conditions (16.2).

Remark. In the case of a hyperbolic system of equations with constant coefficients the properties K and $\bar{K}$ coincide for the difference schemes.

Definition. We will say that the difference scheme (15.1) possesses the properties K_c and K_{-c} if the equations

$$X_c C = 0 \quad \text{and} \quad X_{-c} C = 0.$$

hold, where the vectors

$$X_c := \left(-u + \frac{c}{z}, u^2 - E + \frac{1}{z}(\Theta - uc), 1\right);$$

$$X_{-c} := \left(-u - \frac{c}{z}, u^2 - E + \frac{1}{z}(\Theta + uc), 1\right),$$

are such that

$$X_c A = (u + c) X_c$$

and

$$X_{-c} A = (u - c) X_{-c}.$$

It is not difficult to see that if the difference scheme (15.1) belongs to class Γ, possesses the property K and if

$$r_i = cv_{2i} \quad (r_i = -cv_{2i}), \quad i = 1, 2, 3$$

that then the difference scheme possesses the property $K_c(K_{-c})$.

From this fact the following lemma can be derived.

Lemma 16.4. If the difference scheme belongs to the class $\Gamma \cap K$, satisfying one of the properties K_c or K_{-c} excludes the fulfilment of the other. Or, in other words: then the scheme can belong to the class $\Gamma \cap K \cap K_c$ or to the class $\Gamma \cap K \cap K_{-c}$.

From the results of Sect. 15.1 and Sect. 15.2 it follows that the following statement is true.

. . . .

Theorem 16.1. If the difference scheme (15.1) belongs to class $\Gamma \cap K$ then in addition the difference scheme can possess only one of the following properties: $M, \bar{K}, K_c, K_{-c}$.

. . . .

16.3 Polynomial Form of the Viscosity Matrix

We consider a class of invariant schemes with a viscosity matrix in the first differential approximation in the form of a polynomial of the matrix A. The following lemma holds.

Lemma 16.5. If $\bar{C}$ is a Γ-matrix then the matrix $(A - uI)\bar{C}$ is also a Γ-matrix.

Proof. Let the matrix $\bar{C}$ have the form (16.1). Then for the elements q_{ij} $(i,j = 1, 2, 3)$ of the matrix $(A - uI)\bar{C}$ the following formulas hold:

$$q_{1j} = uq_{2j} + \bar{r}_j;$$
$$q_{2j} = \bar{r}_j;$$
$$q_{3j} = \frac{u^2}{2} q_{2j} + u\bar{r}_j + \bar{k}_j,$$

where

$$\bar{r}_j := (\Theta - \varepsilon z)v_{2j} + \bar{k}_j z;$$
$$\bar{k}_j := (\varepsilon + m)\bar{r}_j;$$
$$\frac{\partial \bar{r}_j}{\partial t} = \frac{\partial \bar{r}_j}{\partial x} = \frac{\partial \bar{r}_j}{\partial u} = 0;$$
$$\frac{\partial \bar{k}_j}{\partial t} = \frac{\partial \bar{k}_j}{\partial x} = \frac{\partial \bar{k}_j}{\partial u} = 0,$$

and, consequently, the matrix $(A - uI)\bar{C}$ is a Γ-matrix. □

Lemma 16.6. The following relations hold:

$$(A - uI)^{2\xi + 2} = c^{2\xi}(A - uI)^2;$$
$$(A - uI)^{2\xi + 1} = c^{2\xi}(A - uI);$$
$$\xi = 0, 1, 2, \ldots$$

These relations can easily be proved by the method of transfinite induction.

Corollary. Any polynomial of the form

$$C = \mathscr{P}(A) = \alpha_0 I + \alpha_1 (A - uI) + \ldots + \alpha_k (A - uI)^k$$

can be represented as a three–term quadratic expression

$$C = \mathscr{P}_2(A) = \beta_0 I + \beta_1 (A - uI) + \beta_2 (A - uI)^2,$$

where $\beta_0 = \alpha_0$; β_1 and β_2 are functions of α_i and c^2 $(i = 1, \ldots, k)$.

. . . .

Theorem 16.2. That the difference scheme (15.1) with the matrix $\Omega = C + \dfrac{\tau}{2} A^2$, where

$$C = \mathscr{P}(A) = \alpha_0 I + \alpha_1 A + \alpha_2 A^2 \qquad (16.3)$$

belongs to the class Γ it is necessary and sufficient that the polynomial $\mathscr{P}(A)$ can be represented in the form

$$C = \mathscr{P}(A) = \beta_0 I + \beta_1 (A - uI) + \beta_2 (A - uI)^2. \qquad (16.4)$$

Here

$$\frac{\partial \beta_i}{\partial t} = \frac{\partial \beta_i}{\partial x} = \frac{\partial \beta_i}{\partial u} = 0.$$

. . . .

Proof. Necessary condition: According to the assumptions the matrix

$$\bar{C} = CH^{-1} = \alpha_0 H^{-1} + \alpha_1 AH^{-1} + \alpha_2 A^2 H^{-1} = \| v_{ij} \|_1^3$$

is a Γ-matrix, i.e.

$$\frac{\partial v_{2j}}{\partial t} = \frac{\partial v_{2j}}{\partial x} = \frac{\partial v_{2j}}{\partial u} = 0;$$

$$v_{1j} = u v_{2j} + r_j;$$

$$v_{3j} = \frac{u^2}{2} v_{2j} + u r_j + k_j.$$

In the case under consideration we have

$$v_{21} = \varrho (\alpha_1 + 2 u \alpha_2);$$

$$v_{22} = \alpha_0 + u \alpha_1 + u^2 \alpha_2;$$

$$v_{23} = \alpha_2.$$

Therefore

$$\alpha_0 = \beta_0 - u \beta_1 + u^2 \beta_2;$$

$$\alpha_1 = \beta_1 - 2 u \beta_2; \qquad (16.5)$$

$$\alpha_2 = \beta_2,$$

where

$$\frac{\partial \beta_i}{\partial t} = \frac{\partial \beta_i}{\partial x} = \frac{\partial \beta_i}{\partial u} = 0.$$

Then

$$C = (\beta_0 - u\beta_1 + u^2\beta_2)I + (\beta_1 - 2u\beta_2)A + \beta_2 A^2$$
$$= \beta_0 I + \beta_1(A - uI) + \beta_2(A - uI)^2.$$

Sufficient condition: Let the matrix C have the form (16.4). Consequently, on the basis of Lemma 16.1 and 16.5 the matrix

$$\bar{C} = \beta_0 H^{-1} + \beta_1(A - uI)H^{-1} + \beta_2(A - uI)^2 H^{-1}$$

is a Γ-matrix and the difference scheme belongs to class Γ. The theorem is proved.   □

We remark that the difference scheme (15.1) with the matrix $\Omega = C + \frac{\tau}{2}A^2$ where the matrix C is a polynomial of the form (16.3) possesses the property K if

$$\alpha_0 + \alpha_1 u + \alpha_2 u^2 = 0.$$

In the case when the difference scheme belongs to class Γ this condition means that $\beta_0 = 0$.

Indeed, from the equation $X_0 C = 0$ it follows that $(\alpha_0 + \alpha_1 u + \alpha_2 u^2)X_0 = 0$ i.e. $\alpha_0 + \alpha_1 u + \alpha_2 u^2 = 0$. If the equations (16.5) are satisfied then

$$\alpha_0 + \alpha_1 u + \alpha_2 u^2 = \beta_0 - u\beta_1 + u^2\beta_2 + (\beta_1 - 2u\beta_2)u + \beta_2 u^2 = \beta_0 = 0.$$

16.4 Numerical Experiments for Equations of Gas Dynamics in Eulerian Coordinates

With invariant difference schemes a series of calculations was performed the results of which were compared with calculations gained by noninvariant schemes [86, 88, 109].

Thus in Fig. 16.1 velocity profiles of a stationary shock wave are shown in three different coordinate systems which differ from each other by a Galilei-transformation. The curves in Fig. 16.1a were calculated by an explicit invariant difference scheme with a viscosity matrix in the first differential approximation of the form

$$C = \mu I, \tag{16.6}$$

where

$$\mu := h\mu_0\varrho\,\frac{|u_x p_x|}{\delta + |u_x p_x|} \tag{16.7}$$

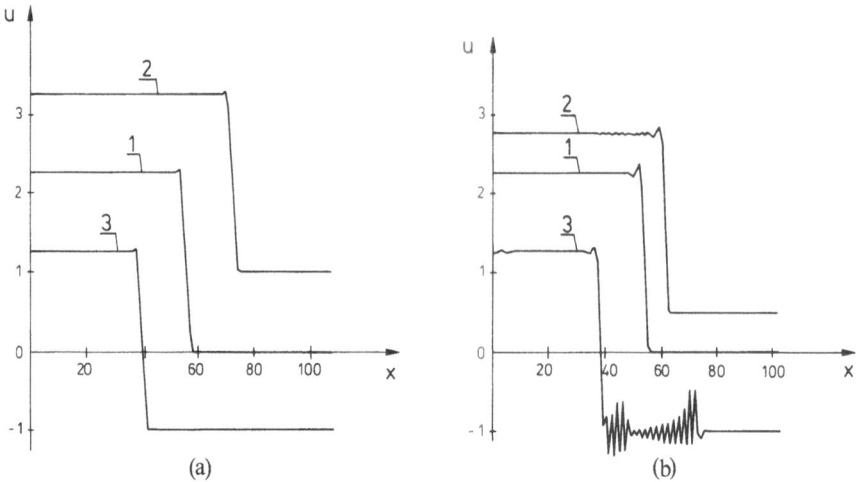

Fig. 16.1. (a) Explicit invariant scheme; (b) Noninvariant Lax-Wendroff-scheme

(μ_0, δ are constants). In this case the difference scheme does not possess the property K. The curves in Fig. 16.1b were found by the help of the noninvariant Lax-Wendroff-scheme.

The invariant difference scheme gives independently of the system of coordinates exactly the same profile for the solution while the corresponding profiles of the solution which were found using a noninvariant difference scheme, are different (see Fig. 16.1). The calculations show a strong dependency of the solution of the noninvariant difference scheme on the choice of the coordinate system.

For example, the calculation of a stationary shock wave with the Lax-Wendroff-scheme

$$\left.\begin{array}{l} u(x,0) = \ \ 2.26 \\ \varrho(x,0) = 43.7 \\ p(x,0) = 76.5 \end{array}\right\}, \quad \text{for} \quad x \leqq 0,$$

$$\left.\begin{array}{l} u(x,0) = \ \ 0 \\ \varrho(x,0) = 10 \\ p(x,0) = \ \ 0 \end{array}\right\}, \quad \text{for} \quad x > 0$$

coincides quite well with the exact solution (see Fig. 16.1b, curve 1), but the calculation of the same problem in other coordinate systems leads to large errors (curves 2 and 3).

In Fig. 16.2 density profiles in the vicinity of a contact discontinuity and a shock wave for the problem of the disintegration of an arbitrary unsteadiness are calculated in three different coordinate systems using an explicit invariant difference scheme with a viscosity matrix in the first differential approximation

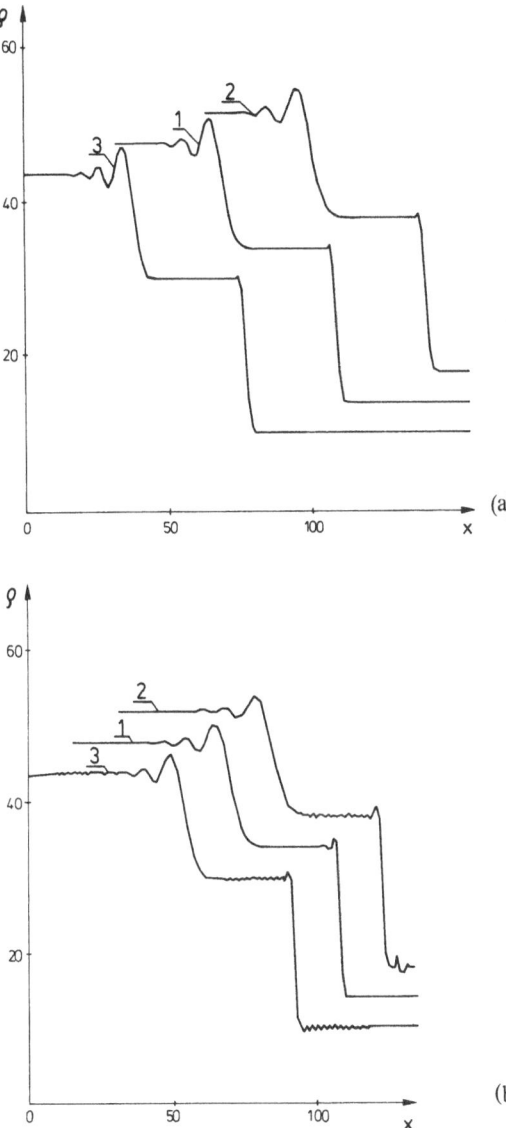

Fig. 16.2. (a) Explicit invariant scheme; (b) Noninvariant Lax-Wendroff-scheme

of the form (16.5) (see Fig. 16.2a) and using the Lax-Wendroff-scheme (see Fig. 16.2b). The coordinate systems differ by Galilei-transformations.

In Figs. 16.3–16.7 results of numerical calculations are presented which were obtained by an explicit invariant difference scheme of the form (15.1) with a viscosity matrix in the first differential approximation

$$C = \frac{h}{2}\,\bar{C}H,$$

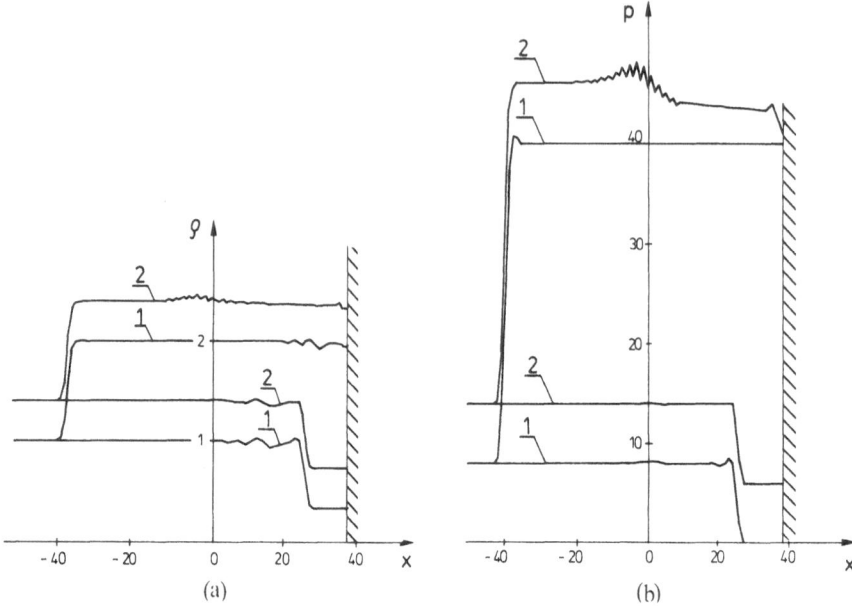

Fig. 16.3. (a) Density distribution (explicit invariant scheme, $v = h\mu_0\varrho$, $\varkappa = 0.05$ (curve 1), $\varkappa = 0.1$ (curve 2)); (b) Pressure distribution (explicit invariant scheme, $v = h\mu_0\varrho$, $\varkappa = 0.05$ (curve 1), $\varkappa = 0.1$ (curve 2))

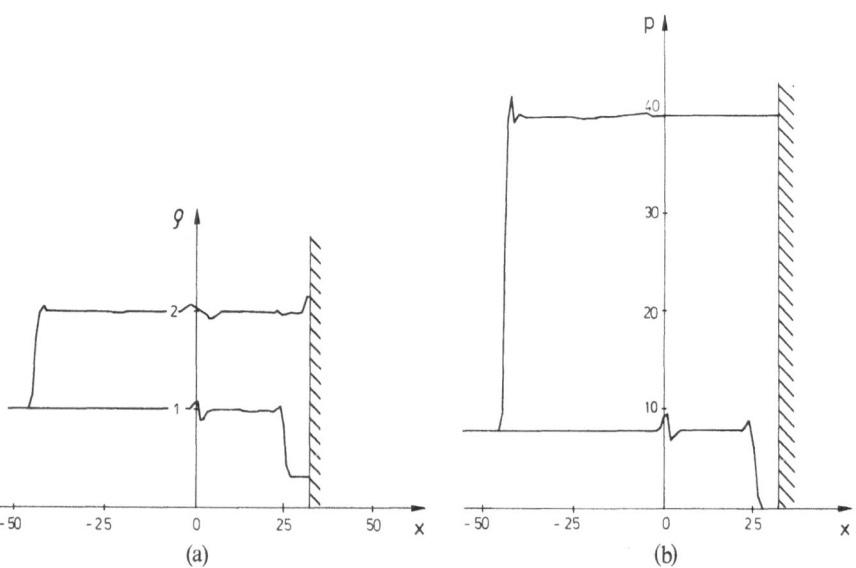

Fig. 16.4. (a) Density distribution (before reflection (curve 1), after reflection (curve 2)); (b) Pressure distribution (before reflection (curve 1), after reflection (curve 2))

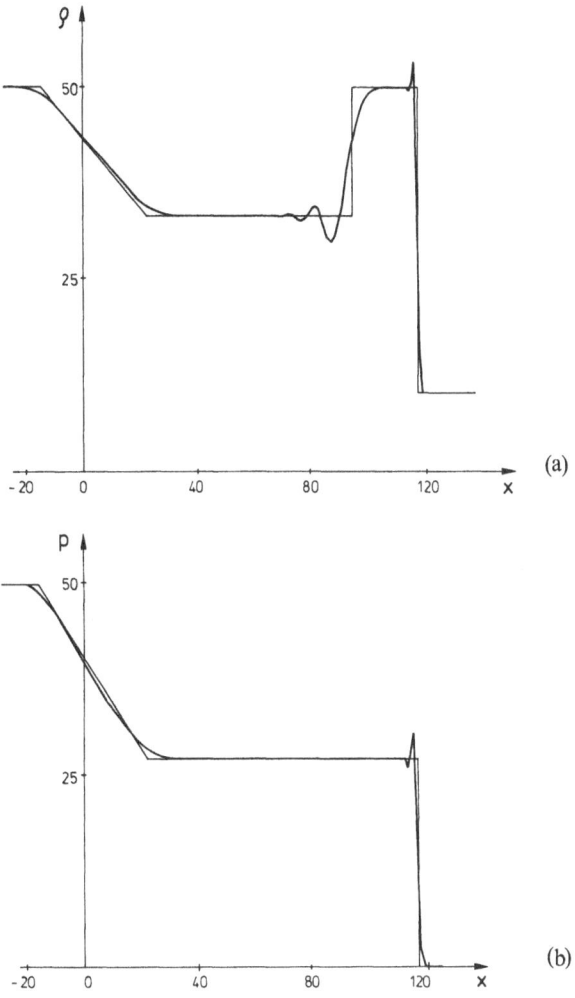

Fig. 16.5. (a) Density distribution of a disintegrating discontinuity; (b) Pressure distribution of a disintegrating discontinuity

where

$$\bar{C} = \begin{pmatrix} v & 0 & 0 \\ 0 & 0 & 0 \\ uv & 0 & 0 \end{pmatrix}; \quad C = \begin{pmatrix} \dfrac{1}{\varrho}v & -\dfrac{u}{\varrho}v & 0 \\ 0 & 0 & 0 \\ \dfrac{1}{\varrho}uv & -\dfrac{u^2}{\varrho}v & 0 \end{pmatrix}, \tag{16.8}$$

and for different choice of the coefficient v. It is not difficult to see that in this case the difference scheme possesses the properties K and M and, moreover, the artificial viscosity enters the first differential approximation in an analogous way to that of the physical viscosity.

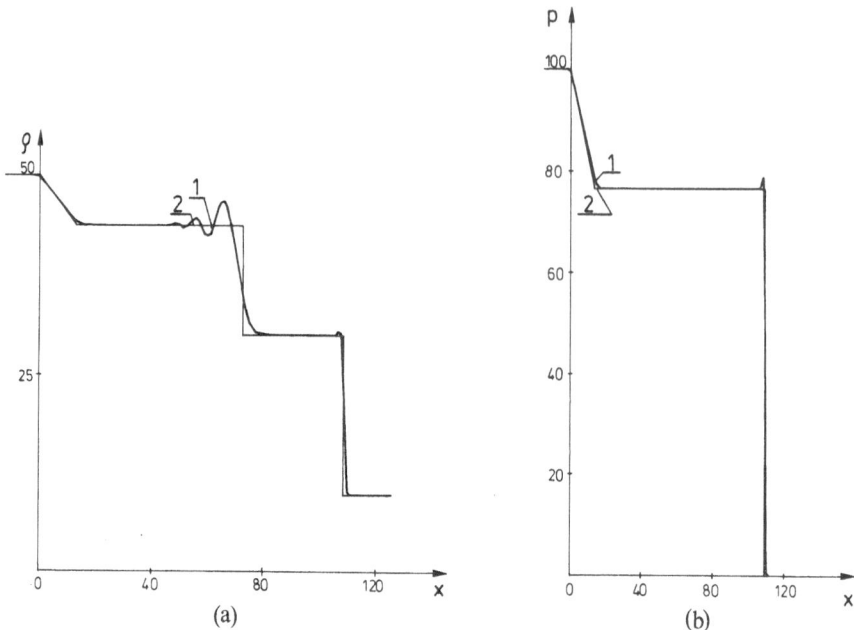

Fig. 16.6. (a) Density distribution of a disintegrating discontinuity; (b) Pressure distribution of a disintegrating discontinuity

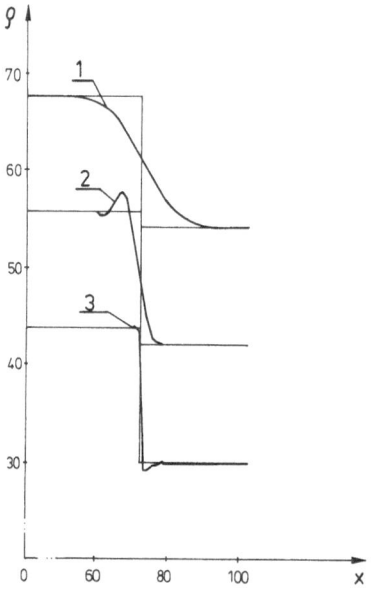

Fig. 16.7. Density distribution (continuity equation with viscous terms (curve 1), third order approximation of the continuity equation (curve 2), Galilei-transformation (curve 3))

The stability analysis of such a class of difference schemes by the method of the differential approximation gave the following bounds:

$$v \geq 0;$$
$$\kappa^2 (u^2 + c^2) + \kappa v \pm \sqrt{4\kappa^2 u^2 c^2 + \kappa^2 v^2} < 1.$$

The numerical calculations have shown the sufficiency of these stability conditions.

In Fig. 16.3 the density profiles (a) and the pressure profiles (b) are shown for the problem of a shock wave running against a rigid wall. The curves correspond to the case where $v = h\mu_0 \varrho$, $\mu_0 = $ const, $\kappa = 0.05$ (curve 1) and $\kappa = 0.1$ (curve 2).

In Fig. 16.4 the density profiles (a) and the pressure profiles (b) are shown for the case $v = \mu$, $\kappa = 0.1$, where μ is defined by formula (16.7). In the Fig. 16.5 and Fig. 16.6 results of calculations are represented for the problem of the disintegration of an arbitrary discontinuity for $v = \mu$, where μ is given by formula (16.7). The rarefaction wave and the shock wave are calculated with good accuracy. But near the contact surface oscillations are generated and a strong smearing occurs. The oscillations can be eliminated by introducing sufficiently viscous terms in the continuity equation (see Fig. 16.7, curve 1) which on the other hand enlarge the zone of smearing-out of the discontinuity. It is possible to diminish the oscillations without enlarging the zone of smearing (see Fig. 16.7, curve 2) by treating the continuity equation with a scheme of third order of approximation (on the whole the difference scheme remains a scheme of first order). The best result (see Fig. 16.7, curve 3) gives a calculation of the contact surface in a coordinate system, where the contact surface is fixed, which is realized by a corresponding Galilei-transformation.

16.5 Damping of Oscillatory Effects

An invariant difference scheme can be constructed also in the following way:

$$\frac{w^*(x) - w^n(x)}{\tau} + \frac{\Delta_1 + \Delta_{-1}}{2h} f^n(x)$$

$$= \frac{\tau}{2h} \left[A^2 \left(x + \frac{h}{2} \right) \frac{\Delta_1}{h} - A^2 \left(x - \frac{h}{2} \right) \frac{\Delta_{-1}}{h} \right] w^n(x),$$

$$w^{n+1}(x) = w^*(x) + \frac{\tau}{h} \left[C \left(x + \frac{h}{2} \right) \frac{\Delta_1}{h} - C \left(x - \frac{h}{2} \right) \frac{\Delta_{-1}}{h} \right] w^*(x).$$

(16.9)

The first differential approximation of the difference scheme (16.9) has the form

$$\frac{\partial w}{\partial t} + \frac{\partial f}{\partial x} = \frac{\partial}{\partial x} \left(C \frac{\partial w}{\partial x} \right).$$

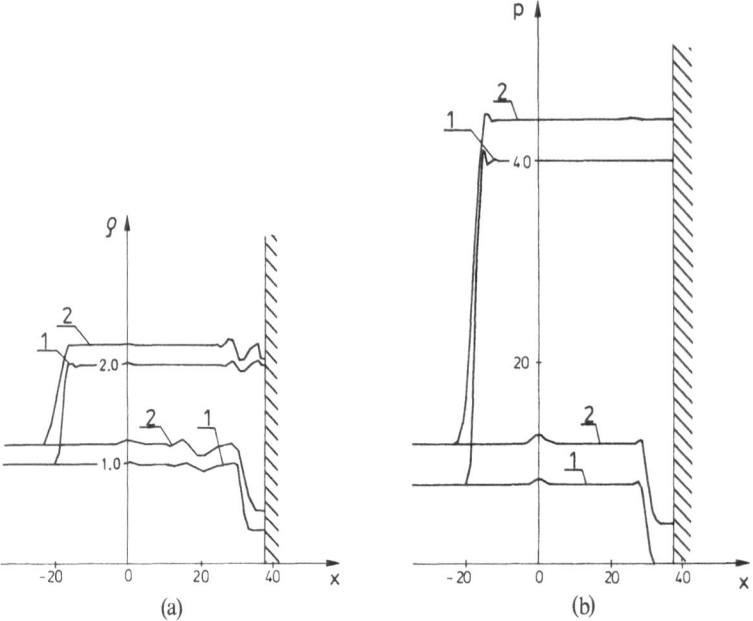

Fig. 16.8. (a) Density distribution; (b) Pressure distribution

If the matrix C is a Γ-matrix then the difference scheme (16.9) belongs to class Γ. The representation of a difference scheme in such a form enables us to diminish the oscillating effects. In Fig. 16.8 the density profiles (a) and pressure profiles (b) are shown for the problem of a shock wave reflecting from a fixed wall. These results obtained by a calculation uing a difference scheme (16.9) with a matrix C of type (16.6), where μ is defined by formula (16.7).

16.6 Γ-matrices in Lagrangean Coordinates

We consider the case of Lagrangean coordinates. The conditions for invariance of a difference scheme (15.28) of the first order of approximation were formulated in Sect. 15.6. From Theorems 15.5 and 15.6 the following lemma can be deduced.

Lemma 16.7. The difference scheme (15.28) belongs to class Γ, if and only if the matrix $\bar{C}$ has the form

$$\bar{C} = \begin{pmatrix} v_{11} & v_{12} & v_{13} \\ v_{21} & v_{22} & v_{23} \\ uv_{11} + k_1 & uv_{12} + k_2 & uv_{13} + k_3 \end{pmatrix}, \tag{16.10}$$

where

$$\frac{\partial v_{ij}}{\partial t} = \frac{\partial v_{ij}}{\partial x} = \frac{\partial v_{ij}}{\partial u} = 0;$$

$$i = 1, 2; \quad j = 1, 2, 3.$$

$$\frac{\partial k_j}{\partial t} = \frac{\partial k_j}{\partial x} = \frac{\partial k_j}{\partial u} = 0;$$

Here

$$\bar{C} = CH^{-1};$$

$$H^{-1} = \begin{pmatrix} 1 & 0 & 0 \\ 0 & 1 & 0 \\ u & -p_v/p_\varepsilon & 1/p_\varepsilon \end{pmatrix}; \quad H = \begin{pmatrix} 1 & 0 & 0 \\ 0 & 1 & 0 \\ up_\varepsilon & p_v & p_\varepsilon \end{pmatrix}; \quad \det\{H\} = p_o.$$

The matrices of type (16.10) will be called, by analogy with the case of Eulerian coordinates, Γ-matrices. In a special problem it is evident which system of coordinates will be used and, consequently, which form of Γ-matrices.

As an example of a Γ-matrix – which can easily be verified – the matrix H^{-1} can serve.

Lemma 16.8.
1) If C_1 and C_2 are Γ-matrices then also their sum is a Γ-matrix.
2) The matrices AH^{-1} and $A^2 H^{-1}$ are Γ-matrices.

Proof. That the first statement is true follows from the definition of Γ-matrices.

For the proof of the second statement we will find the form of the matrices AH^{-1} and $A^2 H^{-1}$. The following representations hold:

$$AH^{-1} = \begin{pmatrix} 0 & 0 & -1 \\ 1 & 0 & 0 \\ -p & 0 & -u \end{pmatrix}; \quad A^2 H^{-1} = \begin{pmatrix} a^2 & 0 & 0 \\ 0 & 0 & -1 \\ ua^2 & 0 & p \end{pmatrix},$$

and, consequently, AH^{-1} and $A^2 H^{-1}$ are Γ-matrices. □

Lemma 16.9. The following formulas can be proved:

$$A^{2k-1} = a^{2k-2} A; \quad A^{2k} = a^{2k-2} A^2; \quad k = 1, 2, \ldots$$

This lemma can easily be proved by application of transfinite induction.

Corollary. Any polynomial of the form

$$\mathscr{P}(A) = \alpha_0 I + \alpha_1 A + \ldots + \alpha_m A^m$$

can be represented as a sum of three terms up to quadratic expressions in A:

$$\mathscr{P}_2(A) = \beta_0 I + \beta_1 A + \beta_2 A^2,$$

where $\beta_0 = \alpha_0$; β_1 and β_2 are functions of α_i and a.

From Lemma 16.7 and 16.8 the following theorem can be deduced.

. . . .

Theorem 16.3. A necessary and sufficient condition that the difference scheme
(15.28) with the matrix $\Omega = C + \dfrac{\tau}{2} A^2$, where

$$C = \alpha_0 I + \alpha_1 A + \alpha_2 A^2,$$

belongs to the class Γ is, that

$$\frac{\partial \alpha_j}{\partial t} = \frac{\partial \alpha_j}{\partial x} = \frac{\partial \alpha_j}{\partial u} = 0; \quad j = 1, 2, 3.$$

. . . .

Remark. If $\alpha_0 = 0$ then the difference scheme possesses also the property K.

Corollary. The Lax' scheme belongs to class Γ and does not possess the property K.

The stability of the difference schemes of the form (15.28), where the matrix $C = \alpha_0 I + \alpha_1 A + \alpha_2 A^2$ can easily be verified by the method of the differential approximation. This method leads to the following bounds:

$$0 \leq \alpha_0 \pm \alpha_1 a + \alpha_2 a^2 \leq \frac{h^2}{2\tau}(1 - \kappa^2 a^2); \quad \alpha_0 \geq 0.$$

It is not difficult to show that under the following relations between the elements of the Γ-matrix (16.9)

$$k_j = - p v_{2j}; \quad j = 1, 2, 3,$$

the difference scheme (15.28) belongs to the class $\Gamma \cap K$.

Definition. We will say that the difference scheme possesses the property $\bar{K}$ if

$$X_0 N_q = 0.$$

The following theorem is true.

. . . .

Theorem 16.4. Let the difference scheme (15.28) possess the property K and

$$N_1 = \phi p_q; \quad N_2 = \phi u_q,$$

where ϕ is an arbitrary function, then the difference scheme possesses also the property $\bar{K}$. If, further,

$$\frac{\partial \phi}{\partial t} = \frac{\partial \phi}{\partial x} = \frac{\partial \phi}{\partial u} = 0, \tag{16.11}$$

then the difference scheme belongs to class Γ.

. . . .

Proof. Let the difference scheme (15.28) have the property K. Then

$$- u N_1 + p N_2 + N_3 = 0.$$

By differentiation of this equation with respect to q we get

$$- u N_{1q} + p N_{2q} + N_{3q} - u_q N_1 + p_q N_2 = X_0 N_q = 0,$$

consequently, the difference scheme possesses the property $\bar{K}$.
The matrix $\bar{C}$ in this case has the form

$$\bar{C} = \begin{pmatrix} 0 & 0 & \phi \\ \phi & 0 & 0 \\ - p\phi & 0 & u\phi \end{pmatrix}$$

and is a Γ-matrix under the conditions (16.11). Thus the theorem is proved. $\square$

When the equations

$$X_a C = 0 \quad \text{and} \quad X_{-a} C = 0$$

are satisfied, then the difference scheme (15.28) possesses the properties K_a and K_{-a}.
Here the vectors

$$X_a = (a - u p_\varepsilon, p_v, p_\varepsilon); \quad X_{-a} = (- a - u p_\varepsilon, p_v, p_\varepsilon)$$

are such that

$$X_a A = a X_a; \quad X_{-a} A = - a X_{-a}.$$

It is not difficult to show that if the difference scheme (15.28) belongs to the class $\Gamma \cap K$ and if

$$v_{1j} = a v_{2j}, \quad (v_{1j} = - a v_{2j}),$$

then the difference scheme possesses the property $K_a (K_{-a})$. From this it follows that the difference scheme (15.28) from the class $\Gamma \cap K$ can not possess at the same time the properties K_a and K_{-a}.
In accordance with the definitions of the properties $\bar{K}$ and M (see Sect. 15.6) the difference scheme (15.28) which belongs to class Γ can not possess at the same time both the properties.
Thus the following theorem holds.

. . . .

Theorem 16.5. If the difference scheme (15.28) belongs to class $\Gamma \cap K$, then in addition it can possess only one of the following properties: M, $\bar{K}$, K_a, K_{-a}.

. . . .

We remark that the fulfilment of the properties K_a and K_{-a} means that in the system of equations of the first differential approximation in the equation for the corresponding invariant a viscosity term appears.

16.7 Width of Shock Smearing

Let us consider difference schemes of the form (15.28) which belong to class $\Gamma \cap K \cap M$. In this case we have

$$
C = \begin{pmatrix} v_{11} - u p_\varepsilon v_{13} & v_{12} + p_v v_{13} & p_\varepsilon v_{13} \\ 0 & 0 & 0 \\ uv_{11} - u^2 p_\varepsilon v_{13} & uv_{12} + u p_v v_{13} & u p_\varepsilon v_{13} \end{pmatrix};
$$

$$
\bar{C} = \begin{pmatrix} v_{11} & v_{12} & v_{13} \\ 0 & 0 & 0 \\ uv_{11} & uv_{12} & uv_{13} \end{pmatrix},
$$

where

$$
\frac{\partial v_{1j}}{\partial t} = \frac{\partial v_{1j}}{\partial x} = \frac{\partial v_{1j}}{\partial u} = 0; \quad j = 1, 2, 3.
$$

The artificial viscosity in the system of equations of the first differential approximation appears in an analogous way to that of the physical viscosity.

We consider for the system of equations of the first differential approximation the problem of a moving stationary shock wave which travels with a constant velocity U. In this case all characteristics of the flow depend only on the variable $y = q - U t$. We will assume that the gas is polytropic. Let

$$
\begin{aligned}
u(+\infty) &= u_0 = 0; & u(-\infty) &= u_1; \\
p(+\infty) &= p_0; & p(-\infty) &= p_1; \\
v(+\infty) &= v_0; & v(-\infty) &= v_1.
\end{aligned}
$$

If we change the variable in the system of equations of the first differential approximation and introduce y we get a system of ordinary differential equations which leads under the assumptions

$$
\omega = v_{11} u_y + v_{12} v_y + v_{13} p_y;
$$
$$
\omega(+\infty) = \omega(-\infty) = 0,
$$

after integration, to the relation

$$
\omega(y) v = \frac{\gamma + 1}{2} U^2 (v_0 - v)(v - v_1). \tag{16.12}
$$

For given values of v_{11}, v_{12}, v_{13} this equation can be solved in a series of cases with respect to v. We assume that

$$
v_{11} = \frac{h \mu_0}{v}; \quad v_{12} = v_{13} = 0; \quad \mu_0 = \text{const}.
$$

Then the equation (16.12) can be integrated and we get the result

$$
v = \frac{v_0 + v_1}{2} + \frac{v_0 - v_1}{2} \cdot \frac{e^{c_1 y + c_2} - 1}{e^{c_1 y + c_2} + 1} = \frac{v_0 + v_1}{2} + \frac{v_0 - v_1}{2} \Phi(y),
$$

where

$$\Phi(y):= \tanh \frac{c_1 y + c_2}{2};$$

$$c_1:= \frac{(\gamma + 1)\, U}{2\,\mu_0\, h};$$

$$\Phi(-\infty) = -1;$$

$$\Phi(+\infty) = 1.$$

The constant of integration c_2 can be determined if the value of v in a point on the interval $(-\infty, +\infty)$ is known.

The effective width of smearing-out the front of the shock wave is given by the formula:

$$\Delta y = \frac{2\, h\, \mu_0}{\Delta v \cdot (\gamma + 1)\, U} \ln \left. \frac{v - v_1}{v_0 - v} \right|_{\tilde{\tilde{v}}}^{\tilde{v}},$$

where $\tilde{v}$, $\tilde{\tilde{v}}$ are some fixed values of v such that

$$|v_1 - \tilde{v}| < \delta; \qquad |v_0 - \tilde{\tilde{v}}| < \delta;$$

and δ is sufficiently small. By reducing μ_0 the profile of the solution can be made steeper and then the effective width of the smeared-out front will be smaller.

16.8 Numerical Experiments on the Equations of Gas Dynamics in Lagrangean Coordinates

To compare properties of invariant and noninvariant difference schemes a great number of calculations with explicit schemes were carried out [86].

The influence of the invariance of a difference scheme on the results of the calculation are shown in Figs. 16.9 and 16.10. We investigated the problem of a shock wave moving with the constant velocity $U = 1$ and the following initial conditions:

$$u = 1, \quad v = 1, \quad p = 1.5714 \quad \text{for} \quad x \leq 0,$$

$$u = 0, \quad v = 2, \quad p = 0.5714 \quad \text{for} \quad x > 0.$$

In Fig. 16.9 profiles of the specific volume v and of the velocity u are shown which were found by an invariant explicit difference scheme of the form (15.28) with the matrix

$$\bar{C} = \begin{pmatrix} v & 0 & 0 \\ 0 & 0 & 0 \\ u\,v & 0 & 0 \end{pmatrix} \tag{16.13}$$

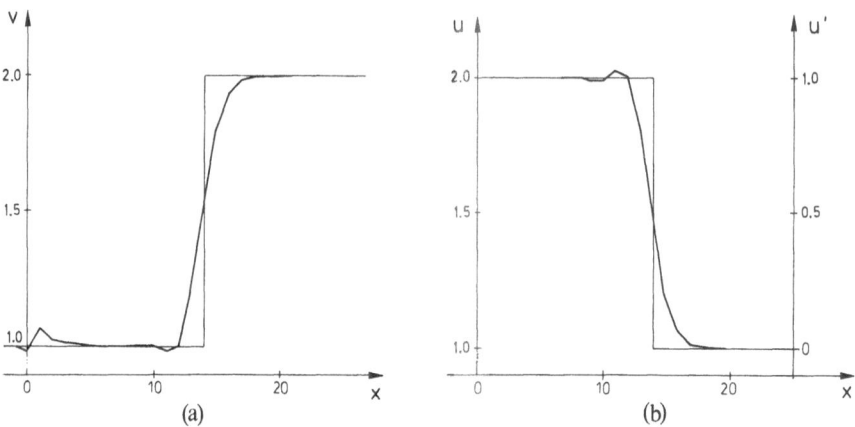

Fig. 16.9. (a) Specific volume (invariant difference scheme); (b) Velocity (invariant difference scheme)

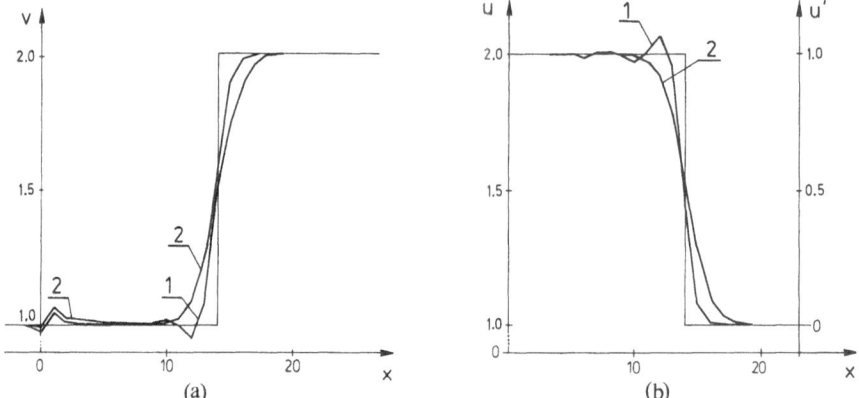

Fig. 16.10. (a) Specific volume (curve 1 and curve 2 belong to different paramters of the Galilei-transformation); (b) Velocity (curve 1 and curve 2 belong to different parameters of the Galilei-transformation)

with respect to two different coordinate systems, which are related by a Galilei-transformation. In the calculations v was chosen according to

$$v = h \mu_0 \frac{|u_q p_q|}{\delta_0 + |u_q p_q|}, \tag{16.14}$$

δ_0 and μ_0 are constants. In this case the difference scheme is invariant and has the properties K and M. The calculations showed that the profiles for u, u' and v, v' coincide with a high degree of accuracy. In Fig. 16.10 the profiles of the specific volume (a) and of the velocity (b), are shown which were found according to an explicit difference scheme of the form (15.28) with a matrix $\bar{C}$ of type

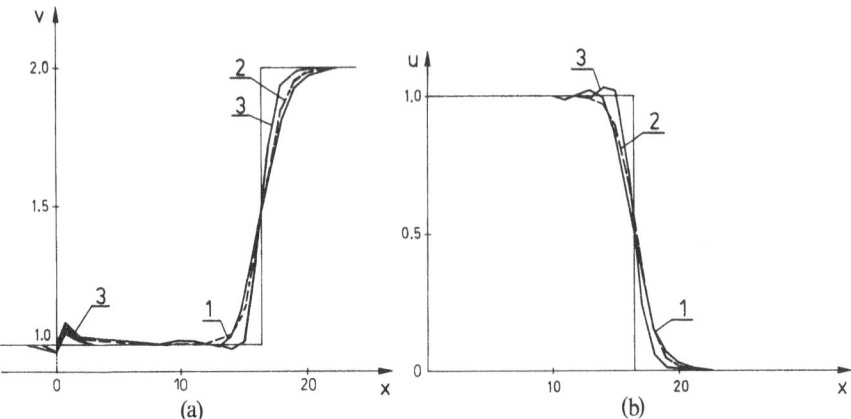

Fig. 16.11. (a) Specific volume ($\varkappa = 0.1$, $\mu_0 = 1$ (curve 1), $\varkappa = 0.4$, $\mu_0 = 0.5$, (curve 2), $\varkappa = 0.4$, $\mu_0 = 1$ (curve 3)); (b) Velocity ($\varkappa = 0.1$, $\mu_0 = 1$ (curve 1), $\varkappa = 0.4$, $\mu_0 = 0.5$, (curve 2), $\varkappa = 0.4$, $\mu_0 = 1$ (curve 3))

(16.13), where

$$v = h\,\mu_0\,u\,\frac{|u_q p_q|}{\delta + |u_q p_q|}.$$

In the given case the difference scheme is not invariant, nevertheless it has the properties K and M. The calculations have shown that the solution of the difference scheme really depends on the parameter of the Galilei-trans-formation.

In Fig. 16.11 the u- and v-profiles are given which were found on the basis of an explicit difference scheme of the form (15.28) with a matrix $\bar{C}$ of type (16.13), where

$$v = \mu_0/v; \qquad \mu_0 = \text{const},$$

and with different values of κ and μ_0;

$$\kappa = 0.1; \qquad \mu_0 = 1, \qquad \text{(curves 1)};$$
$$\kappa = 0.4; \qquad \mu_0 = 0.5, \qquad \text{(curves 2)};$$
$$\kappa = 0.4; \qquad \mu_0 = 1, \qquad \text{(curves 3)};$$

These graphs demonstrate the earlier derived consequences:

1) If the values of the constants are chosen too small, this may lead to an oscillation of the solution,
2) The width of smearing-out the shock can be controlled by choosing an appropriate value for μ_0.

Calculations on the basis of invariant difference schemes were also carried out, where the matrix C is a polynomial of the matrix A. In Fig. 16.12 some results of calculations are plotted which were gained by a difference scheme

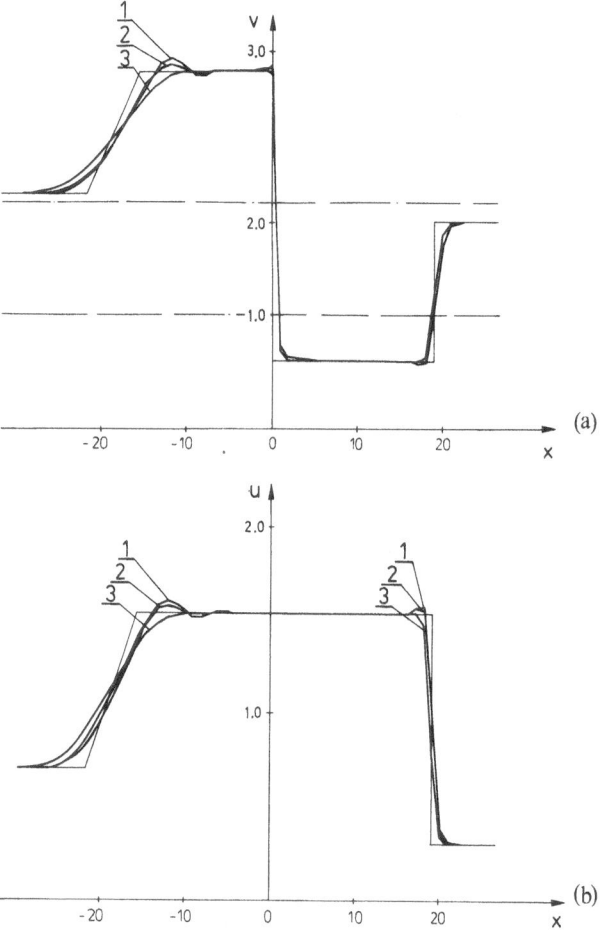

Fig. 16.12. (a) Specific volume ($\lambda = h\mu_0$ (curve 1), $\lambda = h\mu_0 a^2$ (curve 2), $\lambda = h\mu_0 |u_x p_x|/(\delta + |u_x p_x|)$ (curve 3)); (b) Velocity

with the matrix

$$C = \lambda A^2,$$

where

$$\frac{\partial \lambda}{\partial t} = \frac{\partial \lambda}{\partial x} = \frac{\partial \lambda}{\partial u} = 0.$$

Such difference schemes possess the property K and do not possess the property M. For λ we have chosen the following expressions:

$$\lambda_1 = h\mu_0; \qquad\qquad \lambda_2 = h\mu_0 a^2;$$

$$\lambda_3 = h\mu_0 \frac{|u_x p_x|}{\delta + |u_x p_x|}; \qquad \lambda_4 = h\mu_0 \frac{|a_x|}{a^2};$$

$$\lambda_5 = h\mu_0 \frac{|u_x u_{xx}|}{\delta a^2 + |u_x u_{xx}|}; \qquad \lambda_6 = h\mu_0 \frac{|u_x u_{xx}|}{\delta a^2 + \delta_1 |u_x|},$$

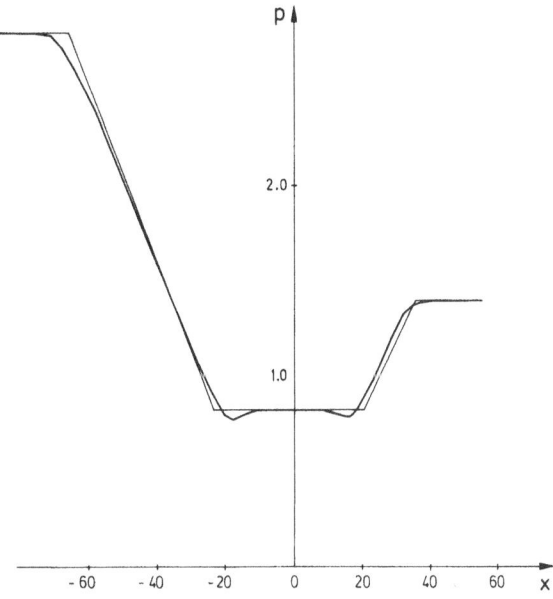

Fig. 16.13. Disintegration of a discontinuity (explicit invariant scheme)

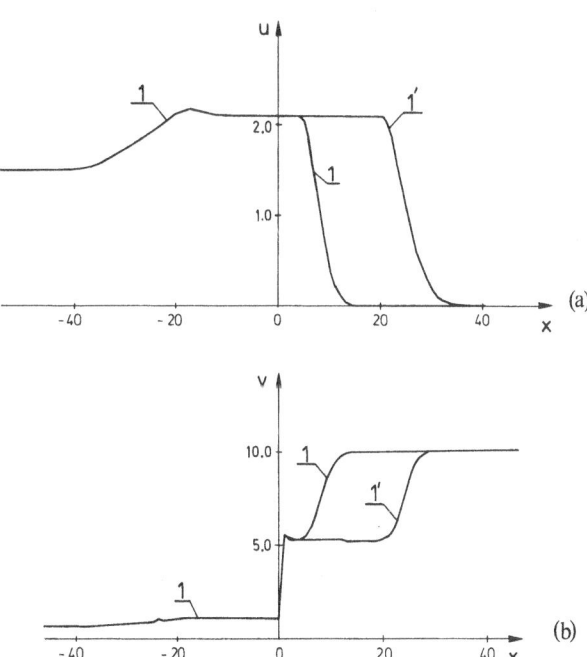

Fig. 16.14. (a) Interaction of shock wave and contact discontinuity, velocity distribution
($n = 42$ (curve 1), $n = 142$ (curve 1')); (b) Specific volume

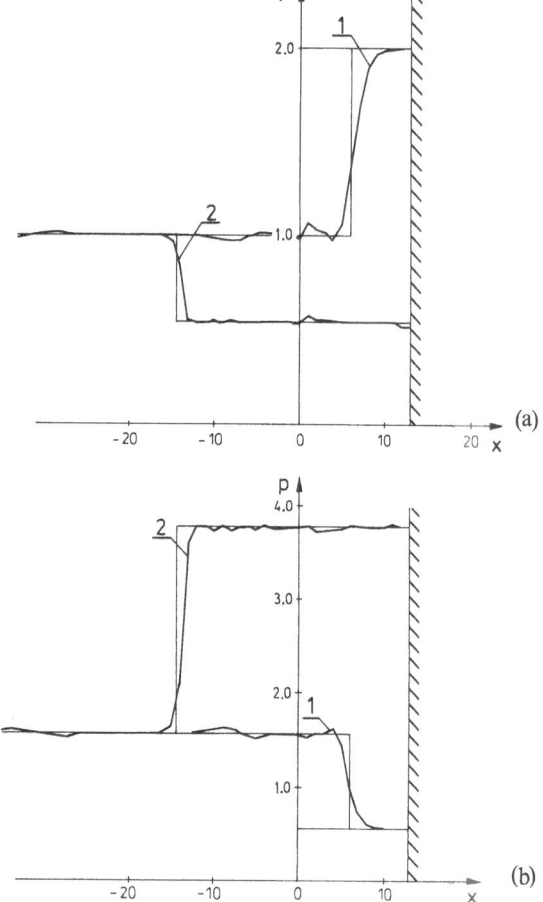

Fig. 16.15. (a) Reflection of a shock wave from a rigid wall, specific volume; (b) Pressure distribution

where μ_0, δ, δ_1 are constants. The numbering of the curves (see Fig. 16.12) corresponds to the indices of the coefficients λ_i; $i = 1, 2, 3$.

In Figs. 16.13–15 numerical results for a number of gasdynamic problems are represented. The results are obtained on the basis of an invariant explicit difference scheme of the form (15.28) with a matrix $\bar{C}$ of type (16.13); the value of v is defined by formula (16.14).

Fig. 16.13 shows the problem of the collapse of a discontinuity, Fig. 16.14 the problem of the interaction of a shock wave with a contact discontinuity (curve 1 corresponds to $n = 42$, curve 1' corresponds to $n = 142$), and Fig. 16.15 shows the problem of the reflection of a shock wave from a rigid wall.

At the end of this section we mention the following properties of the solutions of the considered invariant difference schemes which came out during the calculations:

1. The solution can show oscillatory character only in front of the shock wave but it always remains monotonic behind the shock,

2. With the development of the flow along the characteristic with slope $-a$ a small disturbance is propagated. For the specific volume such a disturbance appears also in the vicinity of the characteristic with the slope zero.

An investigation of the stability of all difference schemes which were used in the calculations described above was carried out on the basis of the method of the differential approximation. Thus for difference schemes with the matrix $C = \lambda A^2$ this analysis leads to the following bounds:

$$\lambda \geq 0; \quad \frac{2\tau}{h^2} \lambda + \kappa^2 a^2 \leq 1.$$

16.9 Conservative and Fully Conservative Schemes

Experience gained from calculations shows that the most important properties which a difference scheme for the equations of gas dynamics should possess are that it should be conservative or fully conservative [13, 194–196]. Later the connection of these properties with properties of the first differential approximation will be pointed out.

Let us consider the family of difference schemes [196]

$$\frac{\Delta_0 u^n(q)}{\tau} + \frac{\Delta_1}{h} p^{\sigma_1}\left(q - \frac{h}{2}\right) = 0; \tag{16.15}$$

$$\frac{\Delta_0 v^n\left(q + \dfrac{h}{2}\right)}{\tau} - \frac{\Delta_1}{h} u^{\sigma_2}(q) = 0; \tag{16.16}$$

$$\frac{\Delta_0}{\tau} \varepsilon^n\left(q + \frac{h}{2}\right) + p^{\sigma_3}\left(q + \frac{h}{2}\right)\frac{\Delta_1}{h} u^{\sigma_4}(q) = 0; \tag{16.17}$$

which approximates the system of equations of gas dynamics in Lagrangean coordinates

$$\frac{\partial u}{\partial t} + \frac{\partial p}{\partial q} = 0; \tag{16.18}$$

$$\frac{\partial v}{\partial t} - \frac{\partial u}{\partial q} = 0; \tag{16.19}$$

$$\frac{\partial \varepsilon}{\partial t} + p \frac{\partial u}{\partial q} = 0. \tag{16.20}$$

Here $f^\sigma := \sigma f^{n+1} + (1-\sigma)f^n$.

The first differential approximation of the scheme (16.15–17) has the form

$$\frac{\partial u}{\partial t} + \frac{\partial p}{\partial q} = \tilde{N}_1; \tag{16.21}$$

$$\frac{\partial v}{\partial t} - \frac{\partial u}{\partial q} = \tilde{N}_2; \tag{16.22}$$

$$\frac{\partial \varepsilon}{\partial t} + p\frac{\partial u}{\partial q} = \tilde{N}_3; \tag{16.23}$$

or, equivalently,

$$\frac{\partial \tilde{w}}{\partial t} + \tilde{A}\frac{\partial \tilde{w}}{\partial q} = \tilde{N}.$$

Here

$$\tilde{w} := \begin{pmatrix} u \\ v \\ \varepsilon \end{pmatrix}; \quad \tilde{A} := \begin{pmatrix} 0 & p_v & p_\varepsilon \\ -1 & 0 & 0 \\ p & 0 & 0 \end{pmatrix}; \quad \tilde{N} := \begin{pmatrix} \tilde{N}_1 \\ \tilde{N}_2 \\ \tilde{N}_3 \end{pmatrix};$$

$$\tilde{N}_1 = -\tau(0.5 - \sigma_1)(a^2 u_q)_q;$$
$$\tilde{N}_2 = \tau(0.5 - \sigma_2)p_{qq}.$$
$$\tilde{N}_3 = -\tau(0.5 - \sigma_3)a^2 u_q^2 - \tau(0.5 - \sigma_4)p\,p_{qq}.$$

Multiplying the system of equations (16.21–23) by the matrix

$$\begin{pmatrix} 1 & 0 & 0 \\ 0 & 1 & 0 \\ u & 0 & 1 \end{pmatrix},$$

we get the following system of equations:

$$\frac{\partial w}{\partial t} + A\frac{\partial w}{\partial q} = N', \tag{16.24}$$

where

$$w := \begin{pmatrix} u \\ v \\ E \end{pmatrix}; \quad A := \begin{pmatrix} -u\,p_\varepsilon & p_v & p_\varepsilon \\ -1 & 0 & 0 \\ -u^2 p_\varepsilon & u\,p_v & u\,p_\varepsilon \end{pmatrix}; \quad N' := \begin{pmatrix} \tilde{N}_1 \\ \tilde{N}_2 \\ \tilde{N}_3 + u\tilde{N}_1 \end{pmatrix}.$$

The system of equations (16.24) can be derived also in a different way. Multiplying equation (16.15) by $u^{0.5}(q)$ and adding the expression to the equation (16.17) we get

$$\frac{\Delta_0 \varepsilon^n \left(q + \dfrac{h}{2}\right)}{\tau} + \frac{\Delta_0}{\tau}\left(\frac{u^2(q)}{2}\right)^n + \frac{\Delta_1}{h} p^{\sigma_1}\left(q - \frac{h}{2}\right) u^{0.5}(q)$$

$$- \tau\left[(\sigma_4 - 0.5) p^{\sigma_1}\left(q + \frac{h}{2}\right)\frac{\Delta_0 \Delta_1}{\tau h} u^n(q) - (\sigma_1 - \sigma_3)\frac{\Delta_0}{\tau} p^n\left(q + \frac{h}{2}\right)\frac{\Delta_1}{h} u^{\sigma_4}(q)\right] = 0.$$

$$(16.25)$$

Equation (16.25) is a consequence of the difference scheme (16.15–17) and approximates the equation for the total energy. When we write down the first differential approximation of the difference scheme (16.15), (16.16), (16.25) which approximates the system of equations (15.28) we get the system of equations (16.24).

. . . .

Theorem 16.6. For the difference scheme (16.15–17) to be fully conservative it is necessary and sufficient that the scheme possesses the property K and that the system of equations of the first differential approximation (16.24) is given in divergence form.

. . . .

Proof. The conditions are necessary:
   Let the difference scheme (16.15–17) be fully conservative. Then as shown in the paper [196] the parameters of the difference scheme are connected by the relations: $\sigma_1 = \sigma_3$, $\sigma_2 = \sigma_4 = 0.5$.
   It is not difficult to see that the system of equations (16.24) for the given $\sigma_i (i = 1, 2, 3, 4)$ has a divergence form

$$\frac{\partial w}{\partial t} + \frac{\partial f}{\partial q} = \frac{\partial N}{\partial q},$$

where

$$f := \begin{pmatrix} p \\ -u \\ up \end{pmatrix}; \quad N := \begin{pmatrix} -\tau(0.5 - \sigma_1)a^2 \dfrac{\partial u}{\partial q} \\ 0 \\ -\tau(0.5 - \sigma_1)u\,a^2 \dfrac{\partial u}{\partial q} \end{pmatrix};$$

and thus the scheme possesses the property K.
   The conditions are sufficient:
   We consider the third equation of the system (16.24) and postulate that it has a divergence form. This is only possible for $\sigma_4 = 0.5$ and $\sigma_1 = \sigma_3$. If we postulate that the scheme has the property K we get $\sigma_2 = 0.5$. Thus $\sigma_2 = \sigma_4 = 0.5$ and $\sigma_1 = \sigma_3$ and as already mentioned under these assumptions the difference scheme (16.15–17) is fully conservative. The theorem is proved.   □

Corollary. When the conditions of the Theorem 16.6 are satisfied the difference scheme (16.15–17) possesses the property M.

For the system of equations (15.28) we consider the following difference scheme:

$$\frac{\Delta_0}{\tau} u^n(q) + \frac{\Delta_1}{h} p^{\sigma_1}\left(q - \frac{h}{2}\right) = 0;$$

$$\frac{\Delta_0}{\tau} v^n\left(q + \frac{h}{2}\right) - \frac{\Delta_1}{h} u^{\sigma_2}(q) = 0; \tag{16.26}$$

$$\frac{\Delta_0}{\tau} \bar{E}^n\left(q + \frac{h}{2}\right) + \frac{\Delta_1}{h} \bar{p}^{\sigma_3}(q) u^{\sigma_4}(q) = 0,$$

where

$$\bar{E}\left(q + \frac{h}{2}\right) := \varepsilon\left(q + \frac{h}{2}\right) + \frac{1}{4}[u^2(q + h) + u^2(q)];$$

$$\bar{p}(q) := \frac{1}{2}\left[p\left(q + \frac{h}{2}\right) + p\left(q - \frac{h}{2}\right)\right].$$

The first differential approximation of the scheme (16.26) has the divergence form

$$\frac{\partial w}{\partial t} + \frac{\partial f}{\partial q} = L_q$$

with

$$L_q := \begin{pmatrix} -\tau(0.5 - \sigma_1) a^2 \dfrac{\partial u}{\partial q} \\[2ex] \tau(0.5 - \sigma_2) \dfrac{\partial p}{\partial q} \\[2ex] -\tau(0.5 - \sigma_3) a^2 u \dfrac{\partial u}{\partial q} - \tau(0.5 - \sigma_4) p \dfrac{\partial p}{\partial q} \end{pmatrix}.$$

. . . .

Theorem 16.7. The difference scheme of the form (16.26) is fully conservative if and only if it has the properties K and M.

. . . .

Proof. We remark that for $\sigma_1 = \sigma_3$ and $\sigma_2 = \sigma_4 = 0.5$ the difference scheme is fully conservative. The conditions mentioned above are necessary and sufficient for the difference scheme (16.26) to possess the properties K and M which follows from the definitions of these properties.   □

The family of the difference schemes (16.15–17) and (16.26) coincide under the condition of being fully conservative [196]. As we have shown fully conservative schemes possess the properties K and M. These difference schemes are also invariant, which can easily be shown when one writes down the matrix of the viscosity in the first differential approximation

$$\frac{\partial w}{\partial t} + \frac{\partial f}{\partial q} = \frac{\partial}{\partial q}\left(C \frac{\partial w}{\partial q}\right).$$

We have

$$C = \begin{pmatrix} -\tau(0.5 - \sigma_1)a^2 & 0 & 0 \\ 0 & 0 & 0 \\ -\tau(0.5 - \sigma_1)u\,a^2 & 0 & 0 \end{pmatrix};$$

i.e. the matrix $\bar{C} = C H^{-1}$ is a Γ-matrix (see Sect. 16.6).

If in the difference scheme of form (16.26) the artificial viscosity is introduced in such a way that it enters the pressure additively and does not depend on x, t, u then the class of difference schemes which is derived under these assumptions is fully conservative and belongs to class $\Gamma \cap K \cap M$.

17. Conditions for the Invariance of Difference Schemes for the Two-dimensional Equations of Gas Dynamics

17.1 Two-level Class of Difference Schemes

We will approximate the system of equations (13.1) by the following difference scheme of first order:

$$\frac{\Delta_0 w^n(x, y)}{\tau} + \frac{T_x - T_{-x}}{2h_1}[\alpha T_0 + (1 - \alpha)I]f^n(x, y)$$

$$+ \frac{T_y - T_{-y}}{2h_2}[\beta T_0 + (1 - \beta)I]g^n(x, y)$$

$$= \left\{ (T_{x/2} - T_{-x/2}) \left[\frac{1}{h_1} \Omega_{11}(T_{x/2} - T_{-x/2}) + \frac{1}{h_2} \Omega_{12}(T_{y/2} - T_{-y/2}) \right] \right.$$

$$\left. + (T_{y/2} - T_{-y/2}) \left[\frac{1}{h_1} \Omega_{21}(T_{x/2} - T_{-x/2}) + \frac{1}{h_2} \Omega_{22}(T_{y/2} - T_{-y/2}) \right] \right\}$$

$$\cdot [\gamma T_0 + (1 - \gamma)I]w^n(x, y). \tag{17.1}$$

Here h_1 and h_2 are the step sizes of the grid in x- and y-direction, respectively. T_x is the shift-operator along the x-axis, T_y the shift-operator along the y-axis. Ω_{ij} is a (4×4)-matrix, the elements of which may depend on t, x, y, w, w_x, w_y, $w_{xx}, w_{xy}, w_{yx}, w_{yy}; 0 \leq \alpha, \beta, \gamma \leq 1$,

$$f^n(x, y) := f(w^n(x, y));$$

$$g^n(x, y) := g(w^n(x, y));$$

$$\tau/h_j =: \kappa_j = \text{const}; \quad i, j = 1, 2.$$

The first differential approximation of the difference scheme (17.1) has the form

$$w_t + f_x + g_y = \frac{\partial}{\partial x}(C_{11}w_x + C_{12}w_y) + \frac{\partial}{\partial y}(C_{21}w_x + C_{22}w_y), \tag{17.2}$$

where

$$C_{1j} = h_1 \Omega_{1j} - dA_1 A_j = \bar{\Omega}_{1j} - dA_1 A_j;$$
$$C_{2j} = h_2 \Omega_{2j} - qA_2 A_j = \bar{\Omega}_{2j} - qA_2 A_j; \qquad j = 1,2$$

$$A_1 := \frac{df}{dw}; \quad A_2 := \frac{dg}{dw};$$

$$d := \frac{\tau}{2}(1 - 2\alpha); \quad q := \frac{\tau}{2}(1 - 2\beta).$$

The system of equations (17.2) can be written in the following form:

$$\mathscr{P}^1 := u_t + uu_x + vu_y + \frac{1}{\varrho} p_x - \frac{1}{\varrho} N_1 + \frac{u}{\varrho} N_3 = 0;$$

$$\mathscr{P}^2 := v_t + uv_x + vv_y + \frac{1}{\varrho} p_y - \frac{1}{\varrho} N_2 + \frac{v}{\varrho} N_3 = 0;$$

$$\mathscr{P}^3 := \varrho_t + u\varrho_x + v\varrho_y + \varrho u_x + \varrho v_y - N_3 \quad = 0; \qquad (17.3)$$

$$\mathscr{P}^4 := p_t + up_x + vp_y + a^2 u_x + a^2 v_y + bN_3$$
$$+ ulN_1 + vlN_2 - lN_4 = 0.$$

Here

$$N := \begin{pmatrix} N_1 \\ N_2 \\ N_3 \\ N_4 \end{pmatrix} = \begin{pmatrix} \Phi_{1x}^{(1)} + \Phi_{1y}^{(2)} \\ \Phi_{2x}^{(1)} + \Phi_{2y}^{(2)} \\ \Phi_{3x}^{(1)} + \Phi_{3y}^{(2)} \\ \Phi_{4x}^{(1)} + \Phi_{4y}^{(2)} \end{pmatrix} = \frac{\partial}{\partial x} \Phi^{(1)} + \frac{\partial}{\partial y} \Phi^{(2)};$$

$$\Phi^{(j)} = C_{j1} \frac{\partial w}{\partial x} + C_{j2} \frac{\partial w}{\partial y};$$

$$\Phi_k^{(j)} = v_{k1}^{j1} u_x + v_{k2}^{j1} v_x + v_{k3}^{j1} \varrho_x + v_{k4}^{j1} p_x + v_{k1}^{j2} u_y + v_{k2}^{j2} v_y + v_{k3}^{j2} \varrho_y + v_{k4}^{j2} p_y,$$

$$v_{k1}^{jm} = \varrho(\omega_{k1}^{jm} + u\omega_{k4}^{jm});$$

$$v_{k2}^{jm} = \varrho(\omega_{k2}^{jm} + v\omega_{k4}^{jm});$$

$$v_{k3}^{jm} = u\omega_{k1}^{jm} + v\omega_{k2}^{jm} + \omega_{k3}^{jm} + E\omega_{k4}^{jm} + \varrho\varepsilon_\varrho \omega_{k4}^{jm},$$

$$v_{k4}^{jm} = \varrho\varepsilon_p \omega_{k4}^{jm};$$

$$C_{jm} = \|\omega_{ki}^{jm}\|_1^4; \quad (j,m = 1,2; \quad k,i = 1,2,3,4);$$

$$a^2 := \frac{p - \varrho^2 \varepsilon_\varrho}{\varrho\varepsilon_p}; \quad b := \frac{E + \varrho\varepsilon_\varrho - u^2 - v^2}{\varrho\varepsilon_p}; \quad l := \frac{1}{\varrho\varepsilon_p}.$$

17.2 Conditions for the Invariance of the Difference Scheme (17.1)

We will find conditions under which the difference scheme (17.1) is invariant with respect to transformation groups, which are admitted by the system of equations (13.1). It is necessary to extend the basis operators (13.2) in the space

$$\mathscr{E}\left(t, x, y, w, w_t, w_x, w_y, w_{xx}, w_{xy}, w_{yx}, w_{yy}, \tau, h_1, h_2\right).$$

The following lemma holds.

Lemma 17.1. The difference scheme (17.1) is invariant if and only if the following relations hold:

$$\frac{\partial}{\partial t} N_j = 0; \qquad \frac{\partial}{\partial x} N_j = 0; \qquad \frac{\partial}{\partial y} N_j = 0. \tag{17.4}$$

$$\left.\begin{aligned}
\frac{\partial}{\partial u} N_1 &= N_3; & \frac{\partial}{\partial u} N_2 &= 0; & \frac{\partial}{\partial u} N_3 &= 0; & \frac{\partial}{\partial u} N_4 &= N_1; \\
\frac{\partial}{\partial v} N_1 &= 0; & \frac{\partial}{\partial v} N_2 &= N_3; & \frac{\partial}{\partial v} N_3 &= 0; & \frac{\partial}{\partial v} N_4 &= N_2;
\end{aligned}\right\} \tag{17.5}$$

$$\left.\begin{aligned}
\bar{L}_6 N_1 &= N_2 - v N_3, \\
\bar{L}_6 N_2 &= u N_3 - N_1, \\
\bar{L}_6 N_3 &= 0, \\
\bar{L}_6 N_4 &= -v N_1 + u N_2,
\end{aligned}\right\} \tag{17.6}$$

$$\bar{L}_7 N_j = 2 N_j; \qquad j = 1, 2, 3, 4, \tag{17.7}$$

where

$$\bar{L}_6 := (v_x + u_y)\left(\frac{\partial}{\partial u_x} - \frac{\partial}{\partial v_y}\right) + (v_y - u_x)\left(\frac{\partial}{\partial u_y} + \frac{\partial}{\partial v_x}\right)$$

$$+ \varrho_y \frac{\partial}{\partial \varrho_x} - \varrho_x \frac{\partial}{\partial \varrho_y} + p_y \frac{\partial}{\partial p_x} - p_x \frac{\partial}{\partial p_y} + (v_{xx} + u_{xy} + u_{yx}) \frac{\partial}{\partial u_{xx}}$$

$$+ (v_{yx} + v_{xy} - u_{xx}) \frac{\partial}{\partial v_{xx}} + (v_{xy} - u_{xx} + u_{yy}) \frac{\partial}{\partial u_{xy}}$$

$$+ (v_{yy} - u_{xy} - v_{xx}) \frac{\partial}{\partial v_{xy}} + (v_{yx} + u_{yy} - u_{xx}) \frac{\partial}{\partial u_{yx}}$$

$$+ (v_{yy} - v_{xx} - u_{yx}) \frac{\partial}{\partial v_{yx}} + (v_{yy} - u_{yx} - u_{xy}) \frac{\partial}{\partial u_{yy}}$$

$$+ (-u_{yy} - v_{yx} - v_{xy}) \frac{\partial}{\partial v_{yy}} + (\varrho_{xy} + \varrho_{yx}) \frac{\partial}{\partial \varrho_{xx}} +$$

$$+ (\varrho_{yy} - \varrho_{xx})\left(\frac{\partial}{\partial\varrho_{xy}} + \frac{\partial}{\partial\varrho_{yx}}\right) - (\varrho_{xy} + \varrho_{yx})\frac{\partial}{\partial\varrho_{yy}}$$

$$+ (p_{xy} + p_{yx})\left(\frac{\partial}{\partial p_{xx}} - \frac{\partial}{\partial p_{yy}}\right) + (p_{yy} - p_{xx})\left(\frac{\partial}{\partial p_{xy}} + \frac{\partial}{\partial p_{yx}}\right);$$

$$\bar{L}_7 := u_x \frac{\partial}{\partial u_x} + u_y \frac{\partial}{\partial u_y} + v_x \frac{\partial}{\partial v_x} + v_y \frac{\partial}{\partial v_y} + \varrho_x \frac{\partial}{\partial\varrho_x} + \varrho_y \frac{\partial}{\partial\varrho_y}$$

$$+ p_x \frac{\partial}{\partial p_x} + p_y \frac{\partial}{\partial p_y} + 2u_{xx}\frac{\partial}{\partial u_{xx}} + 2u_{xy}\frac{\partial}{\partial u_{xy}} + 2u_{yx}\frac{\partial}{\partial u_{yx}}$$

$$+ 2u_{yy}\frac{\partial}{\partial u_{yy}} + 2v_{xx}\frac{\partial}{\partial v_{xx}} + 2v_{xy}\frac{\partial}{\partial v_{xy}} + 2v_{yx}\frac{\partial}{\partial v_{yx}} + 2v_{yy}\frac{\partial}{\partial v_{yy}}$$

$$+ 2\varrho_{xx}\frac{\partial}{\partial\varrho_{xx}} + 2\varrho_{xy}\frac{\partial}{\partial\varrho_{xy}} + 2\varrho_{yx}\frac{\partial}{\partial\varrho_{yx}} + 2\varrho_{yy}\frac{\partial}{\partial\varrho_{yy}} + 2p_{xx}\frac{\partial}{\partial p_{xx}}$$

$$+ 2p_{xy}\frac{\partial}{\partial p_{xy}} + 2p_{yx}\frac{\partial}{\partial p_{yx}} + 2p_{yy}\frac{\partial}{\partial p_{yy}}.$$

Proof. According to the theorem (see Chap. 14) for the invariance of the difference scheme (17.1) it is necessary and sufficient that

$$[\tilde{L}_\alpha \mathscr{F}^\gamma + \tilde{L}_\alpha^{(2)} hR^\gamma]|_{\mathscr{F}^\gamma = 0} = 0; \quad \alpha = 1, \dots, 6; \quad \gamma = 1, 2, 3, 4, \qquad (17.8)$$

where

$$\mathscr{F}^1 := u_t + uu_x + vu_y + \frac{1}{\varrho}p_x = 0;$$

$$\mathscr{F}^2 := v_t + uv_x + vv_y + \frac{1}{\varrho}p_y = 0;$$

$$\mathscr{F}^3 := \varrho_t + u\varrho_x + v\varrho_y + \varrho u_x + \varrho v_y = 0;$$

$$\mathscr{F}^4 := p_t + up_x + vp_y + a^2 u_x + a^2 v_y = 0;$$

$$hR^1 = -\frac{1}{\varrho}N_1 + \frac{u}{\varrho}N_3;$$

$$hR^2 = -\frac{1}{\varrho}N_2 + \frac{v}{\varrho}N_3;$$

$$hR^3 = -N_3;$$

$$hR^4 = bN_3 + ulN_1 + vlN_2 - lN_4.$$

If we write down the relations (17.8) for each value of α and γ then we get, e.g., for $\alpha = 1$ and $\gamma = 3$:

$$\frac{\partial}{\partial t}\mathscr{F}^3 + \frac{\partial}{\partial t}hR^3 = -\frac{\partial}{\partial t}N_3 = 0.$$

In a similar way we get for $\alpha = 2, 3$ and $\gamma = 3$:

$$\frac{\partial N_3}{\partial x} = 0; \quad \frac{\partial N_3}{\partial y} = 0.$$

Then

$$\frac{\partial}{\partial t}(\mathscr{F}^\beta + hR^\beta) = -\frac{1}{\varrho}\delta_\beta^1 \frac{\partial}{\partial t} N_1 - \frac{1}{\varrho}\delta_\beta^2 \frac{\partial}{\partial t} N_2 = 0;$$

$$\frac{\partial}{\partial x}(\mathscr{F}^\beta + hR^\beta) = -\frac{1}{\varrho}\delta_\beta^1 \frac{\partial}{\partial x} N_1 - \frac{1}{\varrho}\delta_\beta^2 \frac{\partial}{\partial x} N_2 = 0;$$

$$\frac{\partial}{\partial y}(\mathscr{F}^\beta + hR^\beta) = -\frac{1}{\varrho}\delta_\beta^1 \frac{\partial}{\partial y} N_1 - \frac{1}{\varrho}\delta_\beta^2 \frac{\partial}{\partial y} N_2 = 0;$$

$$\beta = 1, 2;$$

this means

$$\frac{\partial}{\partial x^k} N_1 = 0; \quad \frac{\partial}{\partial x^k} N_2 = 0; \quad k = 1, 2, 3,$$

where $x^1 = t, x^2 = x, x^3 = y$. Analogously, we get

$$\frac{\partial}{\partial x^k} N_4 = 0.$$

Thus in the case of $\alpha = 1, 2, 3$ the conditions (17.8) are identical with the conditions (17.4).

Further we have

$$\tilde{L}_4 \mathscr{F}^3 + \tilde{L}_4^{(2)} hR^3 = -\frac{\partial}{\partial u} N_3 = 0,$$

$$\tilde{L}_5 \mathscr{F}^3 + \tilde{L}_5^{(2)} hR^3 = -\frac{\partial}{\partial v} N_3 = 0.$$

Then

$$\tilde{L}_4 \mathscr{F}^1 + \tilde{L}_4^{(2)} hR^1 = -\frac{1}{\varrho}\frac{\partial N_1}{\partial u} + \frac{1}{\varrho} N_3 = 0;$$

$$\tilde{L}_4 \mathscr{F}^2 + \tilde{L}_4^{(2)} hR^2 = -\frac{1}{\varrho}\frac{\partial N_2}{\partial u} = 0;$$

$$\tilde{L}_4 \mathscr{F}^4 + \tilde{L}_4^{(2)} hR^4 = lN_1 - l\frac{\partial N_4}{\partial u} = 0;$$

and, consequently,

$$\frac{\partial N_1}{\partial u} = N_3; \quad \frac{\partial N_2}{\partial u} = 0; \quad \frac{\partial N_4}{\partial u} = N_1.$$

Analogous considerations lead to

$$\frac{\partial N_1}{\partial v} = 0; \quad \frac{\partial N_2}{\partial v} = N_3; \quad \frac{\partial N_4}{\partial v} = N_2.$$

Thus the conditions (17.8) for $\alpha = 1, \ldots, 5$ are equivalent to the conditions (17.4, 5).

In a similar way can be proved that under the conditions (17.4, 5) the equations (17.6, 7) are equivalent to the relations (17.8) for $\alpha = 6, 7$.   □

17.3 Theorem on Invariance

. . . .

Theorem 17.1. If the matrices $\Omega_{ij} = \| \Omega_{ij}^{rs} \|_1^4$, $(i, j = 1, 2)$ are chosen such that

$$\frac{\partial \Omega_{ij}^{rs}}{\partial t} = 0, \quad \frac{\partial \Omega_{ij}^{rs}}{\partial x} = 0, \quad \frac{\partial \Omega_{ij}^{rs}}{\partial y} = 0; \tag{17.9}$$

$$\frac{\partial N_3}{\partial u} = 0; \quad \frac{\partial N_3}{\partial v} = 0; \tag{17.10}$$

$$\left. \begin{array}{l} N_1 = u N_3 + R_1, \quad N_2 = v N_3 + R_2, \\[2mm] N_4 = \tfrac{1}{2}(u^2 + v^2) N_3 + u R_1 + v R_2 + R_3, \\[2mm] \dfrac{\partial R_k}{\partial u} = 0; \quad \dfrac{\partial R_k}{\partial v} = 0; \quad (k = 1, 2, 3), \end{array} \right\} \tag{17.11}$$

and besides this, if also the conditions (17.6, 7) are satisfied then the difference scheme (17.1) admits the same group of transformations as the system of equations (13.1).

. . . .

Indeed, from the conditions (17.9) the relations (17.4) follow, and from the conditions (17.10, 11) the equations (17.5) follow. Thus the sufficient conditions of lemma (17.1) are satisfied and, consequently, the difference scheme (17.1) is invariant.

17.4 Property M

Definition: We will say that the difference scheme (17.1) has the property M, if in its first differential approximation the law of conservation of mass is satisfied which means that $N_3 = 0$.

If

$$v_{3k}^{ij} = 0; \quad (i, j = 1, 2; \quad k = 1, 2, 3, 4),$$

then $N_3 = 0$ and, especially, we have

$$\omega_{31}^{ij} + u\omega_{34}^{ij} = 0;$$

$$\omega_{32}^{ij} + v\omega_{34}^{ij} = 0;$$

$$u\omega_{31}^{ij} + v\omega_{32}^{ij} + \omega_{33}^{ij} + (E + \varrho\varepsilon_\varrho)\omega_{34}^{ij} = 0;$$

$$\omega_{34}^{ij} = 0.$$

and, consequently,

$$\omega_{3k}^{ij} = 0. \tag{17.12}$$

In accordance with condition (17.12) we get

$$\bar\Omega_{31}^{11} = du(2 - z);$$
$$\bar\Omega_{32}^{11} = - dvz;$$
$$\bar\Omega_{34}^{11} = dz;$$
$$\bar\Omega_{33}^{11} = d(r - u^2 + zv^2 + u^2 z - Ez);$$
$$\bar\Omega_{31}^{12} = dv;$$
$$\bar\Omega_{32}^{12} = du;$$
$$\bar\Omega_{33}^{12} = - duv;$$
$$\bar\Omega_{34}^{12} = \Omega_{34}^{21} = 0; \tag{17.13}$$
$$\bar\Omega_{31}^{21} = qv;$$
$$\bar\Omega_{32}^{21} = qu;$$
$$\bar\Omega_{33}^{21} = - quv;$$
$$\bar\Omega_{31}^{22} = - quz;$$
$$\bar\Omega_{32}^{22} = qv(2 - z);$$
$$\bar\Omega_{33}^{22} = q(r - v^2 + v^2 z + u^2 z - Ez);$$
$$\bar\Omega_{34}^{22} = qz,$$

where

$$r := p_\varrho; \quad z := \frac{1}{\varrho} p_\varepsilon.$$

The following theorem holds.

. . . .

Theorem 17.2. If in the difference scheme (17.1) the elements of the matrix Ω_{ij} $(i, j = 1, 2)$ are chosen such, that the conditions (17.9), (17.13) are satisfied and further the conditions

$$\frac{\partial N_j}{\partial u} = 0; \qquad \frac{\partial N_j}{\partial v} = 0; \qquad j = 1,2; \tag{17.14}$$

$$N_4 = uN_1 + vN_2 + R;$$

$$\frac{\partial R}{\partial u} = 0; \qquad \frac{\partial R}{\partial v} = 0;$$

$$\bar{L}_6 N_1 = N_2;$$

$$\bar{L}_6 N_2 = -N_1;$$

$$\bar{L}_6 N_4 = -vN_1 + uN_2;$$

$$\bar{L}_7 N_k = 2 N_k; \qquad k = 1,2,3;$$

hold, then the difference scheme is invariant and has the property *M*.

. . . .

Indeed, from the conditions (17.13) it follows that $N_3 = 0$. The conditions of Theorem 17.1 are satisfied and thus the Theorem 17.2 is proved.

Remark. The conditions (17.14) can be replaced by the following equations using the relations (17.13):

$$\bar{\Omega}^{ij}_{3-j3-i} = \mu^{ij}_{3-j3-i} - (v\delta^1_i + u\delta^2_i)\mu^{ij}_{3-j4} + d\delta^1_i(u^2 - v^2 z)$$
$$\qquad + q\delta^2_i(v^2 - u^2 z);$$

$$\bar{\Omega}^{3-jj}_{j3-j} = \mu^{3-jj}_{j3-j} - (v\delta^1_j + u\delta^2_j)\mu^{3-jj}_{j4} + q\delta^1_j(u^2 - v^2 z) + d\delta^2_j(v^2 - u^2 z);$$

$$\bar{\Omega}^{ij}_{3-ji} = \mu^{ij}_{3-ji} - (u\delta^1_i + v\delta^2_i)\mu^{ij}_{3-j4} + (d\delta^1_i + q\delta^2_i)uv(2 - z);$$

$$\bar{\Omega}^{3-jj}_{jj} = \mu^{3-jj}_{jj} - (u\delta^1_j + v\delta^2_j)\mu^{3-jj}_{j4} + (q\delta^1_j + d\delta^2_j)uv(2 - z);$$

$$\bar{\Omega}^{ij}_{3-j3} = \mu^{ij}_{3-j3} + (dv\delta^1_i + qu\delta^2_i)(-2u^2\delta^1_i - 2v^2\delta^2_i + rz + r)$$
$$\qquad - u(\mu^{ij}_{3-j1} - u\mu^{ij}_{3-j4}) - v(\mu^{ij}_{3-j2} - v\mu^{ij}_{3-j4})$$
$$\qquad - \tfrac{1}{2}(u^2 + v^2)\mu^{ij}_{3-j4};$$

$$\bar{\Omega}^{3-jj}_{j3} = \mu^{3-jj}_{j3} + (qv\delta^1_j + du\delta^2_j)(-2u^2\delta^1_j - 2v^2\delta^2_j + rz + r)$$
$$\qquad - u(\mu^{3-jj}_{j1} - u\mu^{3-jj}_{j4}) - v(\mu^{3-jj}_{j2} - v\mu^{3-jj}_{j4}) - \tfrac{1}{2}(u^2 + v^2)\mu^{3-jj}_{j4};$$

$$\bar{\Omega}^{jj}_{jj} = \mu^{jj}_{jj} - (u\delta^1_j + v\delta^2_j)\mu^{jj}_{j4} + 3(du^2\delta^1_j + qv^2\delta^2_j)(1 - z);$$

$$\bar{\Omega}^{jj}_{j3-j} = \mu^{jj}_{j3-j} - (v\delta^1_j + u\delta^2_j)\mu^{jj}_{j4} - 3(d\delta^1_j + q\delta^2_j)uvz;$$

$$\bar{\Omega}^{jj}_{j3} = \mu^{jj}_{j3} - (du\,\delta^1_j + qv\,\delta^2_j)(2u^2\,\delta^1_j + 2v\delta^2_j - 2r - 2rz)$$
$$\qquad - (u\delta^1_j + v\delta^2_j)[\mu^{jj}_{jj} - (u\delta^1_j + v\delta^2_j)\mu^{jj}_{j4} - (d\delta^1_j + q\delta^2_j)r(1 + z)]$$
$$\qquad - \tfrac{1}{2}(u^2 + v^2)\mu^{jj}_{j4} - (v\delta^1_j + u\delta^2_j)[\mu^{jj}_{j3-j} - (v\delta^1_j + u\delta^2_j)\mu^{jj}_{j4}];$$

$$\bar{\Omega}^{il}_{j4} = \mu^{il}_{j4} + (1 + 2\,\delta^l_j)(du\,\delta^1_i + qv\,\delta^2_i)z\delta^l_j + (dv\,\delta^1_i + qu\,\delta^2_i)\delta^{3-j}_i,$$

where

$$\frac{\partial \mu^{ij}_{kl}}{\partial u} = 0; \qquad \frac{\partial \mu^{ij}_{kl}}{\partial v} = 0; \qquad i,j = 1,2; \quad k,l = 1,2,3,4.$$

Indeed, in this case we get:

$$v_{11}^{11} = \varrho\,[\mu_{11}^{11} - d\,(\Theta + mz)]; \qquad v_{11}^{12} = \varrho\mu_{11}^{12};$$
$$v_{12}^{11} = \varrho\mu_{12}^{11}; \qquad v_{12}^{12} = \varrho\,[\mu_{22}^{12} - d\,(\Theta + mz)];$$
$$v_{13}^{11} = \mu_{13}^{11} + \phi\mu_{14}^{11}; \qquad v_{13}^{12} = \mu_{13}^{12} + \phi\mu_{14}^{12};$$
$$v_{14}^{11} = \psi\mu_{14}^{11}; \qquad v_{14}^{12} = \psi\mu_{14}^{12};$$

$$v_{11}^{21} = \varrho\mu_{11}^{21}; \qquad v_{11}^{22} = \varrho\mu_{11}^{22};$$
$$v_{12}^{21} = \varrho\mu_{21}^{21}; \qquad v_{12}^{22} = \varrho\mu_{12}^{22};$$
$$v_{13}^{21} = \mu_{13}^{21} + \phi\mu_{14}^{21}; \qquad v_{13}^{22} = \mu_{13}^{22} + \phi\mu_{14}^{22};$$
$$v_{14}^{21} = \psi\mu_{14}^{21}; \qquad v_{14}^{22} = \psi\mu_{14}^{22};$$

$$v_{21}^{11} = \varrho\mu_{21}^{11}; \qquad v_{21}^{12} = \varrho\mu_{21}^{12};$$
$$v_{22}^{11} = \varrho\mu_{22}^{11}; \qquad v_{22}^{12} = \varrho\mu_{22}^{12};$$
$$v_{23}^{11} = \mu_{23}^{11} + \phi\mu_{24}^{11}; \qquad v_{23}^{12} = \mu_{23}^{12} + \phi\mu_{24}^{12};$$
$$v_{24}^{11} = \psi\mu_{24}^{11}; \qquad v_{24}^{12} = \psi\mu_{24}^{12};$$

$$v_{21}^{21} = \varrho\,[\mu_{21}^{21} - q\,(\Theta + mz)]; \qquad v_{21}^{22} = \varrho\mu_{21}^{22};$$
$$v_{22}^{21} = \varrho\mu_{22}^{21}; \qquad v_{22}^{22} = \varrho\,[\mu_{22}^{22} - q\,(\Theta + mz)];$$
$$v_{23}^{21} = \mu_{23}^{21} + \phi\mu_{24}^{21}; \qquad v_{23}^{22} = \mu_{23}^{22} + \phi\mu_{24}^{22};$$
$$v_{24}^{21} = \psi\mu_{24}^{21}; \qquad v_{24}^{22} = \psi\mu_{24}^{22};$$

where

$$\phi = \varepsilon + \varrho\varepsilon_p; \qquad \psi = \varrho\varepsilon_p.$$

Thus we get

$$\frac{\partial v_{\xi k}^{ij}}{\partial u} = 0; \qquad \frac{\partial v_{\xi k}^{ij}}{\partial v} = 0; \qquad i,j = 1,2; \quad \xi,k = 1,2,3,4;$$

and, consequently, (17.14) is satisfied.

17.5 Property $\bar{K}$

Definition. We will say that the difference scheme (17.1) possesses the property $\bar{K}$ if the equation

$$X\,[(C_{11}w_x + C_{12}w_y)_x + (C_{21}w_x + C_{22}w_y)_y] = 0$$

holds where X is a vector such that

$$XA_1 = uX,$$
$$XA_2 = vX.$$

It is possible to show that if the difference scheme (17.1) possesses the property $\bar{K}$ then the entropy s in the first differential approximation satisfies the equation

$$s_t + us_x + vs_y = 0.$$

This fact can be proved by multiplying the system of equations (17.3) from left by the vector X and using the second law of thermodynamics.

17.6 Some Remarks on Stability

The stability of the considered classes of difference schemes can be investigated by means of the method of the differential approximation. To this end it is necessary to find conditions under which the system of equations (17.2) is not fully parabolic and the region of dependence of the hyperbolic form of the first differential approximation does not overlap the region of dependence of the difference scheme.

In Sect. 17.7 an analysis of specific invariant schemes will be given.

Finally we remark that the group classification proposed above can be carried out in an analogous way for difference schemes of higher order of approximation and also for difference schemes which approximate the equations of gas dynamics in Lagrangean coordinates.

17.7 Γ-matrices

We consider some special classes of invariant difference schemes for the two-dimensional equations of gas dynamics.

Definition. We will say that the difference scheme (17.1) possesses the property Γ (or, equivalently, belongs to class Γ) if it is invariant with respect to a shift-transformation along the axes t, x, y and invariant with respect to a Galilei-transformation along the x- and y-axis.

From Lemma 17.1 it follows that the difference scheme (17.1) belongs to class Γ if and only if the elements $\bar{\omega}_{kl}^{ij}$ ($i, j = 1, 2; k, l = 1, 2, 3, 4$) of the matrices $\bar{C}_{ij} = C_{ij} H^{-1}$, with

$$H^{-1} := \begin{pmatrix} \varrho & 0 & u & 0 \\ 0 & \varrho & v & 0 \\ 0 & 0 & 1 & 0 \\ \varrho u & \varrho v & (E - \varrho p_\varrho/p_\varepsilon) & \varrho/p_\varepsilon \end{pmatrix},$$

have the following form:

$$\bar{\omega}_{11}^{ij} = uv_l^{ij} + r_l^{ij};$$
$$\bar{\omega}_{21}^{ij} = vv_l^{ij} + s_l^{ij};$$
$$\bar{\omega}_{31}^{ij} = v_l^{ij};$$
$$\bar{\omega}_{41}^{ij} = \tfrac{1}{2}(u^2 + v^2)v_l^{ij} + ur_l^{ij} + vs_l^{ij} + q_l^{ij},$$

(17.15)

where

$$\frac{\partial v_l^{ij}}{\partial z} = \frac{\partial r_l^{ij}}{\partial z} = \frac{\partial s_l^{ij}}{\partial z} = \frac{\partial q_l^{ij}}{\partial z} = 0 \quad (z = t, x, y, u, v).$$

Matrices of type $\bar{C}_{ij}$ will be called Γ-matrices.

An example of a Γ-matrix is the matrix H^{-1}.

Taking into account the results of Sects. 17.4/5 it can be shown easily:

1) If in a Γ-matrix $v_l^{ij} = 0$, then the corresponding difference scheme possesses the property M.

2) If in a Γ-matrix $v_l^{ij} = 0$ and $q_l^{ij} = 0$, then the difference scheme possesses the properties K and M.

Lemma 17.2. The sum of two Γ-matrices is also a Γ-matrix.

The proof of this statement is carried out by using the definition of a Γ-matrix.

Lemma 17.3. If the difference scheme (17.1) possesses the property K and for the elements $\bar{\omega}_{kl}^{ij}$ of the matrix $\bar{C}_{ij}$ the relations

$$\bar{\omega}_{11}^{ij} = u\bar{\omega}_{31}^{ij} + r_l^{ij};$$
$$\bar{\omega}_{21}^{ij} = v\bar{\omega}_{3}^{ijl} + s_l^{ij};$$
$$\frac{\partial \bar{\omega}_{31}^{ij}}{\partial z} = \frac{\partial r_l^{ij}}{\partial z} = \frac{\partial s_l^{ij}}{\partial z} = 0; \quad (z = t, x, y, u, v),$$

hold then the difference scheme belongs to class Γ.

Indeed, because of the property K we have

$$\bar{\omega}_{41}^{ij} = -\left(E - \frac{p}{\varrho}\right)\bar{\omega}_{31}^{ij} + u\bar{\omega}_{11}^{ij} + v\bar{\omega}_{21}^{ij},$$

and therefore the matrix $\bar{C}_{ij}$ is a Γ-matrix and, consequently, the difference scheme belongs to class Γ.

Simply by calculating one can verify that if $\bar{C}$ is a Γ-matrix, then also the matrices

$$(A_1 - uI)^{\alpha_1}(A_2 - vI)^{\alpha_2}\bar{C},$$
$$(A_2 - vI)^{\alpha_3}(A_1 - uI)^{\alpha_4}\bar{C},$$

where $\alpha_1, \alpha_2, \alpha_3$ and α_4 are positive integers, are Γ-matrices.

As in the one-dimensional case also in the two-dimensional case those difference schemes are of theoretical and practical interest, in the first differential approximation of which the viscosity matrices are represented by polynomials of the matrices $A_1 - uI$, $A_2 - vI$. As

$$(A_1 - uI)^{2k+1} = c^{2k}(A_1 - uI);$$
$$(A_1 - uI)^{2k+2} = c^{2k}(A_1 - uI)^2;$$
$$(A_2 - vI)^{2k+1} = c^{2k}(A_2 - vI);$$
$$(A_2 - vI)^{2k+2} = c^{2k}(A_2 - vI)^2,$$

which is easily verified by simple matrix multiplication, it is sufficient to consider polynomials of second degree.

On the basis of the statements mentioned above the following theorem can be proved.

. . . .

Theorem 17.3. If

$$\bar{C}_{ij} = \beta_0 I + \sum_{q=1}^{Q} \Phi_{ij}^q,$$

where Q is some integer positive number,

$$\Phi_{ij}^q = \beta_{ij}^q (A_1 - uI)^{\gamma_{ij}^1}(A_2 - vI)^{\gamma_{ij}^2}\ldots(A_1 - uI)^{\gamma_{ij}^q};$$
$$\gamma_{ij}^1, \ldots, \gamma_{ij}^q = 0, 1, 2$$

and

$$\frac{\partial \beta_0}{\partial z} = \frac{\partial \beta_{ij}^q}{\partial z} = 0, \quad (z = t, x, y, u, v);$$

then the difference scheme (17.1) with the matrices

$$\Omega_{ij} = C_{ij} + \frac{\tau}{2} A_i A_j, \quad (i, j = 1, 2)$$

possesses the property Γ. If $\beta_0 = 0$ then the scheme belongs to class K.

. . . .

Corollary. The difference scheme (17.1) where

$$\Omega_{ij} = \phi(A_i - u_i I)(A_j - u_j I) + \frac{\tau}{2} A_i A_j;$$

$$u_1 = u; \quad u_2 = v;$$

$$\frac{\partial \phi}{\partial z} = 0, \quad (z = t, x, y, u, v)$$

belongs to class $\Gamma \cap K$.

We consider difference schemes of the form (17.1) for which

$$\bar{C}_{ii} = \begin{pmatrix} r_i & 0 & 0 & 0 \\ 0 & s_i & 0 & 0 \\ 0 & 0 & 0 & 0 \\ ur_i & vs_i & 0 & 0 \end{pmatrix}; \quad i = 1, 2;$$

$$C_{12} = C_{21} = 0;$$

$$\frac{\partial r_i}{\partial z} = \frac{\partial s_i}{\partial z} = 0; \quad z = t, x, y, u, v.$$

In this case the scheme belongs to class Γ and possesses the properties K and M.

From the Lemma 17.1, it follows that if $r_i = s_i = \phi$ and the function ϕ is invariant with respect to a rotation then the difference scheme also admits a transformation of the kind of a rotation. The function ϕ can be chosen, e.g., as one of the functions

$$\phi = \text{const};$$

$$\phi = \mu \left(\frac{\partial u}{\partial x} + \frac{\partial v}{\partial y} \right);$$
$$\qquad\qquad\qquad \mu = \text{const}$$
$$\phi = \mu \left(\frac{\partial u}{\partial y} - \frac{\partial v}{\partial x} \right);$$

17.8 Numerical Experiments for the Problem of the Interaction of a Shock with Obstacles

The proof of the invariance of an explicit difference scheme from the class (17.1) was performed in the paper [61]. A difference scheme with

$$\bar{C}_{11} = \bar{C}_{22} = \begin{pmatrix} \phi & 0 & 0 & 0 \\ 0 & \phi & 0 & 0 \\ 0 & 0 & 0 & 0 \\ u\phi & v\phi & 0 & 0 \end{pmatrix}; \quad \bar{C}_{12} = \bar{C}_{21} = 0;$$

$$\alpha = \beta = \gamma = 0;$$

was investigated. The calculations showed the effectiveness of invariant difference schemes.

A. G. Marchuk [50] applied an anlogous difference scheme for the calculation of the interaction of shock waves with obstacles. In the calculations such problems were investigated in which the flow of the gas is maintained in regions which are bounded by rigid walls. As the schemes are used without change in the whole flow region, the lines of the shock in the flow region are not taken into account in a special way, and besides the walls only the axis of symmetry plays the role of a boundary.

The conditions on the axis of symmetry for $y = 0$ have the form

$$v = \frac{\partial p}{\partial y} = \frac{\partial \varrho}{\partial y} = \frac{\partial u}{\partial y} = 0.$$

The boundary conditions on the wall are given by

$$- u \sin \theta + v \cos \theta = 0,$$

where θ is the angle of inclination of the wall with respect to the x-axis.

In the calculations based on the difference scheme the computation was carried out within the flow region. At the boundary the values of the gas-dynamic quantities were not calculated but they were assigned in each of the following steps taking into account the fact that the gas can not penetrate the wall. For example, let the point $(x, y) = (kh_1, lh_2)$ belong to the rigid wall which is parallel to the x-axis and let the flow region be situated above the point. Then before the calculation of each following step (with respect to time), we put

$$u(k, l) = u(k, l + 1); \quad v(k, l) = 0; \quad \varrho(k, l) = \varrho(k, l + 1);$$
$$p(k, l) = p(k, l + 1);$$

i.e. the values are taken from the neighbouring points. In an analogous way points are treated which are situated on walls which are parallel to the y-axis. For inclined walls the calculation is somehow more difficult. E.g., let the wall have an angle of inclination of $45°$ with respect to the x-axis. In this case we assume

$$u(k, l) = v(k, l) = \sqrt{\tfrac{1}{2}[u^2(k, l + 1) + v^2(k, l + 1)]},$$

and as one can easily show, the velocity will have the direction parallel to the wall, i.e. the condition of not penetrating the wall is satisfied and the modulus of the velocity is constant. The rest of the quantities are treated in the same way as in the foregoing case. If the wall is inclined at an angle θ to the x-axis in this case we treat the problem in the following way:

$$\begin{pmatrix} h_1 \\ h_2 \end{pmatrix} = h \begin{pmatrix} \cos \theta \\ \sin \theta \end{pmatrix}; \quad \begin{pmatrix} u(k, l) \\ v(k, l) \end{pmatrix} = \sqrt{u^2(k, l + 1) + v^2(k, l + 1)} \begin{pmatrix} \cos \theta \\ \sin \theta \end{pmatrix}.$$

The calculations were carried out for some problems concerning the interaction of shock waves with obstacles of different forms for a polytropic gas ($\varepsilon = p/(\gamma - 1)\varrho$; γ is the adiabatic exponent). The investigated types of problems are schematically shown in Fig. 17.1 where by the dashed lines the initial position of the shock waves is given, and by the heavy lines the shock waves are given which are generated after the interaction. In the legend of the pictures the initial data of the parameters on the shock wave (index 1 means left, index 2 means right side of the shock) and the velocity of the wave are given in tables. In all cases the original shock wave is parallel to the y-axis and the x-axis is the axis of symmetry.

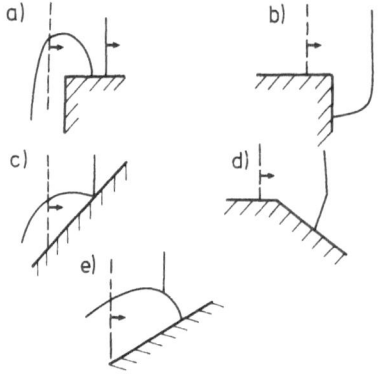

Fig. 17.1 Initial profile of the shock (------), shock front after interaction (————),
a) *Case 1:* $p_1 = 0$, $\varrho_1 = 1$, $u_1 = 0$, $v_1 = 0$, $p_2 = 5$, $\varrho_2 = 5$, $u_2 = 2$, $v_2 = 0$,
   Case 2: $p_1 = 0$, $\varrho_1 = 1$, $u_1 = 0$, $v_1 = 0$, $p_2 = 0.75$, $\varrho_2 = 2.5$, $u_2 = 0.75$, $v_2 = 0$;
b) *Case 1:* $p_1 = 0$, $\varrho_1 = 1$, $u_1 = 0$, $v_1 = 0$, $p_2 = 5$, $\varrho_2 = 5$, $u_2 = 2$, $v_2 = 0$,
   Case 2: $p_1 = 1$, $\varrho_1 = 1$, $u_1 = 0$, $v_1 = 0$, $p_2 = 4$, $\varrho_2 = 2.5$, $u_2 = 1.34$, $v_2 = 0$,
c) $p_1 = 0$, $\varrho_1 = 1$, $u_1 = 0$, $v_1 = 0$, $p_2 = 5$, $\varrho_2 = 5$, $u_2 = 2$, $v_2 = 0$.
d) $p_1 = 0$, $\varrho_1 = 1$, $u_1 = 0$, $v_1 = 0$, $p_2 = 5$, $\varrho_2 = 5$, $u_2 = 2$, $v_2 = 0$.
e) $p_1 = 0$, $\varrho_1 = 1$, $u_1 = 0$, $v_1 = 0$, $h_1 = 0.0868$, $p_2 = 4$, $\varrho_2 = 2.5$, $u_2 = 1.34$, $v_2 = 0$, $h_2 = 0.05$

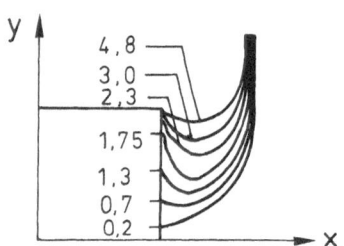

Fig. 17.2 Isobars ($\phi = \mu h$), b) Case 1

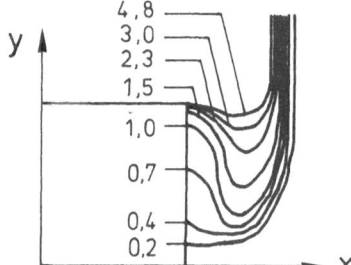

Fig. 17.3 Isobars
($\phi = \mu h^2 \sqrt{u_x^2 p_x^2 + v_y^2 p_y^2} / (\delta + h^2 \sqrt{u_x^2 p_x^2 + v_y^2 p_y^2})$,
b) Case 1

In the calculations the following choice was made:

$$\phi = \mu h; \quad \mu = \text{const}; \quad h = h_1 = h_2 = 0.1.$$

(In the following, if not stated especially to the contrary, these values are chosen.)

In the Figs. 17.2 and 17.3 the results of the numerical calculation of the problem b) are shown with different values of the viscosity ϕ. In the first case a linear viscosity $\phi = \mu h$, $\mu = $ const was chosen, in the second case the viscosity ϕ was chosen according the formula

$$\phi = \mu \frac{h^2 \sqrt{u_x^2 p_x^2 + v_y^2 p_y^2}}{\delta + h^2 \sqrt{u_x^2 p_x^2 + v_y^2 p_y^2}}. \qquad (17.16)$$

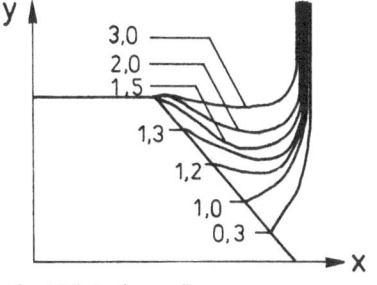

Fig. 17.4 Isobars, d)

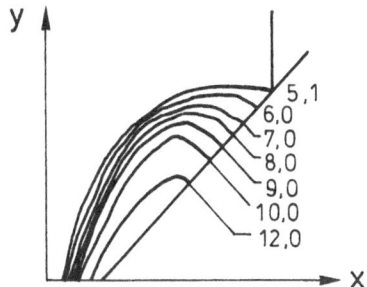

Fig. 17.5 Isobars for simple reflection
$(h_1 = h_2 = 0.1)$, c)

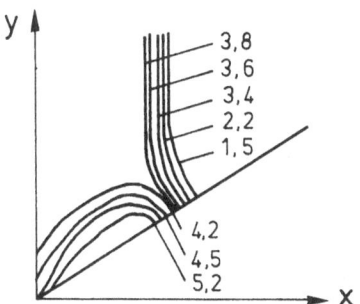

Fig. 17.6. Isobars for Mach-reflection
$(h_1 = 0.0868, h_2 = 0.05)$, e)

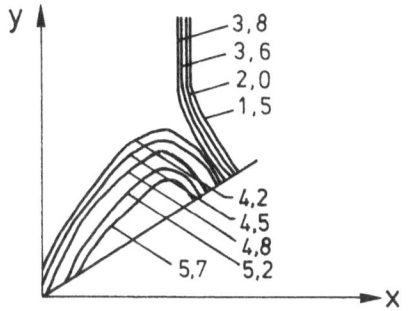

Fig. 17.7. Isobars for Mach-reflection, e)
$(\phi = \mu h^2 \sqrt{u_x^2 p_x^2 + v_y^2 p_y^2} / (\delta + h^2 \sqrt{u_x^2 p_x^2 + v_y^2 p_y^2})$

In the figures lines of constant pressure (isobars) are plotted, where in the Fig. 17.2 the nonuniformity of the shock-smearing along the wall is more pronounced.

In Fig. 17.4 the isobars of problem d) are given.

In the Figs. 17.5 and 17.6 the isobars for the interaction of a shock wave with an inclined solid wall are plotted. Fig. 17.5 corresponds to the simple reflexion (problem c)) with $h_1 = h_2 = 0.1$, and Fig. 17.6 corresponds to a Mach-reflexion (problem e)) with $h_1 = 0.0866$, $h_2 = 0.05$.

In Fig. 17.7 (as in Fig. 17.6) lines of constant pressure are plotted which are the result of the calculation of problem 1 e), but in contrast to Fig. 17.6 in the present calculations the viscosity defined by formula (17.16) was used. Comparing Fig. 17.6 and Fig. 17.7 we recognize that for small gradients the results are nearly identical, but for large gradients the viscosity (17.16) leads to less smearing.

In Figs. 17.8 and 17.9 the results of the calculations of the problems 1 b) (second variant) and 1 a) (second variant) are shown. In Rusanov's paper [204] the results of the calculations according to his scheme are discussed. Our

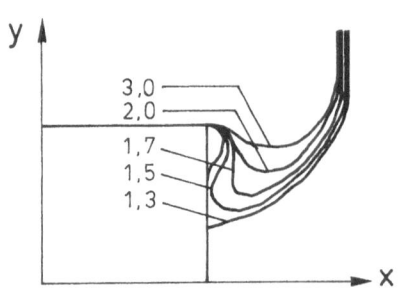

Fig. 17.8. Isobars, b) Case 2

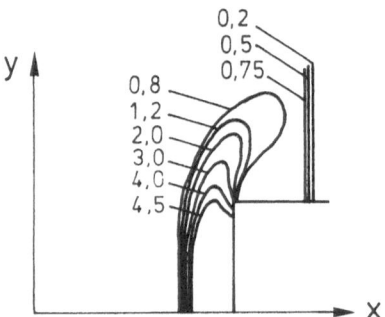

Fig. 17.9. Isobars, a) Case 2

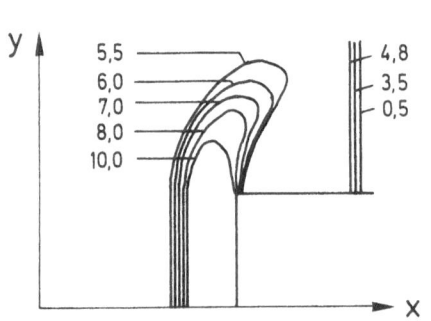

Fig. 17.10. Isobars, a) Case 1

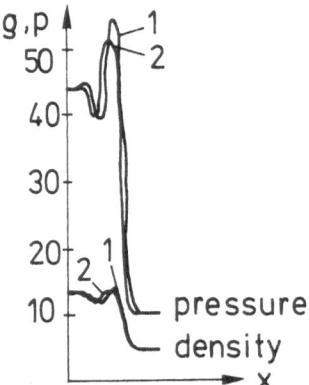

Fig. 17.11 Pressure and density distribution for a shock wave reflected from a wall (Curve 1: Linear Viscosity, curve 2: Viscosity according to (17.16)

calculations have shown a qualitative and also a quantitative coincidence of the results with the results of the paper [204]. The largest difference is encountered in the region where the shock wave is smeared out, and also in the vicinity of the boundary. Apparently the latter is clear from the different kind of realization of the boundary conditions.

In Fig. 17.10 the results of the calculations of problem 1 a) are shown in the plots of isobars.

In Fig. 17.11 plots of the pressure and density change are shown for the problem of a reflected shock wave (with parameters $p_1 = 0$, $\varrho_1 = 1$, $p_2 = 10$; $\varrho_2 = 5$) from a rigid wall. The lines 1 show the pressure and density variation taking into account a linear viscosity, the lines 2 taking into account a viscosity which is defined by (17.16).

The results of the calculation have shown, that in the calculation with the viscosity (17.16) the oscillations at the front of the shock wave are very weak, which leads to the conclusion that the application of viscosity (17.16) is more suitable also for the calculation of more difficult problems (see also Figs. 17.3, 17.7). Thus the construction of invariant difference schemes can be used for the calculation of complicated gasdynamic problems. The numerical results which were gained correspond very well to the real pictures of gas flows.

17.9 Numerical Experiments for the Shallow Water Equations

The system of equations of the shallow water theory which describes the long wave movement of an ideal incompressible fluid will be considered in divergence form

$$w_t + F_x + G_y + Cw = \Phi, \tag{17.17}$$

where

$$w := \begin{pmatrix} \xi \\ U \\ V \end{pmatrix}; \quad F := \begin{pmatrix} U \\ Uu + g\xi^2/2 \\ Uv \end{pmatrix}; \quad G := \begin{pmatrix} V \\ Vu \\ Vv + g\xi^2/2 \end{pmatrix};$$

$$C := \begin{pmatrix} 0 & 0 & 0 \\ 0 & 0 & -f \\ 0 & f & 0 \end{pmatrix}; \quad \Phi := \begin{pmatrix} 0 \\ g\xi \dfrac{\partial h}{\partial x} \\ g\xi \dfrac{\partial h}{\partial y} \end{pmatrix}.$$

$U = u\xi$, $V = v\xi$, $\xi = \eta + h$; u, v are the velocity components in x- and y-direction, $\xi = \xi(x, y, t)$ is the total height of the fluid, $\eta = \eta(x, y, t)$ is the deviation of the surface of the fluid above the equilibrium state, $h = h(x, y)$ is the height of the fluid in the nondisturbed state, g is the acceleration, and f the Coriolis-parameter. We approximate the system of equations (17.17) by the following difference scheme [78], [61]:

$$\frac{\Delta_0 w^n(x, y)}{\tau} + \frac{\Delta_1 + \Delta_{-1}}{2h_1} \tilde{F}^n(x, y) + \frac{\Delta_2 + \Delta_{-2}}{2h_2} \tilde{G}^n(x, y) + \tilde{C}\psi^n(x, y)$$

$$= \left\{ \frac{1}{h_1^2} \bar{\Delta}_1 [\Omega_{11}(x, y) \bar{\Delta}_1] + \frac{1}{h_2^2} \bar{\Delta}_2 [\Omega_{22}(x, y) \bar{\Delta}_2] \right. \tag{17.18}$$

$$\left. + \frac{1}{h_1 h_2} \bar{\Lambda}_1 [\Omega_{12}(x, y) \bar{\Delta}_2] + \frac{1}{h_1 h_2} \bar{\Delta}_2 [\Omega_{21}(x, y) \bar{\Delta}_1] \right\} w^n(x, y).$$

Here $t = n\tau$; τ, h_1, h_2 are the steps of the difference net with respect to time and space in x- and y-direction, respectively:

$$\frac{\tau}{h_i} := \kappa_i = \text{const}, \quad \Delta_i := T_i^1 - E,$$

$$\Delta_{-i} := E - T_i^{-1},$$

$$\bar{\Delta}_i := T_i^{\frac{1}{2}} - T_i^{-\frac{1}{2}}; \quad i = 1, 2,$$

$$\Delta_0 := T_0 - E,$$

$$T_0 \phi(x, y, t) = \phi(x, y, t + \tau),$$

$$T_1^\alpha \phi(x, y, t) = \phi(x + \alpha h_1, y, t),$$

$$T_2^\alpha \phi(x, y, t) = \phi(x, y + \alpha h_2, t),$$

$$E \phi(x, y, t) = \phi(x, y, t),$$

$$\psi^n(x, y) = C w^n(x, y) - \Phi^n(x, y),$$

$$\tilde{C} = I - \frac{\tau}{2} C; \quad I = \text{Identity matrix},$$

$$\tilde{F}^n(x, y) = \tilde{C} F^n(x, y) - \frac{\tau}{2} (A_1 \psi)^n(x, y),$$

$$A_1 = \frac{dF}{dw},$$

$$\tilde{G}^n(x, y) = \tilde{C} G^n(x, y) - \frac{\tau}{2} (A_2 \psi)^n(x, y),$$

$$A_2 = \frac{dG}{dw},$$

$$\Omega_{ij} = \frac{\tau}{2} A_i A_j + C_{ij}; \quad i, j = 1, 2,$$

$$C_{ij} = \text{Matrix of artificial viscosity of order } O(\tau).$$

If $C_{ij} = 0$ ($i, j = 1, 2$) then the difference scheme (17.18) has the order 2, otherwise the order of accuracy is one.

By a special choice of the matrix C_{ij} it is possible to manage that the first differential approximation of the scheme (17.18) admits the same transformations as the system of equations (17.17). In this case we will call the difference scheme invariant [78].

We assume that the function $h = h(x, y)$ is such, that the original system of equation admits a transformation of rotation in the (x, y)-plane. We choose the viscosity matrices in the following way:

$$C_{12} = C_{21} = 0; \quad C_{11} = C_{22} = \phi I,$$

where $\phi = \phi(w, w_x, w_y)$ is a scalar function of order $O(\tau)$, which does not change under a rotation. The choice of such functions is possible. In this case the difference scheme (17.18) will be invariant with respect to the transformation of rotation which will be shown later.

Using the difference scheme described the problem of the motion of a fluid in a paraboloidal basin was solved. The initial state of the fluid is given by a plane deviation of the free surface from the state of equilibrium. This condition together with the given form of the bottom:

$$h(x, y) = h_0 \left[1 - \frac{(x - x_0)^2}{a^2} - \frac{(y - y_0)^2}{a^2} \right],$$

$h_0 = \text{const}$, $a = \text{const}$, $x_0 = \text{const}$, $y_0 = \text{const}$, $h_0 \ll a$, leads to non-linear os-cillations of the fluid. Then the free surface not only does not break, but also stays plane during the time. If $f = 0$ the free surface oscillates such that the normal remains in the same plane. If $f \neq 0$, the free surface rotates around the vertical axis. The frequency of the oscillation of the system is given by the formula

$$\omega_{\pm} = \sqrt{\Omega^2 + \tfrac{1}{4}f^2} \pm \tfrac{1}{2}f, \qquad \Omega := \sqrt{\frac{2\,g h_0}{a^2}}. \tag{17.19}$$

The stated problem and its solution were described for the first time by F. K. Boll in 1962. Sielecki and Wurtele [220] used this problem as a test for the reliability of the computing method for the phenomenon of a wind-driven accumulation of water. According to the paper cited above we choose the following values of the parameters of the problem:

$$a = 14\,\text{km}, \qquad h_0 = 10\,\text{m}, \qquad f = 10^{-3}\,\text{s}^{-1}.$$

The geometry of the problem is shown in Fig. 17.12.

We emphasize that in the given problem one has to calculate the motion of the liquid in a region with a moving boundary which is represented in every moment of time as a contact line of the free surface with the bottom. Let $\gamma(t_\phi; x, y) = 0$ be the equation of the projection of this line in the (x, y)-plane for the time $t = t_\phi$. For the solution of the problem the method of the running calculation was used which does not need complete boundary conditions along the contact line in the problem. If one tries to apply the second order scheme (17.18) for the calculation of the described problem without viscosity matrices (which means by choosing $C_{ij} = 0$; $i,j = 1, 2$), then in the vicinity of $\gamma(t_\phi; x, y) = 0$ small-scale oscillations are generated which spread very fast over the region of computation and destroy the whole picture of the flow. Therefore in the calculations the first order scheme was used.

We mention that Sielecki and Wurtele used a second order scheme without smoothing. To avoid oscillations they determined for each time step the posi-tion of the contact line $\gamma(t_\phi; x, y) = 0$ and calculated along the line the values

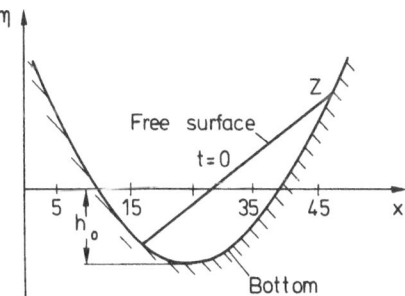

Fig. 17.12. Geometry of the problem of wind-driven accumulation of water

of the function ξ by an extrapolation. Such a procedure is more complicated from the logical point of view than the method of the running calculation. Basically this procedure is useful for smoothing disturbances at the boundary while the calculation in the inner region of the flow is carried out with formulas of higher accuracy.

If in the scheme (17.18) the following matrix is chosen as viscosity matrix:

$$C_{11} = C_{22} = \phi I; \quad C_{12} = C_{21} = 0, \quad \phi = \mu \frac{1}{\alpha(\xi) + 1}, \tag{17.20}$$

where $\mu = \text{const}$, $\alpha = \alpha(\xi)$ is a positive monotonic function, then during the calculation the line $\gamma(t_\phi; x, y) = 0$ is determined automatically. Indeed, along the line $\xi = \xi(x, y, t_\phi) = 0$ also the value of the viscosity coefficient is maximal.

For the numerical calculations the system of equations (17.17) was written in non-dimensional form. As characteristic length-scales the height h_0 and a (the width of the basin in horizontal direction) were chosen. Then we get

$$\bar{x} = \frac{x}{a}; \quad \bar{y} = \frac{y}{a}; \quad \bar{t} = t \frac{\sqrt{gh_0}}{a}, \quad \bar{f} = f \frac{a}{\sqrt{gh_0}},$$

$$\bar{u} = \frac{u}{\sqrt{ga}}, \quad \bar{v} = \frac{v}{\sqrt{ga}}, \quad \bar{\eta} = \frac{\eta}{\eta_0}, \quad \bar{h} = \frac{h}{h_0}.$$

In the transformed system of equation (17.17) the quantity g disappears. The function which describes the geometrical form of the bottom, has the form

$$\bar{h}(\bar{x}, \bar{y}) = 1 - (\bar{x} - \bar{x}_0)^2 - (\bar{y} - \bar{y}_0)^2.$$

In the following the bars above the variables will be omitted.

As a region of computation we choose a square with sidelength $2Nh$, where $h = h_1 = h_2 = 1/14$ for $N = 24$, and $h = 1/7$ for $N = 12$. The initial perturbation is given by

$$\eta(x, y, 0) = 0.968(x - 1.714) - 0.236.$$

Then $\gamma(0; x, y) = 0$ is represented by a unit circle

$$(x - 2.198)^2 + (y - 1.714)^2 = 1. \tag{17.21}$$

For all points of the computing region which are inner points of the unit circle with centre at $O_y = (2.198, 1.714)$ the total height is calculated according to the formula

$$\xi(x, y, 0) = h(x, y) + \eta(x, y, 0);$$

for all other points $\xi(x, y, 0) = 0$. The fluid in the moment $t = 0$ is assumed to be at rest.

When time is running, the line $\gamma(t_\phi, x, y) = 0$ changes its position in the (x, y)-plane but remains a unit circle which osculates the circle Γ of radius 1.484 with the centre $O_\Gamma = (1.714, 1.714)$ which represents the projection of the

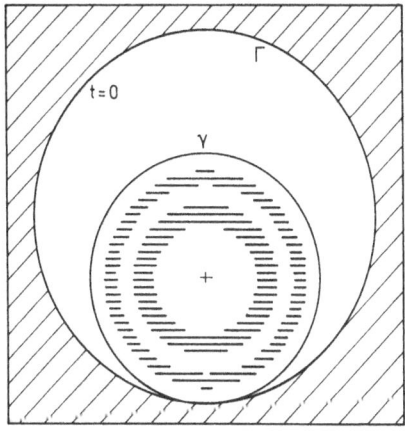

Fig. 17.13. Initial conditions and boundary curve

trajectory of the point z in the (x, y)-plane (see Fig. 17.12). In the hatched region of Fig. 17.13 the height and the velocity for all $t \geqq 0$ are assumed to be zero and they are used as boundary conditions for the difference scheme (17.18).

During the calculation the boundary points were determined as the set of points for which $\xi < \delta$, where δ is a small number. The level lines $\xi(x, y, t_\phi)$ = const were plotted on the printer of the computer.

For comparison of the results gained by different methods besides the scheme (17.18) the well-known Lax' scheme was used which is not invariant with respect to a transformation of rotation.

In the Figs. 17.14 and 17.15 niveau-lines ξ = const are plotted for $t = T/4$ and $t = T/2$ (T is the period calculated by formula (17.19)). The calculation was carried out using the Lax' scheme. From the picture it can be concluded that the level lines are distorted during the time of calculation. By the full line the exact solution is given. The numerical solution is monotonic in any direction but the accuracy of the solution is not very high. For example, the maximal

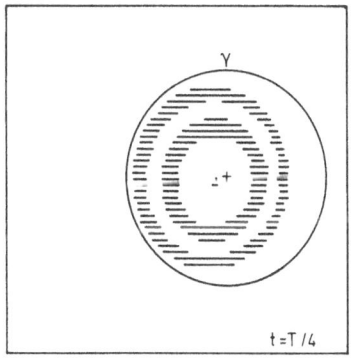

Fig. 17.14

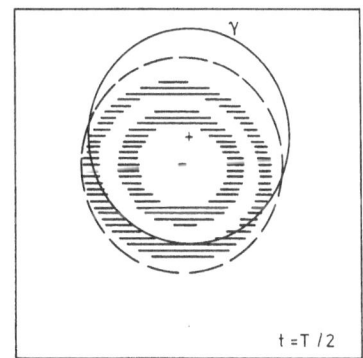

Fig. 17.15

Fig. 17.14. Niveau lines ξ = const (Lax'scheme), $t = T/4$, exact solution (heavy circle)

Fig. 17.15. Niveau lines ξ = const (Lax'scheme), $t = T/2$, exact solution (heavy circle)

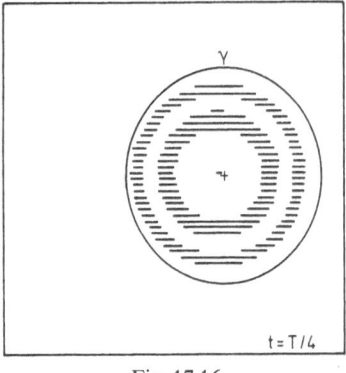
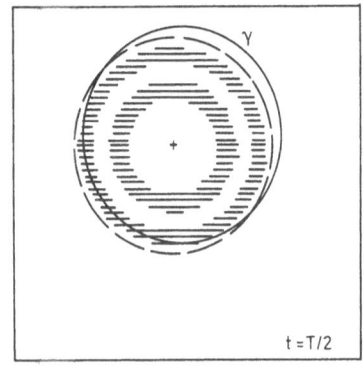

Fig. 17.16   Fig. 17.17

Fig. 17.16. First order invariant scheme (17.18), 49×49 grid points, $t = T/4$

Fig. 17.17. First order invariant scheme (17.18), 49×49 grid points, $t = T/2$

value of ξ in the whole region (it is called Ψ) for $t = T/4$ is 0.861, for $t = T/2$ this value decreases to 0.656. In the exact solution the value of Ψ is 1.

In the Figs. 17.16 and 17.17 the results of calculations are represented which are obtained using the first order invariant scheme (17.18) with

$$C_{11} = C_{22} = \mu_0 \frac{h^2}{2\tau} I, \quad \mu_0 = \text{const}, \quad C_{12} = C_{21} = 0.$$

To perform the calculations the scheme (17.18) was applied to a 10-point star which contained only integer grid points whereas the values at intermediate points were replaced by the formula of the following type:

$$\phi\left(x + \frac{h}{2}\right) = \tfrac{1}{2}[\phi(x + h) + \phi(x)].$$

The level lines $\xi = \text{const}$ which are plotted in Figs. 17.16 and 17.17 are much closer to the form of a circle compared to the Lax' scheme, and although as in the Lax' scheme the difference scheme has first order of accuracy its solution is much closer to the exact one. For example, $\Psi = 0.973$ for $t = T/4$, and $\Psi = 0.853$ for $t = T/2$. The described calculations were performed on a 49×49 grid ($N = 24$, $h = 1/4$).

In the Figs. 17.18–21 level lines ξ are plotted which were calculated on a lattice of 25×25 points $N = 12$, $h = 1/7$).

In the Figs. 17.18 and 17.19 the results are given using the second order scheme (17.18) with $C_{11} = C_{22} = \phi I$, $C_{12} = C_{21} = 0$, $\phi = O(\tau^2)$. The first differential approximation of such a scheme is not invariant with respect to a transformation of rotation.

The function ϕ was chosen in the form (17.20) with $\mu = \mu_0 h^3/\tau$; $\mu_0 = 0.7$, $\alpha(\xi) = 10\,\xi$. We remark that for large values of μ_0 the level lines are less

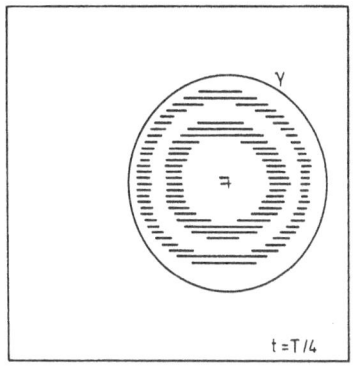

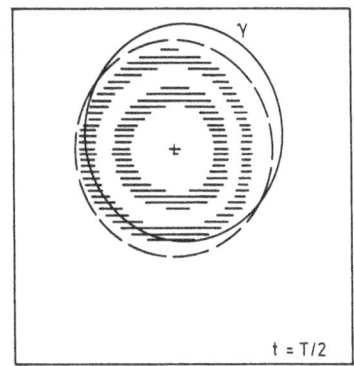

Fig. 17.18   Fig. 17.19

Fig. 17.18. Niveau lines ξ = const, 25×25 grid points (second order scheme (17.18)), $t = T/4$

Fig. 17.19. Niveau lines ξ = const, 25×25 grid points (second order scheme (17.18)), $t = T/2$

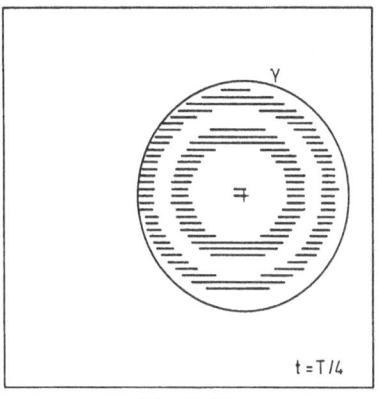

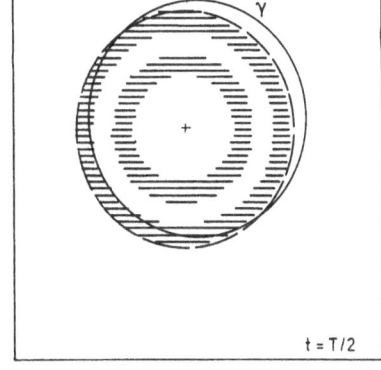

Fig. 17.20   Fig. 17.21

Fig. 17.20. Niveau lines ξ = const, 25×25 grid points (first order invariant scheme (17.18)), $t = T/4$

Fig. 17.21. Niveau lines ξ = const, 25×25 grid points (first order invariant scheme (17.18)), $t = T/2$

distorted. In the Figs. 17.20 and 17.21 level lines are shown which are calculated by a first order invariant scheme of type (17.18) with a viscosity matrix (17.20) with $\alpha(\xi) \equiv 1$, $\phi = \mu_0 h^2/\tau$, $\mu_0 = 0.2$.

In different variants of the calculations μ_0 changed its value between 0.2 and 1, where a larger value of μ_0 gave level lines with a form more similar to a circle. But one has to remember that the investigated schemes are of first order and that an enlargement of μ_0 introduces a strong smearing zone of the contact line of the free surface with the bottom. For the realization of a more accurate calculation of the given problem one has to apply, apparently, invariant schemes of higher order of accuracy.

17.10 Analysis of the Properties of the Artificial Viscosity for Difference Schemes of Two-dimensional Gas Flows

We will compare some of the most commonly used difference schemes of first and second order of approximation for the system of equations (13.1) of gas dynamics with two space variables.

We introduce a computational grid and we are numbering the grid prints on the x-axis by $0, 1, \ldots, J$, and the grid points along the y-axis by $0, 1, \ldots, K$, so that each grid point in the (x, y)-plane is defined by an ordered pair of numbers. The gasdynamic functions will be sought in the grid points and will be denoted by f^n_{jk}, where f represents one of the quantities u, v, ϱ, p, ε. Further we will use the following notations:

τ – time step of the difference scheme,

$\quad t = n\tau;$

h_1, h_2 – step sizes along the x- and y-axis, respectively,

$\quad \kappa_i := \tau/h_i = \text{const}; \quad i = 1, 2.$

First of all we will describe two families of invariant difference schemes of first order of approximation and we will call these schemes of class J_1 and J_2, respectively.

17.10.1 Schemes of Class J_1

We will write these schemes in the following form:

$$\frac{w^{n+1}_{j,k} - w^n_{j,k}}{\tau} + \frac{F^n_{j+1,k} - F^n_{j-1,k}}{2h_1} + \frac{G^n_{j,k+1} - G^n_{j,k-1}}{2h_2}$$

$$= \frac{\kappa_1}{2}\left[(\Omega_{11})^n_{j+\frac{1}{2},k}\frac{w^n_{j+1,k} - w^n_{j,k}}{h_1} - (\Omega_{11})^n_{j-\frac{1}{2},k}\frac{w^n_{j,k} - w^n_{j-1,k}}{h_1}\right.$$

$$\left. + (\Omega_{12})^n_{j+\frac{1}{2},k}\frac{w^n_{j+\frac{1}{2},k+\frac{1}{2}} - w^n_{j+\frac{1}{2},k-\frac{1}{2}}}{h_2} - (\Omega_{12})^n_{j-\frac{1}{2},k}\frac{w^n_{j-\frac{1}{2},k+\frac{1}{2}} - w^n_{j-\frac{1}{2},k-\frac{1}{2}}}{h_2}\right]$$

$$+ \frac{\kappa_2}{2}\left[(\Omega_{22})^n_{j,k+\frac{1}{2}}\frac{w^n_{j,k+1} - w^n_{j,k}}{h_2} - (\Omega_{22})^n_{j,k-\frac{1}{2}}\frac{w^n_{j,k} - w^n_{j,k-1}}{h_2}\right.$$

$$\left. + (\Omega_{21})^n_{j,k+\frac{1}{2}}\frac{w^n_{j+\frac{1}{2},k+\frac{1}{2}} - w^n_{j-\frac{1}{2},k+\frac{1}{2}}}{h_1} - (\Omega_{21})^n_{j,k-\frac{1}{2}}\frac{w^n_{j+\frac{1}{2},k-\frac{1}{2}} - w^n_{j-\frac{1}{2},k-\frac{1}{2}}}{h_1}\right],$$

$$\tag{17.22}$$

$$(\Omega_{ml})^n_{j\pm\frac{1}{2},k} := \Omega_{ml}(w^n_{j\pm\frac{1}{2},k}); \quad (m, l = 1, 2)$$

$$w^n_{j\pm\frac{1}{2},k} := \frac{w^n_{j\pm 1,k} + w^n_{j,k}}{2}.$$

In a similar way the other quantities are defined in points with non-integral indices. The matrices $\Omega_{ml}(m, l = 1, 2)$ are of order $O(1)$, where

$$\Omega_{ml} = A_m A_l + C_{ml}; \quad (m, l = 1, 2).$$

$$A_1 = \frac{dF}{dw}; \quad A_2 = \frac{dG}{dw};$$

(17.23)

here $C_{ml}(m, l = 1, 2)$ are the matrices of the artificial viscosity which are given such, that the difference scheme (17.22) is invariant.

The difference molecule for an interior point consists of nine grid points and is shown in Fig. 17.22.

The amplification matrix of the linearized difference scheme has the form

$$G = I - i(\kappa_1 A_1 \sin\theta_1 + \kappa_2 A_2 \sin\theta_2) - \kappa_1^2 \Omega_{11}(1 - \cos\theta_1)$$

$$- \kappa_2^2 \Omega_{22}(1 - \cos\theta_2) - \frac{\kappa_1 \kappa_2}{2}(\Omega_{12} + \Omega_{21})\sin\theta_1 \sin\theta_2;$$

$$\theta_1 := m_1 h_1, \quad \theta_2 := m_2 h_2,$$

m_1, m_2 are wave numbers.

In [203] a sufficient stability criterion is derived for the case $C_{ml} = 0$ $(m, l = 1, 2)$:

$$\frac{\tau}{h}\sigma^* \leqq \frac{1}{2\sqrt{2}}.$$

(17.24)

Here $h_1 = h_2 = h$, σ^* is the maximum of the spectral radii of the matrices A_1 and A_2. As mentioned in [203] this condition is too restrictive and in some practical cases the time-step can be chosen according to the condition

$$\tau \leqq \frac{h}{\sigma^*\sqrt{2}}$$

(17.25)

and leads to satisfactory results. But in [203] a class of problems is studied for which it is not possible to weaken the condition (17.24) without getting unstable zones.

In the numerical calculations which are described later we used the condition (17.25), and this seemed to be sufficient.

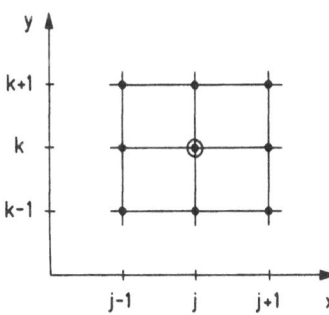

Fig. 17.22.
Difference molecule for a scheme of class J_1 and J_2

If $C_{ml} \neq 0$ our calculations have shown that the condition (17.25) is not sufficient for stability and, instead of it, one has to choose

$$\tau \leqq \frac{\alpha h}{\sigma^* \sqrt{2}}; \quad \alpha < 1,$$

where the quantity α depends on the choice of the matrices $C_{ml}(m, l = 1, 2)$ and must be defined experimentally.

The Π-form of the first differential approximation of the scheme (17.22) has the form

$$w_t + F_x + G_y = \frac{\tau}{2}(C_{11} w_x + C_{12} w_y)_x + \frac{\tau}{2}(C_{21} w_x + C_{22} w_y)_y, \qquad (17.26)$$

where

$$C_{ml} := \Omega_{ml} - A_m A_l; \quad (m, l = 1, 2).$$

17.10.2 Schemes of Class J_2

The schemes of class J_2 are written in the following form: In a first step according to the scheme (17.22) with the matrices $\Omega_{ml} = A_m A_l, (m, l = 1, 2)$ the values $\tilde{w}_{j,k}$ are calculated, and in a second step the final values on the $(n + 1)^{\text{st}}$ layer are found according to the formula

$$w_{j,k}^{n+1} = w_{j,k}^n + \frac{\kappa_1^2}{2}[(\tilde{C}_{11})_{j+\frac{1}{2},k}(\tilde{w}_{j+1,k} - \tilde{w}_{j,k}) - (\tilde{C}_{11})_{j-\frac{1}{2},k}(\tilde{w}_{j,k} - \tilde{w}_{j-1,k})]$$

$$+ \frac{\kappa_1 \kappa_2}{2}[(\tilde{C}_{12})_{j+\frac{1}{2},k}(\tilde{w}_{j+\frac{1}{2},k+\frac{1}{2}} - \tilde{w}_{j+\frac{1}{2},k-\frac{1}{2}}) - (\tilde{C}_{12})_{j-\frac{1}{2},k}(\tilde{w}_{j-\frac{1}{2},k+\frac{1}{2}} - \tilde{w}_{j-\frac{1}{2},k-\frac{1}{2}})]$$

$$+ \frac{\kappa_1 \kappa_2}{2}[(\tilde{C}_{21})_{j,k+\frac{1}{2}}(\tilde{w}_{j+\frac{1}{2},k+\frac{1}{2}} - w_{j-\frac{1}{2},k+\frac{1}{2}}) - (\tilde{C}_{21})_{j,k-\frac{1}{2}}(\tilde{w}_{j+\frac{1}{2},k-\frac{1}{2}} - \tilde{w}_{j-\frac{1}{2},k-\frac{1}{2}})]$$

$$+ \frac{\kappa_2^2}{2}[(\tilde{C}_{22})_{j,k+\frac{1}{2}}(\tilde{w}_{j,k+1} - \tilde{w}_{j,k}) - (\tilde{C}_{22})_{j,k-\frac{1}{2}}(\tilde{w}_{j,k} - \tilde{w}_{j,k-1})];$$

here the values in the points with half integral indices are defined as in the scheme (17.22).

The schemes of class J_2 can be treated as schemes with splitting for the system of equations (17.26) with viscosity. Initially, in the first step we approximate the system of equations of gas dynamics

$$w_t + F_x + G_y = 0$$

by a second order scheme.

In the second step we correct the values of the gasdynamic quantities of the first step using a scheme which approximates the diffusive system of equations:

$$w_t = \frac{\tau}{2}(C_{11} w_x + C_{12} w_y)_x + \frac{\tau}{2}(C_{21} w_x + C_{22} w_y)_y.$$

The amplification matrix of the linearized difference scheme of class J_2 has the following form:

$$G = [I - \kappa_1^2 C_{11}(1 - \cos\theta_1) - \kappa_2^2 C_{22}(1 - \cos\theta_2) - \frac{\kappa_1\kappa_2}{2}(C_{12} + C_{21})\sin\theta_1\sin\theta_2]$$

$$\cdot [I - i(\kappa_1 A_1 \sin\theta_1 + \kappa_2 A_2 \sin\theta_2) - \kappa_1^2 A_1^2(1 - \cos\theta_1) - \kappa_2^2 A_2^2(1 - \cos\theta_2)$$

$$- \frac{\kappa_1\kappa_2}{2}(A_1 A_2 + A_2 A_1)\sin\theta_1\sin\theta_2] = G_1 \cdot G_2.$$

As $\|G\| \leq \|G_1\| \cdot \|G_2\|$, it is sufficient for the stability of the schemes of the considered class that the following inequalities hold:

$$\|G_1\| \leq 1, \quad \|G_2\| \leq 1.$$

As already shown the second inequality is satisfied on the basis of condition (17.24). To satisfy the first condition it is necessary to choose the matrices $C_{ml}(m, l = 1, 2)$ in a suitable manner. We consider the special class of difference schemes with matrices of the artificial viscosity of the following form:

$$C_{12} = C_{21} = 0, \quad C_{11} = C_{22} = \mu I. \tag{17.27}$$

We choose $h_1 = h_2 = h$, $\mu = 2\mu_0 h^2/\tau^2$, $\mu_0 = $ const. Then we get

$$G_1 = \left[1 - 4\mu_0 \sin^2\left(\frac{\theta_1}{2}\right) - 4\mu_0 \sin^2\left(\frac{\theta_2}{2}\right)\right] I,$$

and

$$\|G_1\| \leq 1 \quad \text{for} \quad 0 \leq \mu_0 \leq 0.25. \tag{17.28}$$

Thus a difference scheme of class J_2 with matrices of the artificial viscosity (17.27) is stable if the conditions (17.24) and (17.28) are satisfied.

It is easy to see that for the calculation of the quantities $\tilde{w}_{j,k}$ a scheme was used which coincides for the linear case with the Lax-Wendroff-scheme. To weaken the stability condition (17.24) and to avoid to calculate in each grid point the matrices $A_l A_m$ ($l, m = 1, 2$), one can apply any variant of the Lax-Wendroff-scheme with two steps to determine the values $\tilde{w}_{j,k}$. One can also use the MacCormack-scheme. We attribute such schemes also to the class J_2.

The Π-form of the first differential approximation of schemes from class J_2 has the form (17.26). Thus the features which are defined on the basis of their first differential approximation are in the same way typical for schemes from the class J_1 and J_2. For a more detailed analysis of the considered schemes one has to use differential approximations of higher order.

17.10.3 The Lax' Scheme

The Lax' scheme has the following form:

$$\frac{w_{j,k}^{n+1} - \bar{w}_{j,k}^n}{\tau} + \frac{F_{j+1,k}^n - F_{j-1,k}^n}{2h_1} + \frac{G_{j,k+1}^n - G_{j,k-1}^n}{2h_2} = 0, \tag{17.29}$$

$$\bar{w}_{j,k}^n := \tfrac{1}{4}(w_{j-1,k}^n + w_{j+1,k}^n + w_{j,k-1}^n + w_{j,k+1}^n).$$

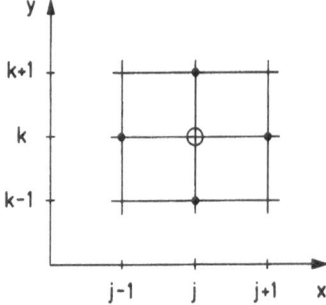

Fig. 17.23. Difference star for the Lax'scheme

To draw the star of this scheme one has to use four points (see Fig. 17.23). The scheme (17.29) can also be written in a different way:

$$\frac{w_{j,k}^{n+1} - w_{j,k}^n}{\tau} + \frac{F_{j+1,k}^n - F_{j-1,k}^n}{2h_1} + \frac{G_{j,k+1}^n - G_{j,k-1}^n}{2h_2}$$

$$= \frac{h_1^2}{4\tau} \frac{w_{j+1,k}^n - 2w_{j,k}^n + w_{j-1,k}^n}{h_1^2} + \frac{h_2^2}{4\tau} \frac{w_{j,k+1}^n - 2w_{j,k}^n + w_{j,k-1}^n}{h_2^2}. \quad (17.30)$$

The scheme (17.30) can be derived from the difference scheme (17.22) if we define

$$\Omega_{12} := \Omega_{21} := 0, \quad \Omega_{11} := \frac{h_1^2}{2\tau^2} I, \quad \Omega_{22} := \frac{h_2^2}{2\tau^2} I.$$

The amplification matrix of the difference scheme has the form

$$G = \frac{\cos\theta_1 + \cos\theta_2}{2} I - i\kappa_1 A_1 \sin\theta_1 - i\kappa_2 A_2 \sin\theta_2.$$

The Lax' scheme is stable, if

$$\kappa_1 \sigma_1^* \leqq \tfrac{1}{2}; \quad \kappa_2 \sigma_2^* \leqq \tfrac{1}{2}.$$

Here σ_1^* and σ_2^* are the moduli of the largest eigenvalues of the matrices A_1 and A_2, respectively.

The Π-form of the first differential approximation of the Lax' scheme has the form:

$$w_t + F_x + G_y = \left[\left(\frac{h_1^2}{4\tau} I - \frac{\tau}{2} A_1^2\right) w_x\right]_x + \left[\left(\frac{h_2^2}{4\tau} I - \frac{\tau}{2} A_2^2\right) w_y\right]_y$$

$$- \frac{\tau}{2}(A_1 A_2 w_y)_x - \frac{\tau}{2}(A_2 A_1 w_x)_y.$$

The considered scheme is not invariant with respect to a Galilei-transformation and a rotation.

17.10.4 The Rusanov-Scheme

The Rusanov-scheme [204] can be written in the form (17.22), if we put

$$\Omega_{12} = \Omega_{21} = 0, \quad \Omega_{11} := \frac{\alpha}{\kappa_1^2} I, \quad \Omega_{22} := \frac{\beta}{\kappa_2^2} I,$$

$$(\Omega_{11})^n_{j+\frac{1}{2},k} = \frac{1}{2\kappa_1^2} (\alpha^n_{j+1,k} + \alpha^n_{j,k}) I,$$

$$(\Omega_{22})^n_{j,k+\frac{1}{2}} = \frac{1}{2\kappa_2^2} (\beta^n_{j,k+1} + \beta^n_{j,k}) I,$$

$$\alpha^n_{j,k} = \omega \kappa (\sqrt{u^2 + v^2} + c)^n_{j,k} \sin^2 \psi,$$

$$\beta^n_{j,k} = \omega \kappa (\sqrt{u^2 + v^2} + c)^n_{j,k} \cos^2 \psi,$$

$$h = \sqrt{h_1^2 + h_2^2}; \quad h_1 = h \cos \psi; \quad h_2 = h \sin \psi;$$

$$\kappa_l = \frac{\tau}{h_l}, (l = 1, 2); \quad \kappa = \sqrt{\kappa_1^2 + \kappa_2^2}; \quad \omega = \text{const.}$$

The scheme can be represented by a star which is shown in Fig. 17.24.
   The transition matrix of the linearized Rusanov-scheme has the form

$$G = I - i(\kappa_1 A_1 \sin \theta_1 + \kappa_2 A_2 \sin \theta_2) - \alpha(1 - \cos \theta_1)I - \beta(1 - \cos \theta_2)I.$$

In [204] it was proved that for stability of the Rusanov-scheme it is necessary to satisfy the conditions

$$\kappa \sigma_0^n \leqq 1, \quad \kappa \sigma_0^n \leqq \omega \leqq \frac{1}{\kappa \sigma_0^n};$$

$$\sigma_0^n := \max_{j,k} \{(\sqrt{u^2 + v^2} + c)^n_{j,k}\}.$$

The Π-form of the first differential approximation of the Rusanov-scheme has the form

$$w_t + F_x + G_y = \left[\left(\frac{h_1 h_2}{2\sqrt{h_1^2 + h_2^2}} \omega \sigma I - \frac{\tau}{2} A_1^2 \right) w_x \right]_x - \frac{\tau}{2} (A_1 A_2 w_y)_x$$

$$- \left[\left(\frac{h_1 h_2}{2\sqrt{h_1^2 + h_2^2}} \omega \sigma I - \frac{\tau}{2} A_2^2 \right) w_y \right]_y - \frac{\tau}{2} (A_2 A_1 w_x)_y,$$

$$\sigma := \sqrt{u^2 + v^2} + c.$$

The difference scheme considered is not invariant with respect to a Galilei-transformation and a rotation.
   In the following we consider some schemes of second order of approximation.

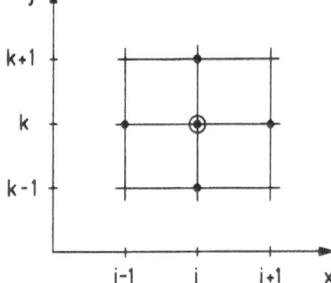

Fig. 17.24. Difference star of the Rusanov-scheme

17.10.5 The Lax-Wendroff-Scheme

The Lax-Wendroff-scheme which was introduced in [205] has the following form:

$$\frac{w_{j,k}^{n+1} - w_{j,k}^{n}}{\tau} + \frac{F_{j+1,k}^{n} - F_{j-1,k}^{n}}{2\,h_1} + \frac{G_{j,k+1}^{n} - G_{j,k-1}^{n}}{2\,h_2}$$

$$= \frac{\kappa_1}{2}\left[(A_1)_{j+\frac{1}{2},k}^{n}\left(\frac{F_{j+1,k}^{n} - F_{j,k}^{n}}{h_1} + \frac{G_{j+\frac{1}{2},k+\frac{1}{2}}^{n} - G_{j+\frac{1}{2},k-\frac{1}{2}}^{n}}{h_2}\right)\right.$$

$$\left. - (A_1)_{j-\frac{1}{2},k}^{n}\left(\frac{F_{j,k}^{n} - F_{j-1,k}^{n}}{h_1} + \frac{G_{j-\frac{1}{2},k+\frac{1}{2}}^{n} - G_{j-\frac{1}{2},k-\frac{1}{2}}^{n}}{h_2}\right)\right]$$

$$+ \frac{\kappa_2}{2}\left[(A_2)_{j,k+\frac{1}{2}}^{n}\left(\frac{F_{j+\frac{1}{2},k+\frac{1}{2}}^{n} - F_{j-\frac{1}{2},k+\frac{1}{2}}^{n}}{h_1} + \frac{G_{j,k+1}^{n} - G_{j,k}^{n}}{h_2}\right)\right.$$

$$\left. - (A_2)_{j,k-\frac{1}{2}}^{n}\left(\frac{F_{j+\frac{1}{2},k-\frac{1}{2}}^{n} - F_{j-\frac{1}{2},k-\frac{1}{2}}^{n}}{h_1} + \frac{G_{j,k}^{n} - G_{j,k-1}^{n}}{h_2}\right)\right], \qquad (17.31)$$

$$(A_l)_{j\pm\frac{1}{2},k}^{n} := A_l\left(\frac{w_{j\pm1,k}^{n} + w_{j,k}^{n}}{2}\right),$$

or

$$(A_l)_{j\pm\frac{1}{2},k}^{n} := \frac{A_l(w_{j\pm1,k}^{n}) + A_l(w_{j,k}^{n})}{2}, \qquad (l = 1, 2).$$

In an analogous way the remaining values in points with half integer indices are defined. The star of the difference scheme is shown in Fig. 17.22. The amplification matrix of the linearized difference scheme can be written in the following way:

$$G = I - i(\kappa_1 A_1 \sin\theta_1 + \kappa_2 A_2 \sin\theta_2)$$

$$- \left[\kappa_1^2 A_1^2(1 - \cos\theta_1) + \kappa_2^2 A_2^2(1 - \cos\theta_2)\right.$$

$$\left. + \frac{\kappa_1 \kappa_2}{2}(A_1 A_2 + A_2 A_1)\sin\theta_1 \sin\theta_2\right].$$

For $h_1 = h_2 = h$ the stability condition was derived in [203] in the form of the inequality (17.24).

The Π-form of the first differential approximation has the form

$$w_t + F_x + G_y = \frac{\tau^2}{6}\{F_{ww}(F_x+G_y)^2 + A_1[A_1(F_x+G_y)]_x + A_1[A_2(F_x+G_y)]_y\}_x$$

$$+ \frac{\tau^2}{6}\{G_{ww}(F_x+G_y)^2 + A_2[A_1(F_x+G_y)]_x$$

$$+ A_2[A_2(F_x+G_y)]_y\}_y$$

$$- \frac{h_1^2}{6}F_{xxx} - \frac{h_2^2}{6}G_{yyy}.$$

17.10.6 Two-step Variant of the Lax-Wendroff-Scheme

The first two-step variant of the Lax-Wendroff-scheme was proposed by Richtmyer and Morton [7]. This method uses a box with sidelengths $2\tau, 2h_1, 2h_2$. Initially intermediate values for $t = (n + 1)\tau$ according to the Lax' scheme are calculated:

$$w_{j,k}^{n+1} = \tfrac{1}{4}(w_{j+1,k}^n + w_{j-1,k}^n + w_{j,k+1}^n + w_{j,k-1}^n)$$

$$- \frac{\kappa_1}{2}(F_{j+1,k}^n - F_{j-1,k}^n) - \frac{\kappa_2}{2}(G_{j,k+1}^n - G_{j,k-1}^n).$$

Then the final values for $t = (n + 2)\tau$ are determined according to the cross-scheme

$$w_{j,k}^{n+2} = w_{j,k}^n - \kappa_1(F_{j+1,k}^{n+1} - F_{j-1,k}^{n+1}) - \kappa_2(G_{j,k+1}^{n+1} - G_{j,k-1}^{n+1}).$$

The star of the Lax-Wendroff-scheme has the form as shown in Fig. 17.25.

The amplification matrix of the linearized version of the difference scheme can be written in the form

$$G = I - i(\cos\theta_1 + \cos\theta_2)(\kappa_1 A_1 \sin\theta_1 + \kappa_2 A_2 \sin\theta_2)$$

$$- 2[\kappa_1 A_1 \sin\theta_1 + \kappa_2 A_2 \sin\theta_2]^2.$$

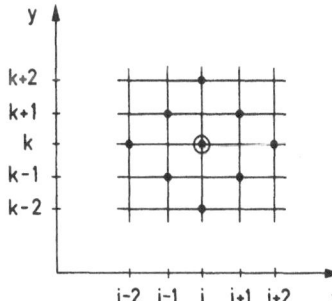

Fig. 17.25. Difference star of the two-step version of the Lax-Wendroff-scheme (Richtmyer-scheme)

The investigation of the eigenvalues of the matrix G shows that for the stability of the considered linearized difference scheme in the case of $h_1 = h_2 = h$ it is necessary and sufficient to satisfy the conditions

$$\frac{\tau \sigma_0^n}{h} \leq \frac{1}{\sqrt{2}}, \qquad \sigma_0^n := \max_{j,k} \{\sqrt{u^2 + v^2} + c\}.$$

The Π-form of the first differential approximation of the Lax-Wendroff-scheme has the form

$$
\begin{aligned}
w_t + F_x + G_y &= \frac{\tau^2}{6} \{[F_{ww}(F_x + G_y)^2]_x + [G_{ww}(F_x + G_y)^2]_y\} \\
&\quad + \tfrac{2}{3}\tau^2 \{A_1[A_1(F_x + G_y)]_x + A_1[A_2(F_x + G_y)]_y\}_x \\
&\quad + \tfrac{2}{3}\tau^2 \{A_2[A_1(F_x + G_y)]_x + A_2[A_2(F_x + G_y)]_y\}_y \\
&\quad - \frac{h_1^2}{6}F_{xxx} - \frac{h_2^2}{6}G_{yyy} - \frac{h_1^2}{4}[A_1(w_{xx} + w_{yy})] \\
&\quad - \frac{h_2^2}{4}[A_2(w_{xx} + w_{yy})]_y.
\end{aligned}
$$

17.10.7 Modification of the Lax-Wendroff-Scheme

We describe now another two-step second-order scheme which represents a modification of the Lax-Wendroff-scheme, and which was proposed by Eilon, Gottlieb, and Zwas [206]. We will call this scheme the EGZLW-scheme. This scheme has the following form:

$$
\begin{aligned}
\tilde{w}_{j+\frac{1}{2},k+\frac{1}{2}} &:= \tfrac{1}{4}(w_{j+1,k+1}^n + w_{j+1,k}^n + w_{j,k+1}^n + w_{j,k}^n) \\
&\quad - \frac{\kappa_1}{2}(F_{j+1,k+\frac{1}{2}}^n - F_{j,k+\frac{1}{2}}^n) - \frac{\kappa_2}{2}(G_{j+\frac{1}{2},k+1}^n - G_{j+\frac{1}{2},k}^n), \quad (17.32)
\end{aligned}
$$

$$
w_{j,k}^{n+1} = w_{j,k}^n - \kappa_1(\tilde{F}_{j+\frac{1}{2},k} - \tilde{F}_{j-\frac{1}{2},k}) - \kappa_2(\tilde{G}_{j,k+\frac{1}{2}} - \tilde{G}_{j,k-\frac{1}{2}});
$$

$$
F_{j+1,k+\frac{1}{2}}^n = F\left(\frac{w_{j+1,k+1}^n + w_{j+1,k}^n}{2}\right), \qquad \text{etc.}
$$

The star of this scheme consists of nine points. Initially in the centres of the boxes auxiliary values $\tilde{w}_{j+\frac{1}{2},k+\frac{1}{2}}$ are calculated using the values $w_{j,k}^n$ in four grid points which are indicated by small open circles in Fig. 17.26. Then the calculated values are used for interpolating the values in the middle of the sides of the calculation boxes (indicated by crosses) and the final value $w_{j,k}^{n+1}$ is determined in the centre of the parallelogram.

The amplification matrix for the linearized version of the difference scheme has the form

$$
\begin{aligned}
G &= I - i\{\kappa_1 A_1[1 - \tfrac{1}{2}(1 - \cos\theta_2)]\sin\theta_1 + \kappa_2 A_2[1 - \tfrac{1}{2}(1 - \cos\theta_1)]\sin\theta_2\} \\
&\quad - \tfrac{1}{2}[\kappa_1^2 A_1^2(1 - \cos\theta_1)(1 + \cos\theta_2) + \kappa_2^2 A_2^2(1 - \cos\theta_2)(1 + \cos\theta_1) \\
&\quad + \kappa_1 \kappa_2(A_1 A_2 + A_2 A_1)\sin\theta_1 \cdot \sin\theta_2].
\end{aligned}
$$

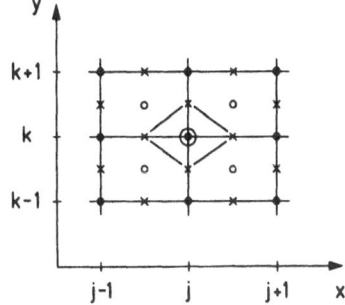

Fig. 17.26. Difference molecule of the modified
Lax-Wendroff-scheme (EGZLW-scheme)

The scheme (17.32) is stable for

$$\frac{\tau}{h}\sigma_0^n \leq 1,$$

$h = h_1 = h_2$, σ_0^n will be defined later.

The Π-form of the first differential approximation of (17.32) for $h_1 = h_2 = h$ has the form

$$w_t + F_x + G_y = \frac{\tau^2}{24}\{[F_{ww}(F_x + G_y)^2]_x + [G_{ww}(F_x + G_y)^2]_y\}$$

$$+ \frac{\tau^2}{6}\{A_1[A_1(F_x + G_y)]_x + A_1[A_2(F_x + G_y)]_y\}_x$$

$$+ \frac{\tau^2}{6}\{A_2[A_1(F_x + G_y)]_x + A_2[A_2(F_x + G_y)]_y\}_y$$

$$- \frac{h^2}{8}[A_1(w_{xx} + w_{yy})]_x - \frac{h^2}{8}[A_1 w_{yy}]_x - \frac{h^2}{24}F_{xxx}$$

$$- \frac{h^2}{8}[A_2(w_{xx} + w_{yy})]_y - \frac{h^2}{8}[A_2 w_{xx}]_y - \frac{h^2}{24}G_{yyy}.$$

17.10.8 The MacCormack-Scheme

The MacCormack-scheme can be written in the form

$$\tilde{w}_{j,k} = w_{j,k}^n - \kappa_1(F_{j+1,k}^n - F_{j,k}^n) - \kappa_2(G_{j,k+1}^n - \tilde{G}_{j,k}^n),$$

$$w_{j,k}^{n+1} = \tfrac{1}{2}(w_{j,k}^n + \tilde{w}_{j,k}^n) - \kappa_1(\tilde{F}_{j,k} - \tilde{F}_{j-1,k}) - \kappa_2(\tilde{G}_{j,k} - \tilde{G}_{j,k-1}).$$

It uses the grid points as shown in Fig. 17.27.

The amplification matrix of the linearized difference scheme has the form

$$G = I - i(\kappa_1 A_1 \sin\theta_1 + \kappa_2 A_2 \sin\theta_2) - \kappa_1^2 A_1^2(1 - \cos\theta_1)$$

$$- \kappa_2^2 A_2^2(1 - \cos\theta_2)$$

$$- 2\kappa_1\kappa_2 \sin\left(\frac{\theta_1}{2}\right)\sin\left(\frac{\theta_2}{2}\right)\left[\cos\left(\frac{\theta_1 - \theta_2}{2}\right)(A_1 A_2 + A_2 A_1)\right]$$

$$- i\sin\left(\frac{\theta_1 - \theta_2}{2}\right)(A_1 A_2 - A_2 A_1).$$

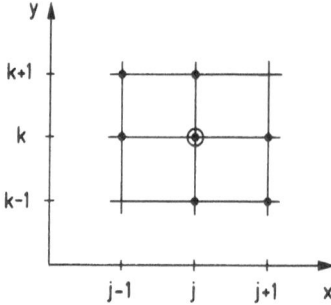

Fig. 17.27. Difference star of the MacCormack-scheme

In [187] it was noticed that for some matrices A_1 and A_2 the MacCormack-scheme is absolutely unstable. If, e.g., $\kappa_1 A_1 = -\kappa_2 A_2$ we get

$$|G(\theta_2 - \theta_1)|^2 = (I + \kappa_1 A_1 \sin \theta_1)^2 > I.$$

For calculations of gasdynamic problems according to the MacCormack-scheme usually the condition $\tilde{\sigma}\tau/h \leq 1$, $\tilde{\sigma} := |u| + |v| + c\sqrt{2}$, $h = h_1 = h_2$, is used. The Π-form of the first differential approximation of the considered scheme is written in the following form:

$$w_t + F_x + G_y = -\frac{\tau^2}{12}\{[F_{ww}(F_x + G_y)^2]_x + [G_{ww}(F_x + G_y)^2]_y\}$$

$$-\frac{\tau^2}{6}\{A_1[A_1(F_x + G_y)]_x + A_1[A_2(F_x + G_y)]_y\}_x$$

$$-\frac{h^2}{6}F_{xxx} - \frac{\tau^2}{6}\{A_2[A_1(F_x + G_y)]_x + A_2[A_2(F_x + G_y)]_y\}_y$$

$$-\frac{h^2}{6}G_{yyy} - \frac{\tau h}{4}\{[A_{1x}(F_x + G_y)]_x + [A_{2y}(F_x + G_y)]_y\};$$

$$h = h_1 = h_2.$$

All schemes described above which are of second order of approximation are not invariant (they do not admit a Galilei-transformation and a rotation).

17.10.9 Comparison of Numerical Results

In the different points above we have discussed a series of wellknown schemes and some of their characteristics. We discuss now the behavior of these schemes for the calculation of a convergent shock wave.

The stated problem is essentially one-dimensional but intentionally we considered this problem as a two-dimensional one in order to show the numerical effects of the difference schemes for two-dimensional problems of gas dynamics (especially the existence of non-symmetry in different directions). The problem is defined as follows:

A diaphragm separates two flow regions: The outer one (1) and the inner one (2), the values of the gasdynamic quantities in region 1 are connected with those of region 2 by the Hugoniot-conditions. For $t = 0$ the diaphragm is suddenly broken and a shock wave is formed which travels to the centre.

As region of computation a square of side-length 2 was chosen. The grid was chosen in two different ways: quadratic ($J = 19$, $K = 19$) and rectangular ($J = 24$, $K = 14$). Because of the symmetry of the flow the calculation was carried out only in a quarter of the whole plane. In Fig. 17.28 the initial position of the shock wave is shown. Along the line AD and AB the boundary

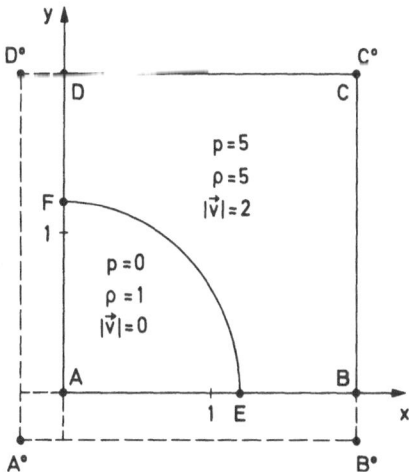

Fig. 17.28. Geometry and initial conditions for a collapsing cylindrical shock wave

conditions are calculated using the original scheme by adding to the grid points of the calculation region ficticious lines A^0D^0 and A^0B^0. The values of the gasdynamic quantities in grid points on the lines A^0D^0 and A^0B^0 are defined from the symmetry conditions of the flow with respect to the lines AD and AB. Along the boundaries CD and CB boundary conditions are given by extrapolations.

The boundary of the shock wave EF is approximated by a polygon. The initial conditions are given in the following way: Let $R = 1.2$ be the radius of the arc S; j, k are the indices of a grid point with respect to the x- and y-axis, $h_1 = 2/J$, $h_2 = 2/K$. If

$$\sqrt{j^2 h_1^2 + k^2 h_2^2} > R,$$

then

$$p(jh_1, k h_2, 0) = 5; \quad \varrho(j h_1, k h_2, 0) = 5; \quad q(j h_1, k h_2, 0) = 2.$$

If

$$\sqrt{j^2 h_1^2 + k^2 h_2^2} \leqq R,$$

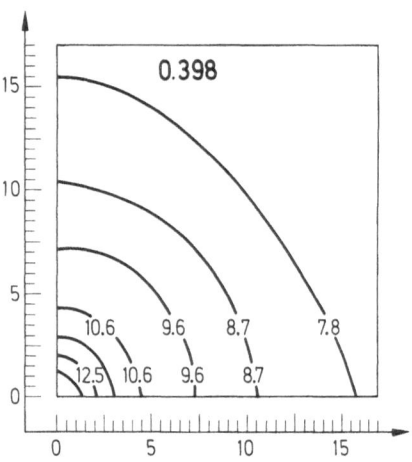

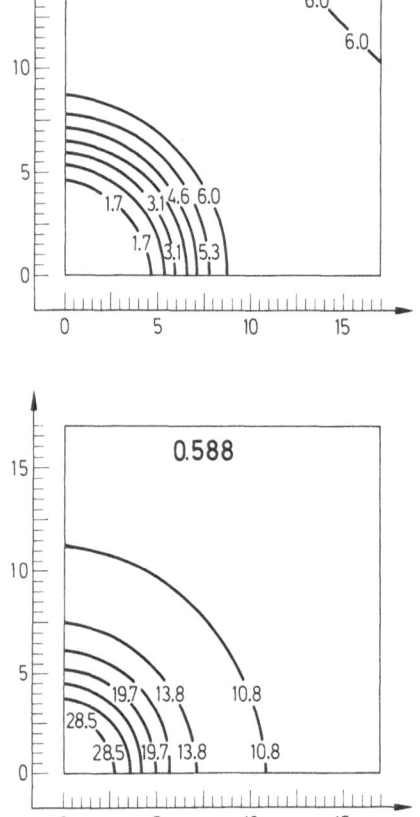

Fig. 17.29. Lines of constant density for different time steps (invariant, monotone difference scheme of class J_2, $\tilde{w}_{j,k}$ are calculated by EGZLW-scheme)

then

$$p(jh_1, kh_2, 0) = 0; \quad \varrho(jh_1, kh_2, 0) = 1; \quad q(jh_1, kh_2, 0) = 0.$$
$$q := \sqrt{u^2 + v^2}; \quad \varepsilon := p/\varrho(\gamma - 1); \quad \gamma = 1.5.$$

According to the invariance of the gasdynamic problem with respect to the transformation of rotation, the level-lines of the two-dimensional convergent (and outgoing) shock wave are represented by circles. Nevertheless the points where the grid functions have constant values and which represent the shock wave given by the difference calculation–because of the noninvariance–are situated on curves which can strongly deviate from a circle if a rectangular grid is introduced.

Now we discuss the numerical results: In the Figs. 17.29–41 lines of constant density ϱ are given which are calculated by the schemes mentioned above in the moments of time $t_k (k = 1, 2, 3)$, whereas t_k is chosen according to the

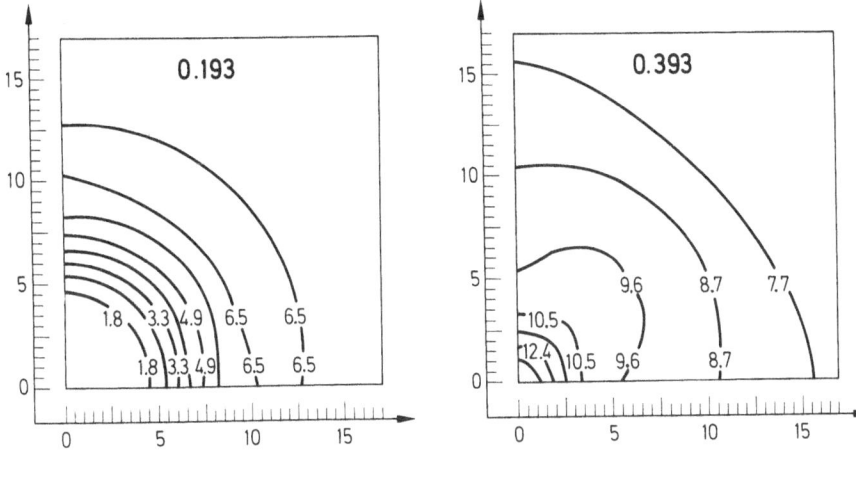

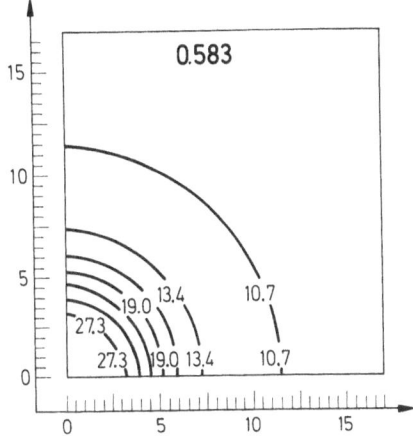

Fig. 17.30. Lines of constant density for different time steps (invariant, monotone difference scheme of class J_2, $\tilde{w}_{j,k}$ are calculated by Lax-Wendroff-scheme)

conditions

$$- \tau^n + k\,t^* \leq t_k \leq \tau^n + k\,t^*, \quad t^* = 0.2.$$

For each time step τ^n is chosen according to the stability condition of the scheme. In each figure lines of constant density are represented which correspond to seven values of ϱ:

$$\varrho_1 = \min_{j,k} \{\varrho_{j,k}\},$$

$$\varrho_7 = \max_{j,k} \{\varrho_{j,k}\},$$

$$\varrho_i = \varrho_1 + \tfrac{1}{6}(\varrho_7 - \varrho_1)(i-1); \quad i = 2,3,4,5,6.$$

Figure 17.29: In this figure lines of constant density are shown which are calculated using an invariant difference scheme from class J_2, whereas for the

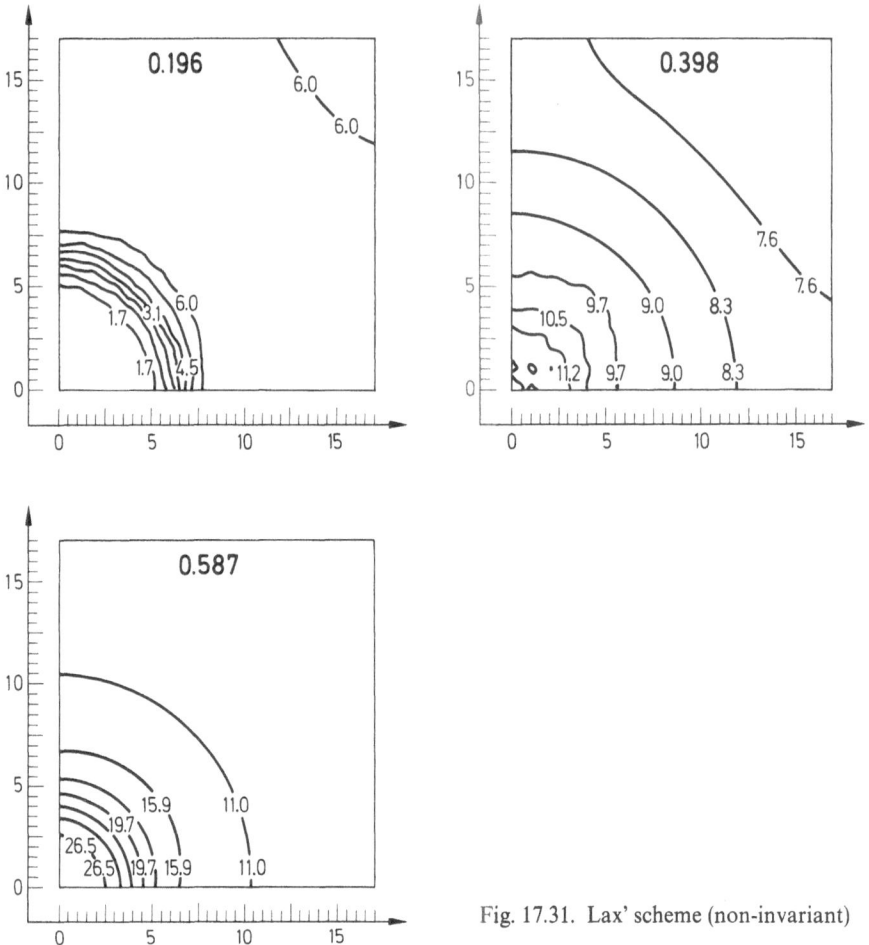

Fig. 17.31. Lax' scheme (non-invariant)

calculation of the values $\tilde{w}_{j,k}$ the EGZLW-scheme was used. The matrices of the artificial viscosity were chosen in the following manner:

$$C_{12} = C_{21} = 0, \qquad C_{11} = C_{22} = \mu I,$$

$$\mu = \frac{2h^2}{\tau^2}\, w_0\, \frac{(u_x + v_y)^2}{w_1 + (u_x + v_y)^2},$$

$$w_0 = \text{const} = 0.2, \qquad w_1 = 0.1, \qquad h = h_1 = h_2.$$

As can be seen from the graphs, the scheme is monotonic and the lines of constant density are represented by circles with sufficient accuracy.

Figure 17.30: Contrary to the foregoing case the values $\tilde{w}_{j,k}$ are calculated on the basis of the Lax-Wendroff-scheme (which was chosen as that variant where on the right hand side the vector $(A^2 w_x)_x$ and not the vector $(A F_x)_x$ is approximated, as usually. Here the isolines of the density, especially in the

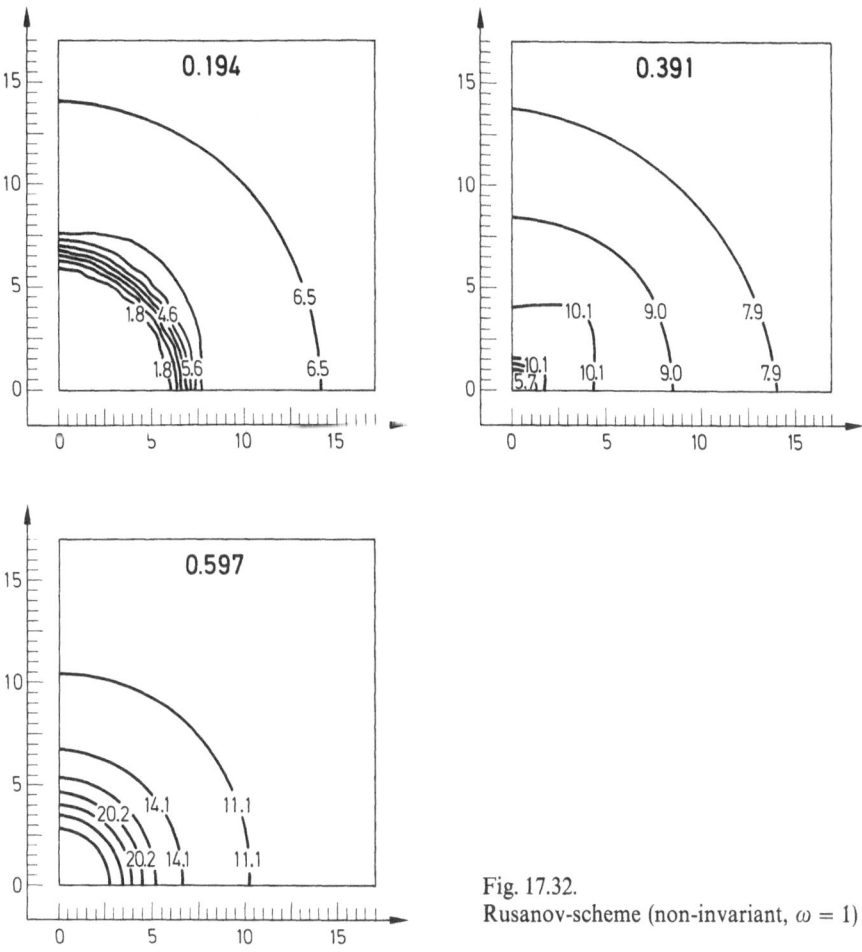

Fig. 17.32.
Rusanov-scheme (non-invariant, $\omega = 1$)

moment when the shock wave meets the centre ($t \approx t_2$), deviate considerably from a circle. In the following modifications of this scheme will be proposed. The matrices of the artificial viscosity in this case are chosen as in the former case.

Figures 17.31 and 17.32: Here the solutions are shown which are gained by the Lax' and the Rusanov-scheme, respectively. For the Rusanov-scheme $\omega = 1$. The time step τ^n in both the cases is chosen according to the stability conditions.

Figures 17.33 and 17.34: Contrary to the cases which precede and succeed the cases here a rectangular grid is chosen instead of a quadratic one, where along the y-axis 14 grid points were chosen. The calculations were carried out on the basis of a scheme from class J_2 which was already used to derive the isolines of Fig. 17.29, and using the Rusanov-scheme. The Rusanov-scheme gives closed isolines.

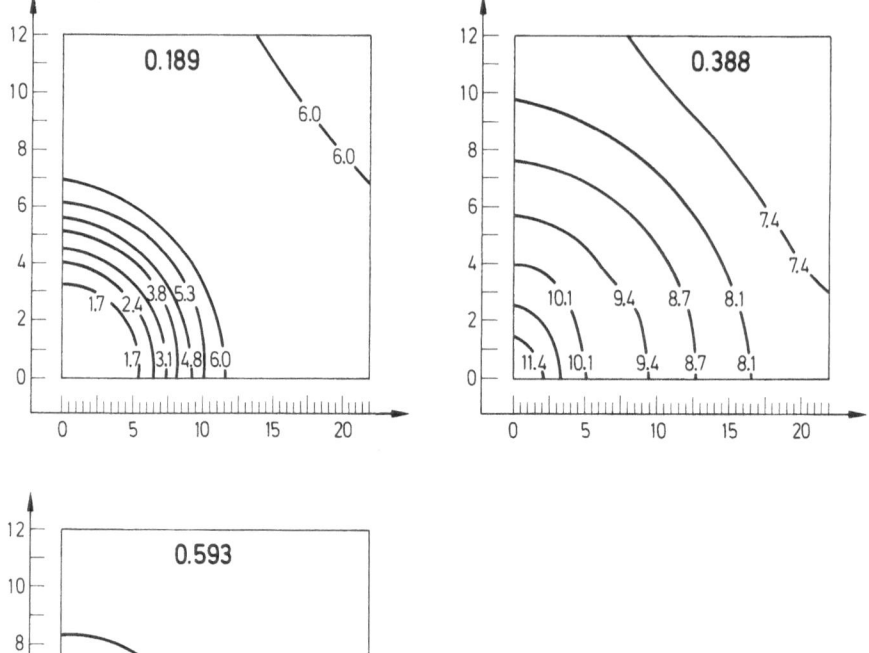

Fig. 17.33. Invariant scheme of class J_2 (rectangular grid)

Figures 17.35–41: In these figures results are represented which were derived with second-order difference schemes. The following schemes were tested (all of them were described above):

Lax-Wendroff-scheme (Figs. 17.35, 40, 41),
MacCormack-scheme (Fig. 17.36),
EGZLW-scheme (Fig. 17.37),
Two-step Richtmyer-scheme (Fig. 17.38),
Splitting-scheme, where as a one-dimensional operator the MacCormack-scheme was chosen (Fig. 17.39).

All schemes considered are noninvariant with respect to the transformation of rotation and are nonmonotone.

The calculations performed make clear how careful one has to be when a difference scheme for the calculation of two-dimensional problems must be chosen. The effects solely due to the difference approximation are so large for

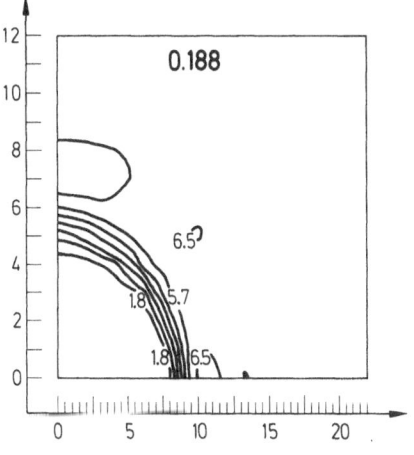

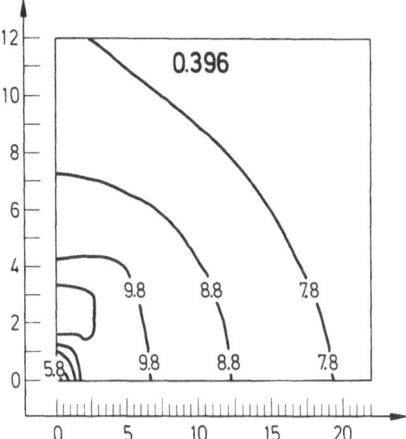

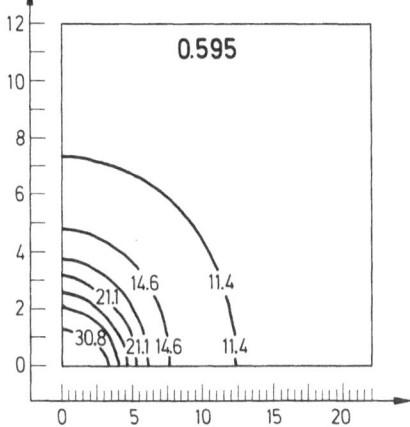

Fig. 17.34. Non-invariant Rusanov-scheme (rectangular grid)

some schemes that they lead to wrong qualitative pictures, not to mention the quantitative character.

We describe the first differential approximation of some schemes considered above which have second order of approximation for the continuity equation, where we will not replace the derivative by spatial derivatives. For the EGZLW-scheme we get:

$$\varrho_t + (\varrho\, u)_x + (\varrho\, v)_y + \frac{\tau^2}{6}\varrho_{ttt}$$

$$= -\frac{h^2}{6}[(\varrho\, u)_{xxx} + (\varrho\, v)_{yyy}] - \frac{h^2}{4}[(\varrho\, u)_{yyx} + (\varrho\, v)_{xxy}].$$

The MacCormack-scheme leads to the equation

$$\varrho_t + (\varrho\, u)_x + (\varrho\, v)_y + \frac{\tau^2}{6}\varrho_{ttt} = -\frac{h^2}{6}[(\varrho\, u)_{xxx} + (\varrho\, v)_{yyy}].$$

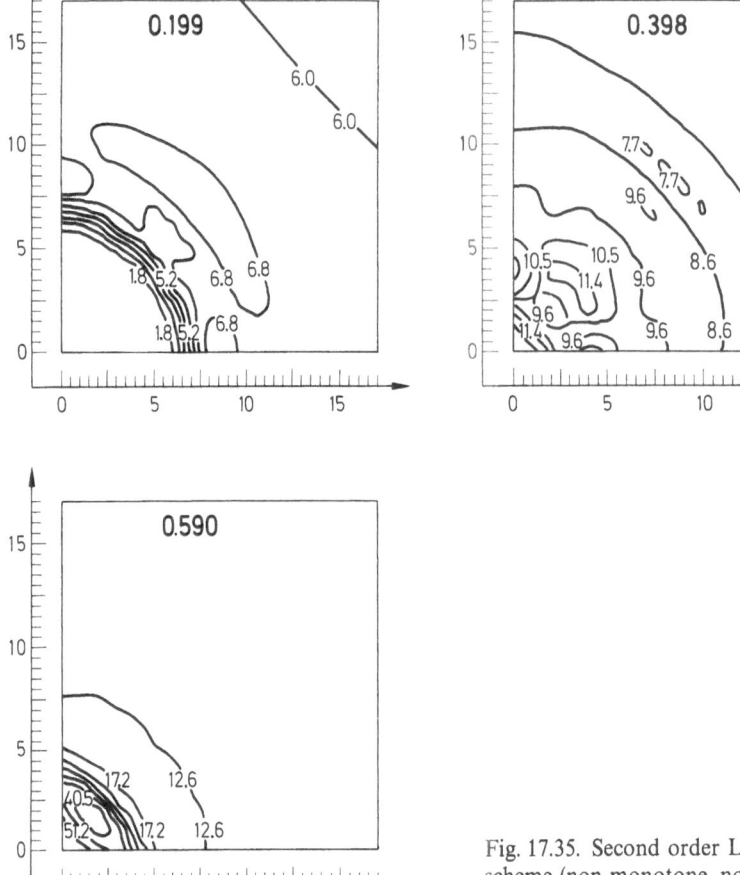

Fig. 17.35. Second order Lax-Wendroff-scheme (non-monotone, non-invariant)

For the Lax-Wendroff-scheme we get the same as for the MacCormack-scheme.

For the two-step Richtmyer method the first differential approximation reads

$$\varrho_t + (\varrho\, u)_x + (\varrho\, v)_y + \frac{\tau^2}{6}\, \varrho_{ttt} = -\frac{5\, h^2}{12}\, [(\varrho\, u)_{xxx} + (\varrho\, v)_{yyy}]$$

$$-\frac{h^2}{4}\, [(\varrho\, u)_{yyx} + (\varrho\, v)_{xxy}].$$

It is easy to see that the viscous terms on the right-hand side of the equations above are not invariant with respect to a rotation in contrast to the left sides which admit this type of transformation.

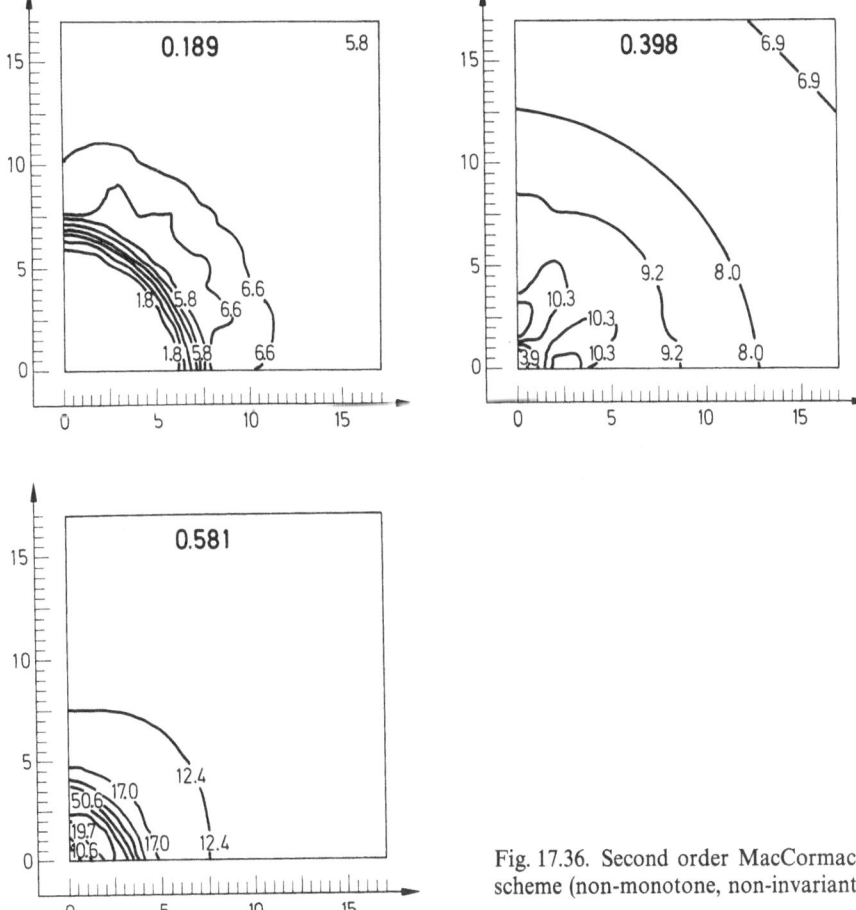

Fig. 17.36. Second order MacCormack-scheme (non-monotone, non-invariant)

We modify the first differential approximation for the continuity equation in the following way:

$$\varrho_t + (\varrho u)_x + (\varrho v)_y + \frac{\tau^2}{6}\varrho_{ttt} = -\frac{h^2}{6}[(\varrho u)_{xxx} + (\varrho v)_{yyy} + (\varrho u)_{yyx} + (\varrho v)_{xxy}].$$

(17.33)

The described first differential approximations of the EGZLW-scheme and the two-step Richtmyer method are similar to this equation. In the first differential approximations of the other schemes the terms with mixed derivatives generally vanish. If one adds to the right-hand side of the Lax-Wendroff-scheme the finite difference approximation of the term

$$-\frac{h^2}{6}(F_{yyx} + G_{xxy}),$$

(17.34)

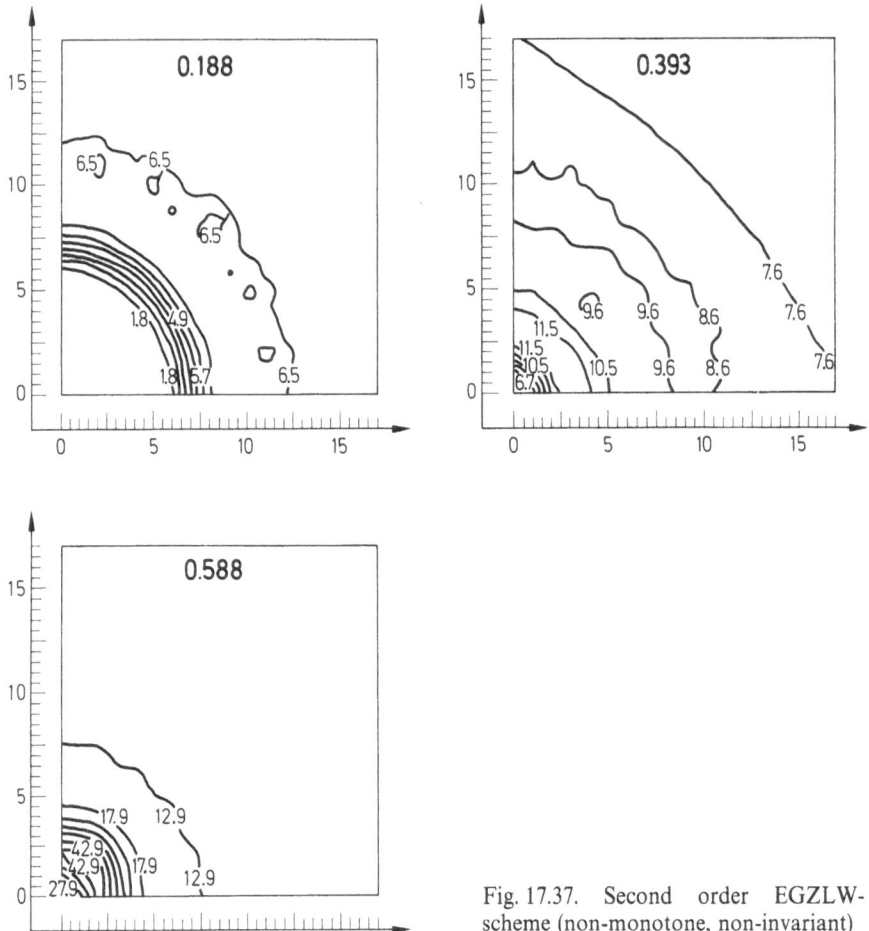

Fig. 17.37. Second order EGZLW-scheme (non-monotone, non-invariant)

the derived scheme will have the equation (17.35) as the first differential approximation for the continuity equation. On the basis of the modified scheme two invariant difference schemes were constructed (they differ by introducing different finite difference analogies of the expression (17.36) into the original scheme).

In the Figs. 17.40 and 17.41 isolines of the density are plotted which were calculated by these schemes. In comparison with those cases which were shown in Fig. 17.30, the isolines are more similar to a circle.

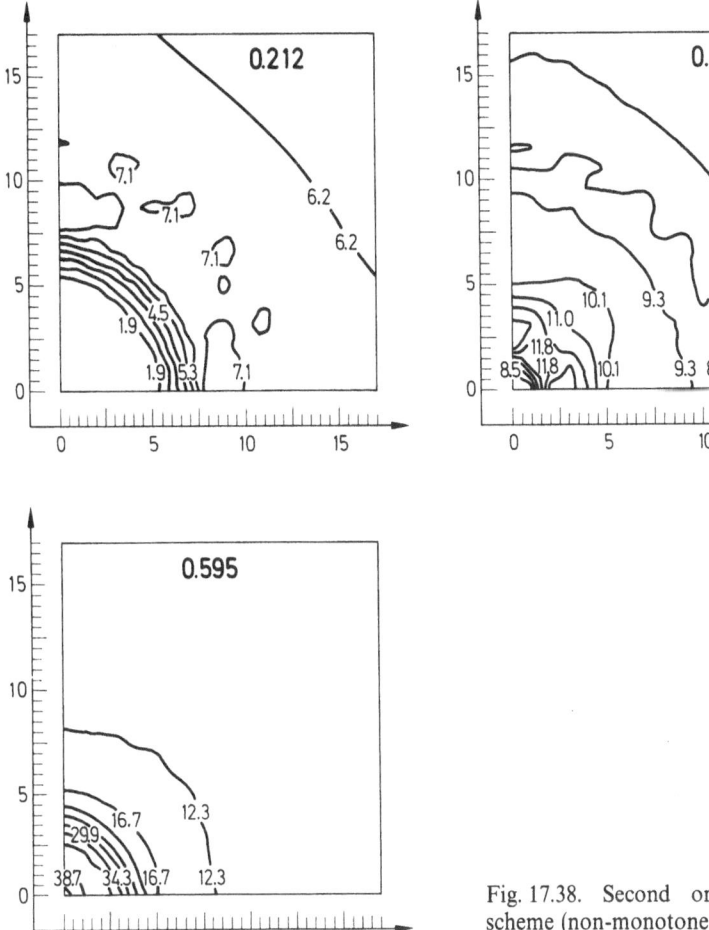

Fig. 17.38. Second order Richtmyer-scheme (non-monotone, non-invariant)

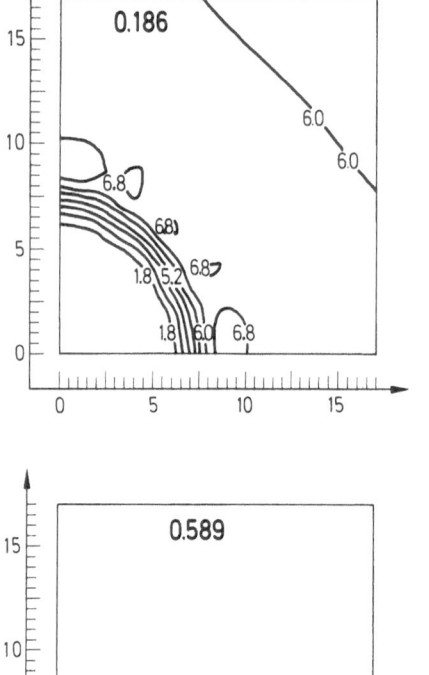

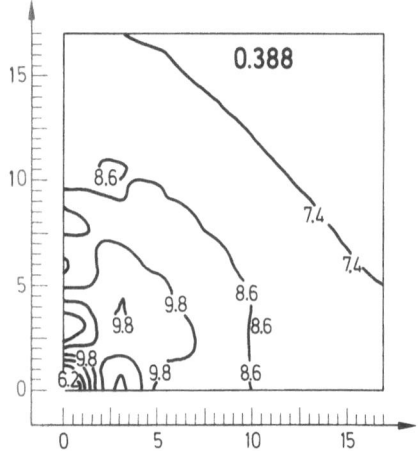

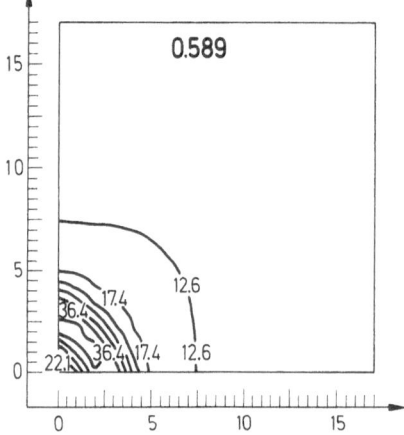

Fig. 17.39. Second order MacCormack-scheme (splitting method), (non-monotone, non-invariant)

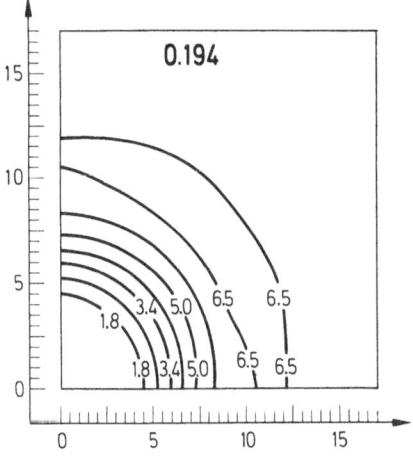

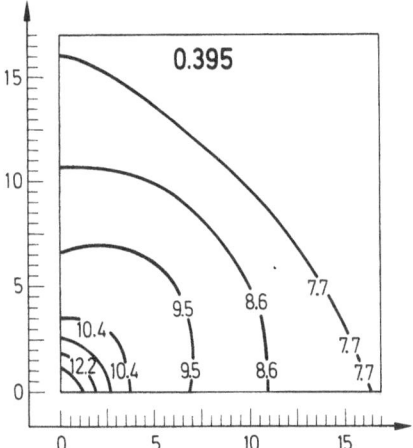

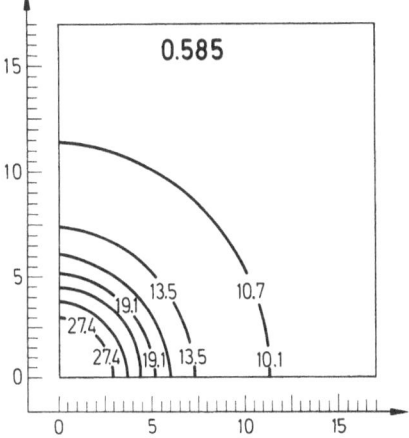

Fig. 17.40. Second order modified Lax-Wendroff-scheme (invariant), special choice of difference approximation of (17.36)

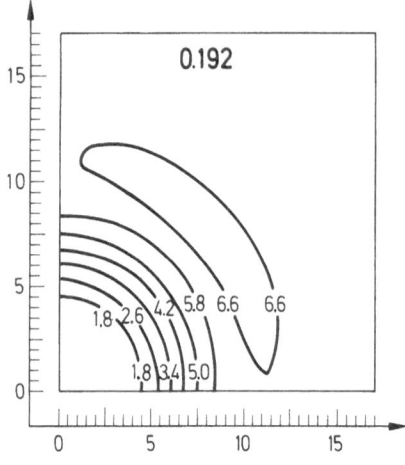

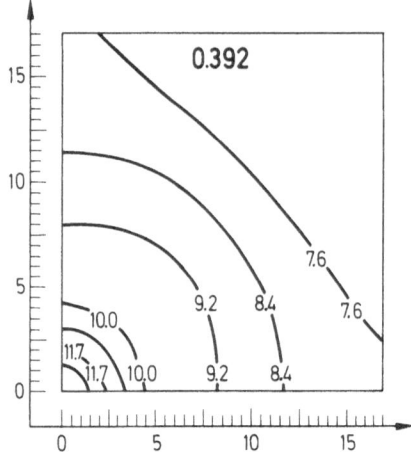

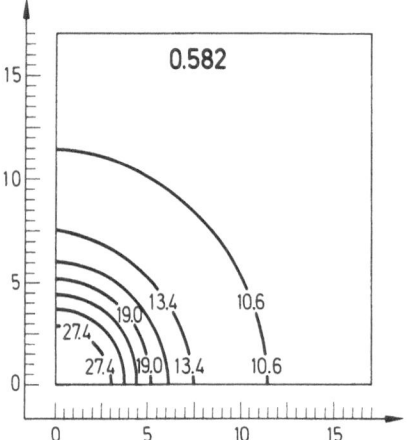

Fig. 17.41. Second order modified Lax-Wendroff-scheme (invariant), difference approximation of (17.36) different to Fig. 17.40

18. Investigation of Difference Schemes with Time-splitting Using the Theory of Groups

18.1 Two First-order Schemes with Time-splitting

For the system of differential equations of gas dynamics in Eulerian coordinates (13.1) we consider two classes of first-order difference schemes:

$$\left.\begin{aligned}
\frac{w^{n+\frac{1}{2}}(x,y) - w^n(x,y)}{\tau} + \frac{f^n(x+h_1,y) - f^n(x-h_1,y)}{2h_1} &= \Lambda^1 w^n(x,y), \\
\frac{w^{n+1}(x,y) - w^{n+\frac{1}{2}}(x,y)}{\tau} + \frac{g^{n+\frac{1}{2}}(x,y+h_2) - g^{n+\frac{1}{2}}(x,y-h_2)}{2h_2} & \\
= \Lambda^2 w^{n+\frac{1}{2}}(x,y);
\end{aligned}\right\} \tag{18.1}$$

$$\left.\begin{aligned}
\frac{w^{n+\frac{1}{2}}(x,y) - w^n(x,y)}{\tau} + \frac{f^{(1)^n}(x+h_1,y) - f^{(1)^n}(x-h_1,y)}{2h_1} & \\
+ \frac{g^{(1)^n}(x,y+h_2) - g^{(1)^n}(x,y-h_2)}{2h_2} = L^1 w^n(x,y), \\
\frac{w^{n+1}(x,y) - w^{n+\frac{1}{2}}(x,y)}{\tau} + \frac{f^{(2)^{n+\frac{1}{2}}}(x+h_1,y) - f^{(2)^{n+\frac{1}{2}}}(x-h_1,y)}{2h_1} & \\
+ \frac{g^{(2)^{n+\frac{1}{2}}}(x,y+h_2) - g^{(2)^{n+\frac{1}{2}}}(x,y-h_2)}{2h_2} = L^2 w^{n+\frac{1}{2}}(x,y).
\end{aligned}\right\} \tag{18.2}$$

In the following we will call these schemes for simplicity scheme (18.1) and scheme (18.2). The following abbreviations are used:

$$\Lambda^k := M^{k(2)}; \quad L^k := M^{k(3)};$$

$$M^{k(l+1)} := \frac{\tau}{h_1}(T_{x/2} - T_{-x/2}) \cdot \left[\frac{1}{h_1}\Lambda_{11}^{k(l+1)}(T_{x/2} - T_{-x/2}) + \frac{1}{h_2}\Lambda_{12}^{k(l+1)}(T_{y/2} - T_{-y/2})\right]$$

$$+ \frac{\tau}{h_2}(T_{y/2} - T_{-y/2}) \cdot \left[\frac{1}{h_1}\Lambda_{21}^{k(l+1)}(T_{x/2} - T_{-x/2}) + \frac{1}{h_2}\Lambda_{22}^{k(l+1)}(T_{y/2} - T_{-y/2})\right];$$

τ, h_1, h_2 are the step sizes of the grid along the t-, x-, and y-axis, respectively. $\Lambda_{rs}^{k(l+1)}$ are (4×4)-matrices, the elements of which may depend on t, x, y, w, w_x, $w_y, w_{xx}, w_{xy}, w_{yx}, w_{yy}$.

$$T_{\pm x/2}\,\phi\,(x, y) := \phi\left(x \pm \frac{h_1}{2}, y\right); \quad T_{\pm y/2}\,\phi\,(x, y) := \phi\left(x, y \pm \frac{h_2}{2}\right);$$

$$f^{n+\frac{1}{2}}(x \pm h_1, y \pm h_2) = f(w^{n+\frac{1}{2}}(x \pm h_1, y \pm h_2));$$

$$g^{n+\frac{1}{2}}(x \pm h_1, y \pm h_2) = g(w^{n+\frac{1}{2}}(x \pm h_1), y \pm h_2));$$

$$f^{(k)n+\frac{1}{2}}(x \pm h_1, y \pm h_2) = f^{(k)}(w^{n+\frac{1}{2}}(x \pm h_1, y \pm h_2));$$

$$g^{(k)n+\frac{1}{2}}(x \pm h_1, y \pm h_2) = g^{(k)}(w^{n+\frac{1}{2}}(x \pm h_1, y \pm h_2));$$

$$(r, s, k, l = 1, 2);$$

$$f^{(1)} := \begin{pmatrix} p \\ 0 \\ 0 \\ up \end{pmatrix}; \quad f^{(2)} := \begin{pmatrix} \varrho\,u^2 \\ \varrho\,u\,v \\ \varrho\,u \\ \varrho\,u\,E \end{pmatrix}; \quad g^{(1)} := \begin{pmatrix} 0 \\ p \\ 0 \\ v\,p \end{pmatrix}; \quad g^{(2)} := \begin{pmatrix} \varrho\,u\,v \\ \varrho\,v^2 \\ \varrho\,v \\ \varrho\,v\,E \end{pmatrix}.$$

It is assumed that the equation of state of the gas has the form

$$p = p\,(\varepsilon, \varrho).$$

We remark that the scheme (18.2) is an asymptotic representation of the particle-in-cell method [95]. We are led to this conclusion by an investigation of the given method. There exists a close connection between the particle-in-cell method and the method of splitting, and in fact the particle-in-cell method is a specific modification of the splitting method. The splitting of the algorithm into two steps in the particle-in-cell method is nothing else than the splitting of the scheme into two intermediate steps. The interpretation of the particle-in-cell method in terms of the splitting method makes easier the mathematical formalization and investigation of such a complicated algorithm as the particle-in-cell method (for a more detailed discussion see [95]).

18.2 Group Properties of the Schemes (18.1) and (18.2)

The first differential approximation of the schemes (18.1) and (18.2) has the form:

$$\frac{\partial w}{\partial t} + \frac{\partial f}{\partial x} + \frac{\partial g}{\partial y} = \frac{\partial}{\partial x}\left(C_{11}^k \frac{\partial w}{\partial x} + C_{12}^k \frac{\partial w}{\partial y}\right) + \frac{\partial}{\partial y}\left(C_{21}^k \frac{\partial w}{\partial x} + C_{22}^k \frac{\partial w}{\partial y}\right),$$

where for the scheme (18.1) $k = 1$ and for the scheme (18.2) $k = 2$;

$$C_{rs}^1 := \tau (\Lambda_{rs}^{1\,(2)} + \Lambda_{rs}^{2\,(2)}) - \frac{\tau}{2}(-1)_r^{\delta_s+1} A_r A_s;$$

$$C_{rs}^2 := \tau (\Lambda_{rs}^{1\,(3)} + \Lambda_{rs}^{2\,(3)}) + \tau (\delta_r^1 B_2 + \delta_r^2 B_4)(\delta_s^1 B_1 + \delta_s^2 B_3) - \frac{\tau}{2} A_r A_s;$$

$$A_1 = \frac{df}{dw}; \quad A_2 = \frac{dg}{dw}; \quad B_r = \frac{df^{(r)}}{dw}; \quad B_{r+2} = \frac{dg^{(r)}}{dw};$$

δ_r^s is the Kronecker-symbol $(r, s = 1, 2)$.

The following theorem holds.

. . . .

Theorem 18.1.

1) If the matrices $\Lambda_{rs}^{k\,(2)} = \| \lambda_{rs}^{k\,\xi\eta} \|_1^4$ are chosen such that

$$\frac{\partial \lambda_{rs}^{k\,\xi\eta}}{\partial x^\gamma} = 0 \quad (k, r, s = 1, 2; \quad \gamma = 1, 2, 3),$$

$$\frac{\partial N^1}{\partial u} = \begin{pmatrix} N_3 \\ 0 \\ 0 \\ N_1^1 \end{pmatrix}; \quad \frac{\partial N^1}{\partial v} = \begin{pmatrix} 0 \\ N_3^1 \\ 0 \\ N_2^1 \end{pmatrix}; \quad L_6^1 N^1 = \begin{pmatrix} N_2^1 \\ -N_1^1 \\ 0 \\ 0 \end{pmatrix};$$

$$\bar{L}_7 N^1 = 2 N^1,$$

then the difference scheme (18.1) admits the same group of transformations which is admitted by the two-dimensional system of the equations of gas dynamics (13.1).

2) If the matrices $\Lambda_{rs}^{k\,(3)} = \| \mu_{rs}^{k\,\xi\eta} \|_1^4$ are chosen such that

$$\frac{\partial (\mu_{rs}^{1\,\xi\eta} + \mu_{rs}^{2\,\xi\eta})}{\partial x^\gamma} = 0 \quad (r, s = 1, 2; \quad \gamma = 1, 2, 3);$$

$$\frac{\partial N^2}{\partial u} = \begin{pmatrix} N_3^2 \\ 0 \\ 0 \\ N_1^2 \end{pmatrix}; \quad \frac{\partial N^2}{\partial v} = \begin{pmatrix} 0 \\ N_3^2 \\ 0 \\ N_2^2 \end{pmatrix}; \quad L_6^1 N^2 = \begin{pmatrix} N_2^2 \\ -N_1^2 \\ 0 \\ 0 \end{pmatrix};$$

$$\bar{L}_7 N^2 = 2 N^2, \tag{18.3}$$

then the difference scheme (18.2) is invariant with respect to the group of transformations which is admitted by the system of equations (13.1).

Here we define:

$$x^1 = t; \quad x^2 = x; \quad x^3 = y;$$

$$L_6^1 = v \frac{\partial}{\partial u} - u \frac{\partial}{\partial v} + \bar{L}_6 ;$$

$$N^k = \begin{pmatrix} N_1^k \\ N_2^k \\ N_3^k \\ N_4^k \end{pmatrix} = \frac{\partial}{\partial x}\left(C_{11}^k \frac{\partial w}{\partial x} + C_{12}^k \frac{\partial w}{\partial y} \right) + \frac{\partial}{\partial y}\left(C_{21}^k \frac{\partial w}{\partial x} + C_{22}^k \frac{\partial w}{\partial y} \right) ;$$

The operators $\bar{L}_6$ and $\bar{L}_7$ are defined in Sect. 17.2.

. . . .

From the Theorem 17.1 both the statements can be derived.

18.3 Conditions for Invariance for a Polytropic Gas

In the case of a specific equation of state the conditions for the invariance can be simplified. Thus for a polytropic gas, for example, the conditions (18.3) for invariance of the scheme (18.2) read

$$\frac{\partial R_1}{\partial u} - R_3 = \tau \left[\frac{\partial^2 (u^2 \varrho)}{\partial x^2} + \frac{\partial^2 (u v \varrho)}{\partial x \partial y} \right],$$

$$\frac{\partial R_2}{\partial u} = \tau \left[\frac{\partial^2 (u v \varrho)}{\partial x^2} + \frac{\partial^2 (v^2 \varrho)}{\partial x \partial y} \right],$$

$$\frac{\partial R_3}{\partial u} = \tau \left[\frac{\partial^2 (u \varrho)}{\partial x^2} + \frac{\partial^2 (v \varrho)}{\partial x \partial y} \right],$$

$$\frac{\partial R_4}{\partial u} - R_1 = \frac{\tau}{2} \left[\frac{\partial^2 U}{\partial x^2} + \frac{\partial^2 V}{\partial x \partial y} \right],$$

$$\frac{\partial R_1}{\partial v} = \tau \left[\frac{\partial^2 (u^2 \varrho)}{\partial x \partial y} + \frac{\partial^2 (u v \varrho)}{\partial y^2} \right],$$

$$\frac{\partial R_2}{\partial v} - R_3 = \tau \left[\frac{\partial^2 (u v \varrho)}{\partial x \partial y} + \frac{\partial^2 (v^2 \varrho)}{\partial y^2} \right],$$

$$\frac{\partial R_3}{\partial v} = \tau \left[\frac{\partial^2 (u \varrho)}{\partial x \partial y} + \frac{\partial^2 (v \varrho)}{\partial y^2} \right],$$

$$\frac{\partial R_4}{\partial v} - R_2 = \tau \left[\frac{\partial^2 U}{\partial x \partial y} + \frac{\partial^2 V}{\partial y^2} \right] ;$$

$$L_6^1 R_1 = R_2, \quad L_6^1 R_2 = - R_1, \quad L_6^1 R_3 = 0, \quad L_6^1 R_4 = 0 ;$$

$$\bar{L}_7 R = 2 R - S ,$$

where

$$U := \frac{2}{\gamma - 1} u p + u^3 p + u v^2 \varrho; \quad V := \frac{2}{\gamma - 1} v p + v^3 p + u^2 v \varrho;$$

$$R = \begin{pmatrix} R_1 \\ R_2 \\ R_3 \\ R_4 \end{pmatrix} = \tau \frac{\partial}{\partial x} \left[(\Lambda_{11}^{1\,(3)} + \Lambda_{11}^{2\,(3)}) \frac{\partial w}{\partial x} + (\Lambda_{12}^{1\,(3)} + \Lambda_{12}^{2\,(3)}) \frac{\partial w}{\partial y} \right]$$
$$+ \tau \frac{\partial}{\partial y} \left[(\Lambda_{21}^{1\,(3)} + \Lambda_{21}^{2\,(3)}) \frac{\partial w}{\partial x} + (\Lambda_{22}^{1\,(3)} + \Lambda_{22}^{2\,(3)}) \frac{\partial w}{\partial y} \right];$$

$$S = \tau \frac{\partial}{\partial x} \left[2 B_2 \left(\frac{\partial f^{(1)}}{\partial x} + \frac{\partial g^{(1)}}{\partial y} \right) - A_1 \left(\frac{\partial f}{\partial x} + \frac{\partial g}{\partial y} \right) \right]$$
$$+ \tau \frac{\partial}{\partial y} \left[2 B_4 \left(\frac{\partial f^{(1)}}{\partial x} + \frac{\partial g^{(1)}}{\partial y} \right) - A_2 \left(\frac{\partial f}{\partial x} + \frac{\partial g}{\partial y} \right) \right].$$

In an analogous way the conditions for invariance of difference schemes with time splitting and with second order of approximation can be derived.

18.4 Comparison of Invariant and Noninvariant Schemes

The typical properties of invariant and noninvariant difference schemes with splitting were compared on the basis of calculations concerning the following model equation [83]:

$$\frac{\partial u}{\partial t} = \alpha y \frac{\partial u}{\partial x} - \alpha x \frac{\partial u}{\partial y}, \tag{18.4}$$

which admits a rotation with the infinitesimal operator $L := y \frac{\partial}{\partial x} - x \frac{\partial}{\partial y}$.

For the equation (18.4) Cauchy's problem was considered using the following initial conditions:

$$u(0, x, y) = \begin{cases} 1 - \dfrac{1}{u_0} z, & z^2 = (x - a)^2 + (y - b)^2 \leq u_0^2 \\ 0, & z^2 > u_0^2. \end{cases} \tag{18.5}$$

The problem (18.4), (18.5) describes the rotation of a circular cone around the origin of the coordinate system with a height of 1 and a radius of its basis equal to u_0. The period of the rotation is chosen to be $2\pi/\alpha$. For $t = 0$ the centre of the basis is situated in the point (a, b).

The exact solution of the problem (18,4, 5) has the following form:

$$u(t, x, y) = \begin{cases} 1 - \dfrac{1}{u_0} r, & r^2 \leq u_0^2, \\ 0, & r^2 > u_0^2, \end{cases}$$

where

$$r^2 := [x \cos(\alpha t) + y \sin(\alpha t) - a]^2 + [-x \sin(\alpha t) + y \cos(\alpha t) - b]^2.$$

The calculations were carried out using the following schemes:

Lax-Wendroff-scheme using the modification proposed by Richtmyer [7]:

$$u^{n+\frac{1}{2}}(x, y) = [\tfrac{1}{2}(\mu_x + \mu_y) + \tfrac{1}{2}\kappa_1 \alpha y \delta_x - \tfrac{1}{2}\kappa_2 \alpha x \delta_y]u^n(x, y),$$

$$u^{n+1}(x, y) = u^n(x, y) + [\kappa_1 \alpha y \delta_x - \kappa_2 \alpha x \delta_y]u^{n+\frac{1}{2}}(x, y). \tag{18.6}$$

MacCormack-scheme [207]:

$$\left.\begin{aligned}
u^{n+\frac{1}{4}}(x, y) &= [I + \kappa_1 \alpha y \Delta_{-x}]u^n(x, y), \\
u^{n+\frac{1}{2}}(x, y) &= \tfrac{1}{2}[u^n(x, y) + u^{n+\frac{1}{4}}(x, y)] + \tfrac{1}{2}\kappa_1 \alpha y \Delta_x u^{n+\frac{1}{4}}(x, y), \\
u^{n+\frac{3}{4}}(x, y) &= [I - \kappa_2 \alpha x \Delta_{-y}]u^{n+\frac{1}{2}}(x, y), \\
u^{n+1}(x, y) &= \tfrac{1}{2}[u^{n+\frac{1}{2}}(x, y) + u^{n+\frac{3}{4}}(x, y)] - \tfrac{1}{2}\kappa_2 \alpha x \Delta_y u^{n+\frac{3}{4}}(x, y).
\end{aligned}\right\} \tag{18.7}$$

Invariant scheme:

$$\left.\begin{aligned}
u^{n+\frac{1}{5}}(x, y) &= [I - \tfrac{1}{6}\kappa_1 \alpha y(1 - \kappa_1^2 \alpha^2 y^2)\delta_x \Delta_x \Delta_{-x} - \alpha^2 y^2 \kappa_1^2 \Delta_x^2 \Delta_{-x}^2] \\
&\quad \cdot [I + \kappa_1 \alpha y \delta_x + (\tfrac{1}{2} - \kappa_1^2 \alpha^2 y^2 \Delta_x \Delta_{-x})\kappa_1^2 y^2 \alpha^2 \Delta_x \Delta_{-x}] \\
&\quad \cdot u^n(x, y), \\
u^{n+\frac{2}{5}}(x, y) &= [I - \tfrac{1}{6}\kappa_1 \alpha^3 y \tau^2 \delta_x + \kappa_1^2 \alpha^6 y^2 \tau^4 \Delta_x \Delta_{-x}] \\
&\quad \cdot [I - \tfrac{1}{2}\alpha^2 x \tau \kappa_1 \delta_x + \alpha^4 \tau^2 x^2 \kappa_1^2 \Delta_x \Delta_{-x}]u^{n+\frac{1}{5}}(x, y), \\
u^{n+\frac{3}{5}}(x, y) &= [I + \tfrac{1}{6}\kappa_2 \alpha x(1 - \kappa_2^2 \alpha^2 x^2)\delta_y \Delta_y \Delta_{-y} - \alpha^2 x^2 \kappa_2^2 \Delta_y^2 \Delta_{-y}^2] \\
&\quad \cdot [I - \alpha x \kappa_2 \delta_y \\
&\quad + (\tfrac{1}{2} - \kappa_2^2 \alpha^2 x^2 \Delta_y \Delta_{-y})\kappa_2^2 x^2 \alpha^2 \Delta_y \Delta_{-y}]u^{n+\frac{2}{5}}(x, y), \\
u^{n+\frac{4}{5}}(x, y) &= [I + \tfrac{1}{6}\kappa_2 \alpha^3 x \tau^2 \delta_y + \kappa_2^2 \alpha^6 x^2 \tau^4 \Delta_y \Delta_{-y}] \\
&\quad \cdot [I - \tfrac{1}{2}\alpha^2 y \tau \kappa_2 \delta_y + \alpha^4 \tau^2 y^2 \kappa_2^2 \Delta_y \Delta_{-y}]u^{n+\frac{3}{5}}(x, y), \\
&\quad [I - \tau^3 \phi + \tau \psi(\kappa_1^2 \Delta_x \Delta_{-x} + (\kappa_2^2 \Delta_y \Delta_{-y})]u^{n+\frac{4}{5}}(x, y),
\end{aligned}\right\} \tag{18.8}$$

Here the functions ϕ and ψ depend only on the argument $z = x^2 + y^2$,

$$\Delta_x := T_x - I; \quad \Delta_{-x} := I - T_{-x}; \quad \Delta_y := T_y - I; \quad \Delta_{-y} := I - T_{-y};$$

$$\mu_x := \tfrac{1}{2}(T_x + T_{-x}); \quad \mu_y := \tfrac{1}{2}(T_y + T_{-y});$$

$$\delta_x := \tfrac{1}{2}(\Delta_x + \Delta_{-x}); \quad \delta_y := \tfrac{1}{2}(\Delta_y + \Delta_{-y}).$$

I is the identity-operator, T_x the shift-operator along the x-, and T_y the shift-operator along the y-axis, $\kappa_1 = \tau/h_1$, $\kappa_2 = \tau/h_2$, h_1 and h_2 are the step sizes in x- and y-direction, respectively. The schemes (18.6) and (18.7) do not admit the transformation of a rotation.

The calculations which were carried out by A.I. Urusov have shown that the invariant schemes with splitting reflect the special character of the exact solution in a better way, qualitatively as well as quantitatively, compared with the noninvariant schemes. The noninvariant schemes lead to distortions. Especially, as it can be seen from the figures given below, the solutions gained by noninvariant schemes differ strongly from the exact solution.

The results of the calculations according to the schemes (18.6–8) are plotted in the Figs. 18.1–18. Results due to the Lax-Wendroff-scheme are shown in the Figs. 18.5, 6, 11, 12, 17, 18. The results according to the MacCormack-scheme are given in the Figs. 18.1, 2, 7, 8, 13, 14; results gained on the basis of the scheme (18.8) are plotted in the Fig. 18.3, 4, 9, 10, 15, 16. In the calculations the following values for the parameters were chosen:

$$ a = b = 2, \quad \alpha = \pi/2, \quad u_0 = 5, \quad \tau = 0.01, \quad \phi(z) = \psi(z) = z. $$

In all figures lines $u(t, x, y) = \text{const}$ were plotted (const $= 0.2, 0.4, 0.6, 0.8$) as from the exact solution (circles) as from the differences solution for $t = 1$ (Figs. 18.1–6), $t = 3$ (Figs. 18.7–12), and $t = 5$ (Figs. 18.13–18). In the pictures with odd numbers lines are plotted which are calculated by schemes on a quadratic grid $h_1 = h_2 = 1$, but in the figures with even numbers a rectangular grid was chosen $h_1 = 1$, $h_2 = 0.5$. Comparing the results of the calculations we see that an invariant scheme allows us to find the solution better qualitatively, and, consequently, using coarse grids (i.e. with sufficient large values of τ, h_1, h_2) it is useful to carry out the calculations applying invariant schemes.

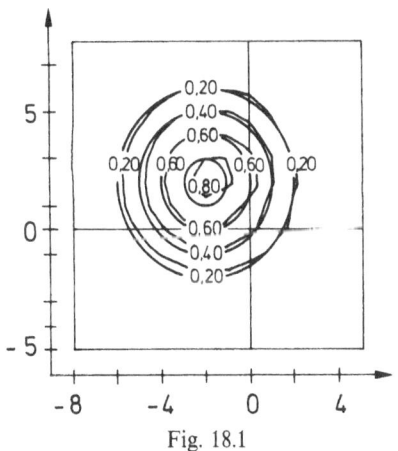
Fig. 18.1

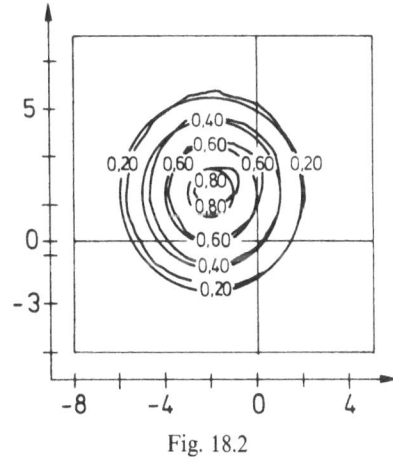
Fig. 18.2

Fig. 18.1. MacCormack-scheme ($\alpha = 1.5708$, $m = 1$, $h_x = 1$, $h_y = 1$, $\tau = 0.01$), $t = 1$

Fig. 18.2. MacCormack-scheme ($\alpha = 1.5708$, $m = 1$, $h_x = 1$, $h_y = 0.5$, $\tau = 0.01$), $t = 1$

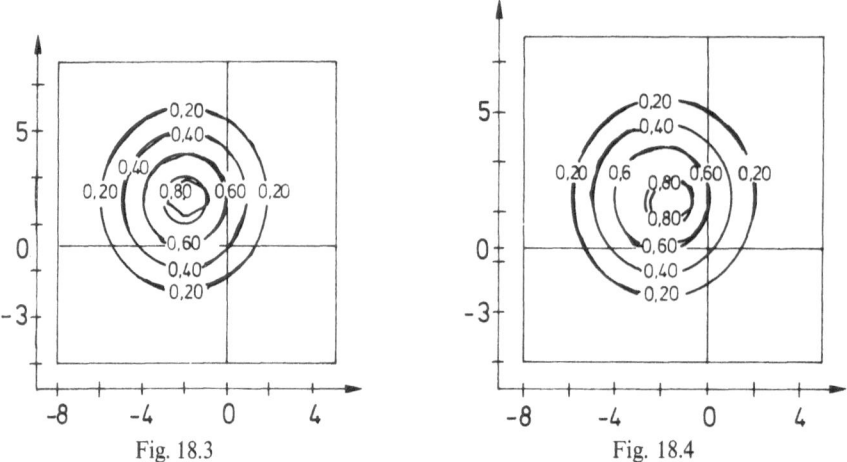

Fig. 18.3   Fig. 18.4

Fig. 18.3. Invariant scheme (18.8); ($\alpha = 1.5708$, $m = 1$, $h_x = 1$, $h_y = 1$, $\tau = 0.01$), $t = 1$

Fig. 18.4. Invariant scheme (18.8); ($\alpha = 1.5708$, $m = 1$, $h_x = 1$, $h_y = 0.5$, $\tau = 0.01$) $t = 1$

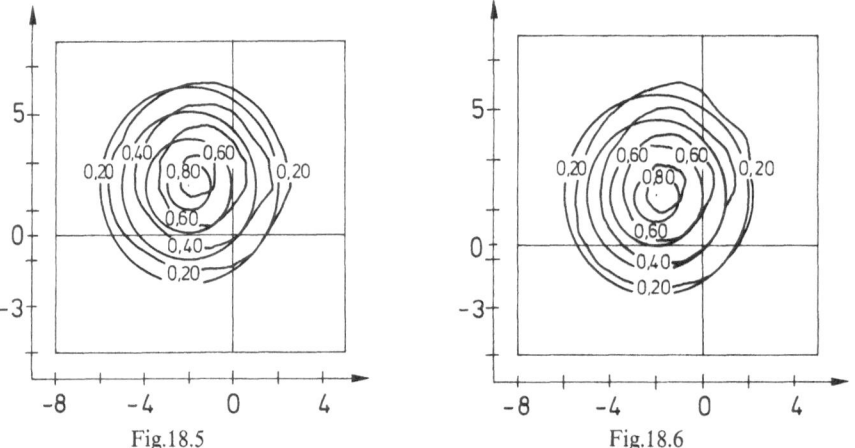

Fig.18.5   Fig.18.6

Fig. 18.5. Lax-Wendroff-scheme ($\alpha = 1.5708$, $m = 1$, $h_x = 1$, $h_y = 1$, $\tau = 0.01$), $t = 1$

Fig. 18.6. Lax-Wendroff-scheme ($\alpha = 1.5708$, $m = 1$, $h_x = 1$, $h_y = 0.5$, $\tau = 0.01$), $t = 1$

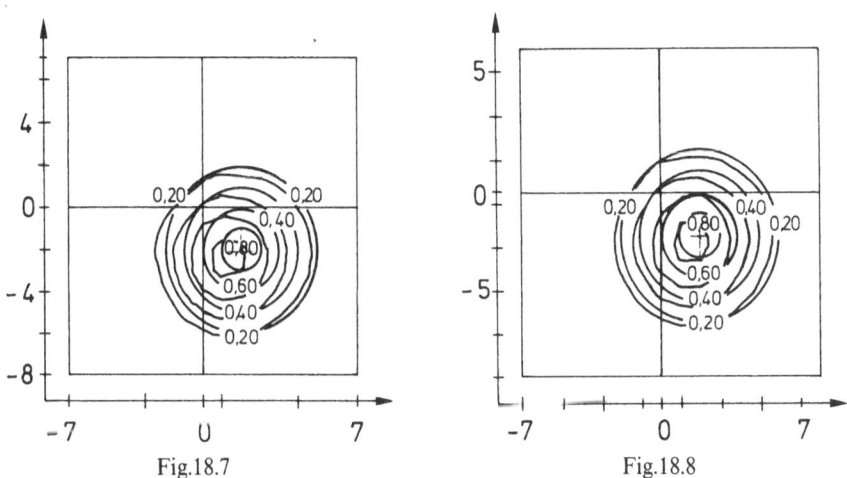

Fig.18.7   Fig.18.8

Fig. 18.7. MacCormack-scheme ($\alpha = 1.5708$, $m = 3$, $h_x = 1$, $h_y = 1$, $\tau = 0.01$), $t = 3$

Fig. 18.8. MacCormack-scheme ($\alpha = 1.5708$, $m = 3$, $h_x = 1$, $h_y = 0.5$, $\tau = 0.01$), $t = 3$

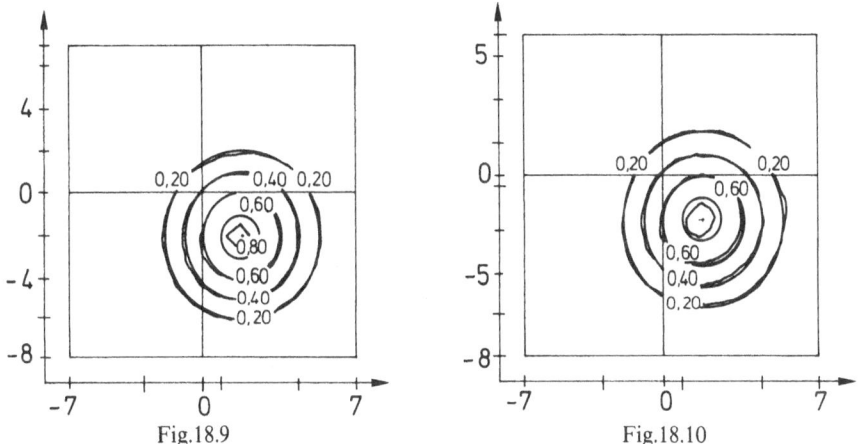

Fig.18.9   Fig.18.10

Fig. 18.9. Invariant scheme (18.8); ($\alpha = 1.5708$, $m = 3$, $h_x = 1$, $h_y = 1$, $\tau = 0.01$), $t = 3$

Fig. 18.10. Invariant scheme (18.8); ($\alpha = 1.5708$, $m = 3$, $h_x = 1$, $h_y = 0.5$, $\tau = 0.01$), $t = 3$

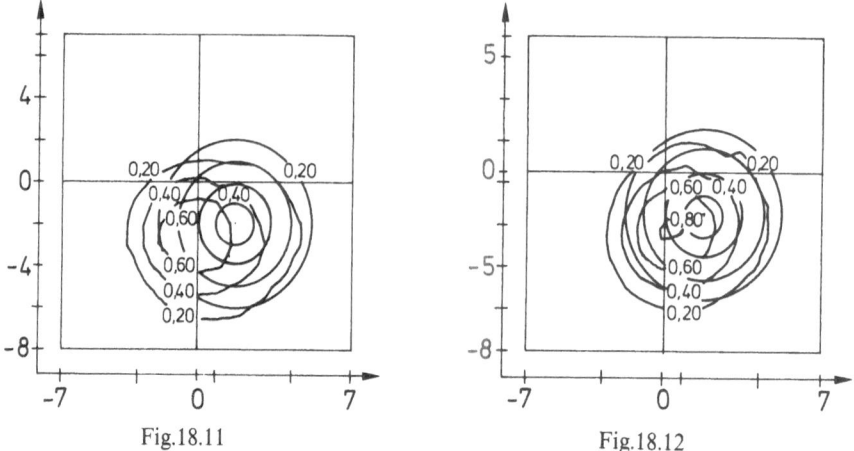

Fig.18.11   Fig.18.12

Fig. 18.11. Lax-Wendroff-scheme ($\alpha = 1.5708$, $m = 3$, $h_x = 1$, $h_y = 1$, $\tau = 0.01$), $t = 3$

Fig. 18.12. Lax-Wendroff-scheme ($\alpha = 1.5708$, $m = 3$, $h_x = 1$, $h_y = 0.5$, $\tau = 0.01$), $t = 3$

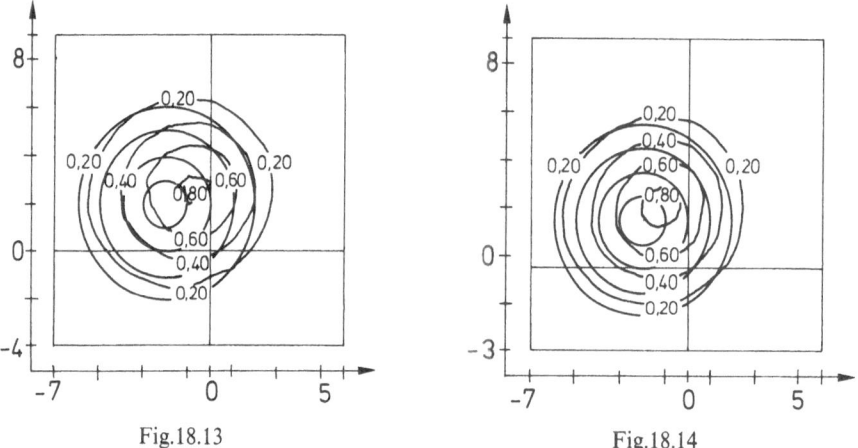

Fig.18.13   Fig.18.14

Fig. 18.13. MacCormack-scheme ($\alpha = 1.5708$, $m = 5$, $h_x = 1$, $h_y = 1$, $\tau = 0.01$), $t = 5$

Fig. 18.14. MacCormack-scheme ($\alpha = 1.5708$, $m = 5$, $h_x = 1$, $h_y = 0.5$, $\tau = 0.01$), $t = 5$

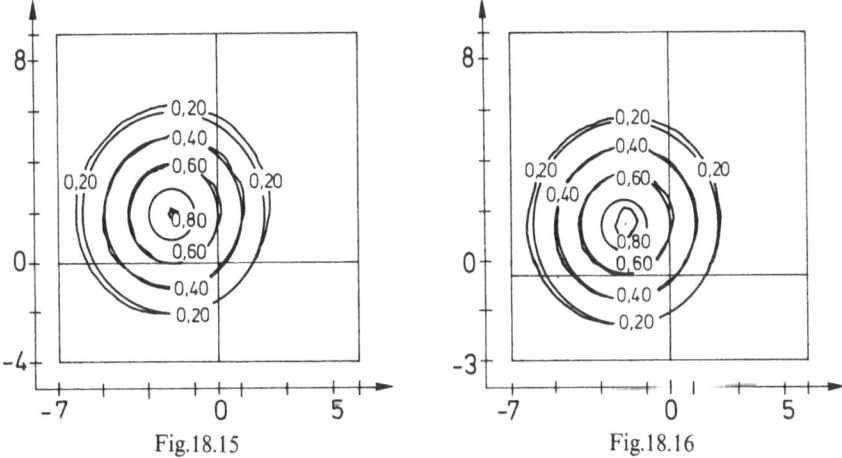

Fig.18.15   Fig.18.16

Fig. 18.15. Invariant scheme (18.8); ($\alpha = 1.5708$, $m = 5$, $h_x = 1$, $h_y = 1$, $\tau = 0.01$), $t = 5$

Fig. 18.16. Invariant scheme (18.8); ($\alpha = 1.5708$, $m = 5$, $h_x = 1$, $h_y = 0.5$, $\tau = 0.01$), $t = 5$

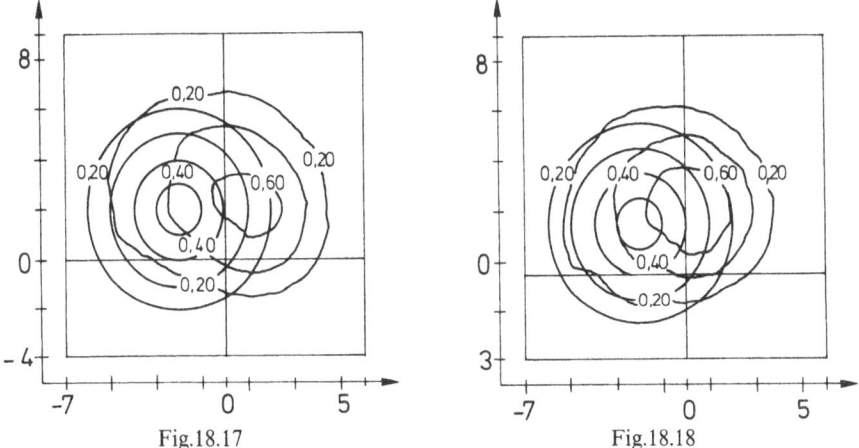

Fig.18.17   Fig.18.18

Fig. 18.17. Lax-Wendroff-scheme ($\alpha = 1.5708$, $m = 5$, $h_x = 1$, $h_y = 1$, $\tau = 0.01$), $t = 5$

Fig. 18.18. Lax-Wendroff-scheme ($\alpha = 1.5708$, $m = 5$, $h_x = 1$, $h_y = 0.5$, $\tau = 0.01$), $t = 5$

Part IV

Appendix

In this appendix the theory of first differential approximation is applied to the classification of difference schemes which are used at the present time for the numerical solution of gasdynamic problems. The contents of this appendix corresponds to results of Yanenko et al. [212].

A.1 Introduction

The aim of the present work is to describe difference schemes for the solution of initial value problems which are used in the present time for the numerical solution of gasdynamic problems. Our classification is based on the properties of difference schemes which are defined by the corresponding characteristics of their differential approximations. The method of differential approximation which is an important tool to analyse properties of difference schemes, the kind of their construction and classification in the present time has found a wide application, and their effectiveness was underlined by practical calculations.

The possibility of the use of differential approximation for the investigation of difference schemes was pointed out the first time in the fiftieth by Zhukov for the case of a very simple equation of propagation type with constant coefficients. In 1968 in the publications of Yanenko and Shokin the notation of the first differential approximation was formulated for an arbitrary difference scheme with constant coefficients, and theorems were given about the connection between the stability of simple and majorant schemes and the correctness of their first differential approximation for hyperbolic systems of differential equations of first order. Later the idea of using the differential approximation for the analysis of properties of difference schemes was developed in an extensive manner by the work of Yanenko and Shokin and also in the work of other mathematicians.

The literature on the method of differential approximation counts about more than 150 references at the present time. For a more detailed information on the results presented here the reader is refered to the references.

In describing the difference schemes we tried to characterize their most important properties in such a way that it would be possible to compare and select difference schemes with properties which are important for the user when calculating given problems. Performing such a classification for difference schemes of gas dynamics is a very complicated problem and it is divided into a set of smaller steps. First of all analogies of difference schemes for gasdynamic problems are considered in the case of an equation of propagation type with constant coefficients:

$$u_t + au_x = 0, \tag{A1}$$

then for the nonlinear equation of propagation type

$$u_t + f_x = 0; \quad f = f(u) = u^2/2, \tag{A2}$$

or, equivalently,

$$u_t + a(u)u_x = 0; \quad a(u) := u = f_u, \tag{A3}$$

and after that difference schemes especially used for the equations of gas dynamics are described.

Such a procedure – in our opinion – reflects in a natural way the kind of procedure that a numerical mathematician would follow for a special difference scheme for equations of gas dynamics. It is helpful to be convinced of the power of a scheme, to emphasize one or the other property (and it is even possible to exclude a scheme) and to point out ways how to polish up schemes for practical problems. The difference schemes and their properties are listed in tables which will be especially useful for the user and which reflect the kind of treatment which is applied.

The references in the field of difference schemes of gas dynamics are very numerous, including hundreds of articles; therefore we refer the reader to basic monographs [1–13] and the original articles.

For convenience we introduce the following frequently used notations:

$$\mu f_i := \tfrac{1}{2}(f_{i+\frac{1}{2}} + f_{i-\frac{1}{2}}),$$
$$\delta f_i := f_{i+\frac{1}{2}} - f_{i-\frac{1}{2}},$$
$$\Delta_1 = T_1 - E,$$
$$\Delta_{-1} = E - T_{-1},$$
$$\Delta_0 = T_0 - E.$$

E = identy operator, T_1 = shift-operator with respect to x, T_0 = shift-operator with respect to t, $u_i^n = u^n(x)$, $x = ih \in \mathbb{R} = (-\infty, +\infty)$, χ_h = dissipation of the scheme, $\Delta \Phi_h$ = dispersion of the scheme, $\xi = kh$, k = wave number, h = step of the scheme in x-direction, τ = step of the scheme in t-direction, $\kappa = \tau/h$ = const, $\sigma = \kappa a$.

A.2 Difference Schemes for the Equation of Propagation

1) First of all we consider the following family of difference schemes:

$$\frac{\Delta_0 w_i^n}{\tau} + a\frac{\Delta_1 + \Delta_{-1}}{2h}u_i^\gamma = \frac{h^2 q}{2\tau}\frac{\Delta_1 \Delta_{-1}}{h^2}u_i^\gamma, \tag{A4}$$

which approximates equation (A1). Here

$$u_i^\gamma := \gamma u_i^{n+1} + (1 - \gamma)u_i^n.$$

In the class of difference schemes (A4) several well-known schemes include the parameters which are given in Table 1, [8], [78], [80], [105], [110], [190], [191], [199], [204], [208].

2) The corresponding family of difference schemes for equation (A2) can be written in the form

$$\frac{\Delta_0 u_i^n}{\tau} + \frac{\Delta_1 + \Delta_{-1}}{2h}f_i^\gamma = \frac{h^2}{2\tau}\frac{\delta q_i^n \delta\phi_i^n}{h^2}. \tag{A5}$$

The parameters of special schemes from the family (A5) are listed in Table 2. We remark that in the linear case the Godunov-scheme coincides with the explicit scheme for initial value problems [9], and with which in the linear case also Rusanov's scheme coincides for $a \geq 0$, $\omega = 1$. The implicit scheme with central approximation is applied using a sweep method which leads to bounds

Table 1

Scheme		w_i^n	γ	q	Order of approximation
Lax	[208]	u_i^n	0	1	$O\left(\tau, h^2, \dfrac{h^2}{\tau}\right)$
Godunov	[191]	u_i^n	0	$\kappa\, a\, \text{sign}\,\{a\}$	$O(\tau, h)$
Rusanov	[204]	u_i^n	0	$\kappa\,\omega\,|a|$	$O(\tau, h)$
Unstable ("Tripod")	[8]	u_i^n	0	0	$O(\tau, h^2)$
Invariant	[78], [80], [105], [110]	u_i^n	0	$\kappa^2 a^2 + \alpha$ $\alpha = 0(\tau, h)$	$O(\tau, h)$
Lax-Wendroff	[199]	u_i^n	0	$\kappa^2 a^2$	$O(\tau^2, h^2)$
Implicit with central approximation	[8]	u_i^n	± 0	0	$O(\tau, h^2), \gamma \neq \frac{1}{2}$ $O(\tau^2, h^2), \gamma = \frac{1}{2}$
Implicit with one-sided approximation	[8]	u_i^n	± 0	$\kappa\, a\, \text{sign}\,\{a\}$	$O(\tau, h), \gamma \neq \frac{1}{2}$ $O(\tau^2, h), \gamma = \frac{1}{2}$
With skew derivative	[190]	$\left(E + \dfrac{\tau a}{2}\dfrac{\Delta_{-1}}{h}\right)u_i^n$	0	0	$O(\tau^2, h^2)$

Table 2

Scheme	q_i	γ	ϕ_i	Order of approximation
Lax	1	0	u_i	$O\left(\tau, h, \dfrac{h^2}{2\tau}\right)$
Godunov	$\kappa\,\text{sign}\,\{u_{i-\frac{1}{2}}\}\,\mu\,u_i$	0	u_i	$O(\tau, h)$
Rusanov	$\alpha\,\omega\,\lvert\mu\,u_i\rvert$	0	u_i	$O(\tau, h)$
Unstable ("Tripod")	0	0	–	$O(\tau, h^2)$
Invariant	$\kappa^2\,\mu\,u_i^2 + \alpha,\ \alpha = O(\tau, h),\ \partial\alpha/\partial u = 0$	0	u_i	$O(\tau, h)$
Lax-Wendroff	$\kappa^2\,\mu\,u_i$	0	f_i	$O(\tau^2, h^2)$
Implicit with central approximation	0	$\neq 0$	–	$O(\tau, h^2),\ \gamma \neq \tfrac{1}{2}$ $O(\tau^2, h^2),\ \gamma = \tfrac{1}{2}$
Implicit with one-sided approximation	$\kappa\,u_i\,\text{sign}\,\{u_i\}$	$\neq 0$	f_i	$O(\tau, h),\ \gamma \neq \tfrac{1}{2}$ $O(\tau^2, h),\ \gamma = \tfrac{1}{2}$

for τ which are connected with the stability of the sweep method. The explicit scheme with one-sided approximation is free from this drawback.

A basic profitable feature of implicit schemes is their absolute stability for $\gamma \geq 1/2$. But for a realization of these schemes a linearization is necessary followed by an iteration because of the nonlinearity. This makes the algorithm complicated. Schemes of the predictor-corrector type are much simpler for the realization [9], [62].

In the linear case the predictor-corrector scheme can be written in such a form:

$$\frac{u_i^* - u_i^n}{\gamma\tau} + a\,\frac{u_{i+1}^* - u_{i-1}^*}{2h} = \alpha\,\frac{h}{2}\,a\,\frac{\Delta_1\Delta_{-1}}{h^2}\,u_i^*,$$

$$\frac{\Delta_0 u_i^n}{\tau} + a\,\frac{\Delta_1 + \Delta_{-1}}{2h}\,u_i^* = 0,$$

(A6)

where $\alpha = \text{sign}\,\{a\}$ or $\alpha = 0$.

In the nonlinear case the predictor-corrector scheme has the form:

$$\frac{u_i^* - u_i^n}{\gamma\tau} + u_i^n\,\frac{\Delta_1 + \Delta_{-1}}{2h}\,u_i^* = \alpha\,\frac{h}{2}\,u_i^n\,\frac{\Delta_1\Delta_{-1}}{h^2}\,u_i^*,$$

$$\frac{\Delta_0 u_i^n}{\tau} + \frac{\Delta_1 + \Delta_{-1}}{2h}\,f_i^* = 0,$$

(A7)

where $f_i^* = f(u_i^*)$, $\alpha = 0$ or $\alpha = \text{sign}\,\{u\}$. For $\alpha = 0$ the scheme is realized by a sweep method, for $\alpha = \text{sign}\,\{u\}$ the predictor is an implicit scheme of the type for initial value problems.

3) Many authors have suggested several two-step difference schemes of second order of approximation with three-point formulas with respect to the space variable. These schemes can be included in a three-parameter family $\mathscr{L}^{\alpha}_{\beta,\varepsilon}$:

$$\frac{\tilde{u}_i - u_i^n}{\tau} + \alpha \frac{\Delta_1 f_i^n}{h} = \frac{\beta h}{\tau} \cdot \frac{\Delta_1}{h} u_i^n + \alpha h \varepsilon \frac{\Delta_1 \Delta_{-1}}{h^2} f_i^n,$$

$$\frac{\Delta_0 u_i^n}{\tau} + \frac{\Delta_1 + \Delta_{-1}}{2h} f_i^n = \frac{1}{2\alpha} \frac{\Delta_{-1}}{h} f_i^n - \frac{1}{2\alpha} \frac{\Delta_{-1}}{h} \tilde{f}_i$$

$$+ \frac{\varepsilon + \beta}{2\alpha} h \frac{\Delta_1 \Delta_{-1}}{h^2} f_i^n - \frac{\varepsilon}{2\alpha} h \frac{\Delta_1 \Delta_{-1}}{h^2} \tilde{f}_i.$$

(A8)

Here $\beta = 0$ or $\varepsilon - 0$, α is an arbitrary parameter. The values of the parameters of special schemes from family (A8) are given in Table 3, [7], [143], [158], [209–211].

In the linear case the schemes of family $\mathscr{L}^{\alpha}_{\beta,\varepsilon}$ coincide with the linear analogue to the Lax-Wendroff-scheme. We remark that $\mathscr{L}^1_{1,0} = \mathscr{L}^1_{0,1}$.

4) We will consider now a set of schemes which can be grouped together in a family of schemes but which leads to very lengthy terms and therefore they are written down separately.

Kolgan's scheme [221] is using a variable star containing four variants. For $a > 0$ the scheme has the form:

1) $\dfrac{\Delta_0 u_i^n}{\tau} + a \dfrac{\Delta_1 + \Delta_{-1}}{2h} u_i^n = 0;$ $|\Delta_1 u_{i-2}| \geq |\Delta_1 u_{i-1}| \geq |\Delta_1 u_i|;$

2) $\dfrac{\Delta_0 u_i^n}{\tau} + a \dfrac{\Delta_{-1}}{h} u_i^n = -\dfrac{ah}{2} \dfrac{\Delta_1 \Delta_{-1}}{h^2} u_{i-1}^n;$ $|\Delta_1 u_{i-2}| \leq |\Delta_1 u_{i-1}| \leq |\Delta_1 u_i|;$

3) $\dfrac{\Delta_0 u_i^n}{\tau} + a \dfrac{\Delta_{-1}}{h} u_i^n = 0;$ $|\Delta_1 u_{i-2}| \geq |\Delta_1 u_{i-1}|;$ $|\Delta_1 u_i| \geq |\Delta_1 u_{i-1}|;$

4) $\dfrac{\Delta_0 u_i^n}{\tau} + a \dfrac{\Delta_1 + \Delta_{-1}}{2h} u_i^n = -\dfrac{ah}{2} \dfrac{\Delta_1 \Delta_{-1}}{h^2} u_{i-1}^n;$ $|\Delta_1 u_{i-2}| \leq |\Delta_1 u_{i-1}|;$ $|\Delta_1 u_i| \leq |\Delta_1 u_{i-1}|.$

(A9)

Table 3

Scheme		Class	Order of approximation	α	β	ε
Richtmyer	[7]	$\mathscr{L}^{1/2}_{1/2,0}$	$O(\tau^2, h^2)$	$\frac{1}{2}$	$\frac{1}{2}$	0
Rubin-Burstein	[211]	$\mathscr{L}^1_{1/2,0}$	$O(\tau^2, h^2)$	1	$\frac{1}{2}$	0
Gourlay-Morris	[210]	$\mathscr{L}^{\alpha}_{1/2,0}$	$O(\tau^2, h^2)$	α	$\frac{1}{2}$	0
Lerat-Peyret	[143]	$\mathscr{L}^{\alpha}_{\beta,0}$	$O(\tau^2, h^2)$	α	β	0
Warming-Kutler-Lomax	[158]	$\mathscr{L}^{\alpha}_{0,\varepsilon}$	$O(\tau^2, h^2)$	α	0	ε
MacCormack	[209]	$\mathscr{L}^1_{0,1}$	$O(\tau^2, h^2)$	1	0	1
MacCormack	[209]	$\mathscr{L}^1_{0,0}$	$O(\tau^2, h^2)$	1	0	0

The three-level leap-frog-scheme has the form

$$\frac{u_i^{n+1} - u_i^{n-1}}{2\tau} + a \frac{\Delta_1 + \Delta_{-1}}{2h} u_i^n = 0,$$

$$\frac{u_i^{n+1} - u_i^{n-1}}{2\tau} + \frac{\Delta_1 + \Delta_{-1}}{2h} f_i^n = 0, \tag{A10}$$

in the linear and nonlinear case, respectively. Iteration schemes which are based on the Lax-Wendroff-scheme [222, 233] have the form:

$$u_i^{n+1, 0} = u_i^n,$$

$$u_i^{n+1, s+1} = u_i^n + (1 - \Theta)\left[-\tau \frac{\Delta_1 + \Delta_{-1}}{2h} f_i^n\right.$$

$$+ \frac{\kappa^2}{2}(a_{i+\frac{1}{2}}^n \Delta_1 f_i^n - a_{i-\frac{1}{2}}^n \Delta_{-1} f_i^n)\right]$$

$$+ \Theta\left[-\tau \frac{\Delta_1 + \Delta_{-1}}{2h} f_i^{n+1, s}\right. \tag{A11}$$

$$+ \frac{\kappa^2}{2}(a_{i+\frac{1}{2}}^{n+1, s} \Delta_1 f_i^{n+1, s} - a_{i-\frac{1}{2}}^{n+1, s} \Delta_{-1} f_i^{n+1, s})\right];$$

$$a_{i+\frac{1}{2}} = \mu a_i,$$

s is the iteration index, and Θ is a parameter. With a growing number of iterations the star increases. In the linear case these systems look such:

$$\frac{\Delta_0 u_i^n}{\tau} = (1 + \eta) \sum_{r=1}^{s+1} \left(-\frac{a \Delta_1}{2h} + \frac{\tau a^2}{2} \frac{\Delta_1 \Delta_{-1}}{h^2}\right)^r u_i^n; \quad \eta := \frac{1 - \Theta}{\Theta}. \tag{A12}$$

The scheme of the Shasta-method in the linear case [224] has the form:

$$u_i^{n+\frac{1}{2}} = (\tfrac{3}{4} - \kappa^2 a^2) u_i^n + \left(\frac{1}{8} + \frac{\kappa^2 a^2}{2} - \frac{\kappa a}{2}\right) u_{i+1}^n + \left(\frac{1}{8} + \frac{\kappa^2 a^2}{2} + \frac{\kappa a}{2}\right) u_{i-1}^n,$$

$$u_i^{n+1} = u_i^{n+\frac{1}{2}} - \tfrac{1}{8}\Delta_1 \Delta_{-1} u_i^{n+\frac{1}{2}}. \tag{A13}$$

Here in the first step the calculations are carried out on the basis of a stable scheme of first order of approximation, but in the second step with a non-

Table 4

Scheme		Class	Order of approximation	α	ω	β	ε
Rusanov	[192]	$\mathscr{L}_{1/2, 0}^{1/3, \omega}$	$O\left(\tau^3, \frac{h^4}{\tau}\right)$	$\frac{1}{3}$	ω	$\frac{1}{2}$	0
Burstein-Mirin [225]		$\mathscr{L}_{1/2, 0}^{\alpha, \omega}$	$O\left(\tau^3, \frac{h^4}{\tau}\right)$	α	ω	$\frac{1}{2}$	0
Warming-Kutler-Lomax [158]		$\mathscr{L}_{0, \varepsilon}^{\alpha, \omega}$	$O\left(\tau^3, \frac{h^4}{\tau}\right)$	α	ω	0	ε

correct scheme. On the whole, the scheme is stable and has second order of approximation.

5) The well-known schemes of third order of approximation based on a five-point star can be unified in a four-parameter family $\mathscr{L}_{\beta,\,\varepsilon}^{\alpha,\,\omega}$:

$$u_i^{[0]} = u_i^n,$$

$$u_i^{[1]} = (1 - \beta)\,u_i^{[0]} + \beta u_{i+1}^{[0]} - \alpha\kappa\,(\Delta_1 f_i^{[0]} - \varepsilon\Delta_1\Delta_{-1}f_i^{[0]}),$$

$$u_i^{[2]} = u_i^{[0]} - \frac{\kappa}{9\alpha}\,[3\,\alpha\,(\Delta_1 + \Delta_{-1})\,f_i^{[0]} - 2\Delta_{-1}f_i^{[0]} - 2\,(\varepsilon + \beta)\,\Delta_1\,\Delta_{-1}f_i^{[0]}]$$

$$- \frac{2}{9\alpha}\,\kappa\,(\Delta_{-1}f_i^{[1]} + \varepsilon\Delta_1\,\Delta_{-1}f_i^{[1]}),$$

$$u_i^{n+1} = u_i^{[0]} - \frac{\omega}{24}\,\delta^4\,u_i^{[0]} - \frac{\kappa}{4}\,[(E - \tfrac{2}{3}\delta^2)\,\mu\delta f_i^{[0]}] - \tfrac{3}{4}\kappa\mu\delta f_i^{[2]}.$$

Here $\beta = 0$ or $\varepsilon = 0$; ω, α are parameters. The term $(\omega/24)\,\delta^4\,u_i^{[0]}$ is introduced for stability. The values for the parameters in case of special schemes are given in the Table 4 [192], [225], [158].

Balakin's scheme [226] of third order of approximation has the form:

$$u_i^{[0]} = u_i^n,$$

$$u_i^{[1]} = \mu u_i^{[0]} - \tfrac{1}{3}\kappa\delta f_i^{[0]}, \tag{A15}$$

$$u_i^{[2]} = u_i^{[0]} - \tfrac{2}{3}\kappa\delta f_i^{[1]},$$

$$u_i^{n+1} = u_i^{[3]} = \tfrac{1}{4}\mu u_i^{[0]} + (\tfrac{3}{4}E - \tfrac{1}{8}\delta^2)\,u_i^{[1]} - \tfrac{3}{4}\kappa\delta f_i^{[2]}.$$

In the linear case this scheme coincides with Strang's scheme [227] of maximal order of approximation for a four-point star:

$$u_i^{n+1} = \frac{1}{16}(18\,\mu u_i^n - u_{i-\frac{3}{2}}^n - u_{i+\frac{3}{2}}^n) - \frac{\kappa a}{24}\,(27\,\delta u_i^n - u_{i+\frac{3}{2}}^n + u_{i-\frac{3}{2}}^n)$$

$$+ \frac{\kappa^2 a^2}{4}\,\delta\,(\Delta_1 + \Delta_{-1})\,u_i^n - \frac{\kappa^3 \alpha^3}{6}\,\delta\Delta_1\,\Delta_{-1}\,u_i^n. \tag{A16}$$

If one finds u_i on the layer $n + \tfrac{2}{3}$ according to Balakin's scheme of third order of approximation, and if one performs in the defined way a further step, one gets Balakin's scheme of second order of approximation [226]:

$$u_i^{[0]} = u_i^n,$$

$$u_{i+\frac{1}{2}}^{[1]} = \mu u_{i+\frac{1}{2}}^{[0]} - \frac{\kappa}{4}\,\delta f_{i+\frac{1}{2}}^{[0]},$$

$$u_i^{[2]} = u_i^{[0]} - \frac{\kappa}{2}\,\delta f_i^{[1]}, \tag{A17}$$

$$u_{i+\frac{1}{2}}^{[3]} = \tfrac{1}{4}\mu u_{i+\frac{1}{2}}^{[0]} + (\tfrac{3}{4}E - \tfrac{1}{8}\delta^2)\,u_{i+\frac{1}{2}}^{[1]} - \tfrac{9}{16}\kappa\delta f_{i+\frac{1}{2}}^{[2]},$$

$$u_i^{[4]} = u_i^{[2]} - \frac{16}{27}\kappa\delta f_i^{[3]} + \frac{5\kappa}{54}\,\delta^3 f_i^{[1]} + \frac{5\kappa}{18}\,\delta f_i^{[1]} - \frac{5\kappa}{27}\,\mu\delta f_i^{[0]},$$

Table 5

Scheme	Star	Scheme	Star
1	2	1	2
Lax		Implicit with one-sided approximation	$a > 0$
Godunov	$a > 0$ $a < 0$		$a < 0$
		With skew derivative	
Rusanov		Predictor-corrector with central differences in the predictor	
Invariant		Predictor-corrector with one-sided differences in the predictor	
Lax-Wendroff		Three-parameter family $\mathscr{L}^{\alpha}_{\beta,\varepsilon}$	

Table 5 (continued)

Scheme	Star	Scheme	Star
1	2	1	2

Scheme	Star	Scheme	Star
Implicit with central approximation		Four-parameter family of schemes with third-order of approximation $\mathscr{L}^{\alpha,\omega}_{\beta,\varepsilon}$	$\mathscr{L}^{\alpha,\omega}_{\frac{1}{2},0}$
Kolgan	Type 1 Type 2 Type 3 Type 4		$\mathscr{L}^{\alpha,\omega}_{0,0}$ $\mathscr{L}^{\alpha,\omega}_{0,1}$
"leap-frog"		Balakin's scheme of third order of approximation	
Iterated scheme with s iterations		Balakin's scheme of fourth order of approximation	

Table 5 (continued)

Scheme	Star	Scheme	Star
1	2	1	2
"Shasta"		Tusheva	

which has fourth order of approximation in the linear case and looks such:

$$u_i^{n+1} = \left(1 + \frac{\kappa^4 a^4}{4} - \frac{5}{4}\kappa^2 a^2\right)u_i^n - \tfrac{1}{6}\kappa a(1-\kappa a)(4-\kappa^2 a^2)u_{i+1}^n$$

$$+ \tfrac{1}{6}\kappa a(\kappa a + 1)(4 - \kappa^2 a^2)u_{i-1}^n + \frac{\kappa a}{24}(\kappa^2 a^2 - 1)(\kappa a - 2)u_{i+2}^n$$

$$+ \tfrac{1}{24}\kappa a(\kappa^2 a^2 - 1)(\kappa a + 2)u_{i-2}^n. \tag{A18}$$

Tusheva's scheme [57] is based on the Simpson's rule:

$$-\kappa f_{i+1}^{n+1} - u_{i+1}^{n+1} - 4 u_i^{n+1} - u_{i-1}^{n+1} + \kappa f_{i-1}^{n+1} - 4\kappa f_{i+1}^n + 4\kappa f_{i-1}^n$$

$$-\kappa f_{i+1}^{n-1} + u_{i+1}^{n-1} + 4 u_i^{n-1} + u_{i-1}^{n-1} + \kappa f_{i-1}^{n-1} = 0. \tag{A19}$$

In the linear case it has the following form:

$$-(1 + \kappa a)u_{i+1}^{n+1} - 4 u_i^{n+1} - (1 - \kappa a)u_{i-1}^{n+1} - 4\kappa a u_{i+1}^n$$

$$+ 4\kappa a u_{i-1}^n + (1 + \kappa a)u_{i-1}^{n-1} + 4 u_i^{n-1} + (1 - \kappa a)u_{i+1}^{n-1} = 0. \tag{A20}$$

6) In [228] a class of absolutely stable implicit schemes of maximal order of accuracy on the basis of a given star is discussed for equation (A1):

$$\sum_{k=-q}^{p} [b_k u^{n+1}(x + kh) + c_k u^n(x + kh)] = 0.$$

The coefficients b_k and c_k; ($k = -q, \ldots, p$) are defined by the demand to get the best approximation for both the variables, which leads to a system of compatibility conditions:

$$\sum_{k=-q}^{p} [b_k(k - \kappa)^l + c_k k^l] = 0; \quad \kappa := \frac{a\tau}{h}; \quad l = 0, \ldots, L; \; L = 2(p + q),$$

the order of approximation of this scheme is $L = 2(p + q)$, if these conditions are satisfied. The author proves that the schemes constructed by these conditions are absolutely stable for arbitrary p and q.

Table 6

No.	Scheme	Order of approximation	ϱ
1	Lax	$O\left(\dfrac{h^2}{2\tau}, h^2\right)$	$\cos\xi - i\sigma\sin\xi$
2	Godunov	$O(\tau, h)$	$1 - i\sigma\sin\xi - |\sigma|(1 - \cos\xi)$
3	Rusanov	$O(\tau, h)$	$1 - i\sigma\sin\xi - \omega|\sigma|(1 - \cos\xi)$
4	Invariant	$O(\tau, h)$	$1 - i\sigma\sin\xi - (\sigma^2 + \alpha_0\kappa)(1 - \cos\xi), \; \alpha_0 = \dfrac{\alpha}{\kappa}$
5	Lax-Wendroff	$O(\tau^2, h^2)$	$1 - i\sigma\sin\xi - \sigma^2(1 - \cos\xi)$
6	Implicit with central approximation	$O(\tau, h^2), \gamma \neq \frac{1}{2}$; $O(\tau^2, \dot{h}^2), \gamma = \frac{1}{2}$	$[1 - i(1 - \gamma)\sigma\sin\xi]/(1 + i\gamma\sigma\sin\xi)$
7	Implicit with one-sided approximation	$O(\tau, h), \gamma \neq \frac{1}{2}$; $O(\tau^2, h), \gamma = \frac{1}{2}$	$\dfrac{1 - 2(1 - \gamma)|\sigma|\sin^2\left(\dfrac{\xi}{2}\right) - i(1 - \gamma)\sigma\sin\xi}{1 + 2\gamma|\sigma|\sin^2\left(\dfrac{\xi}{2}\right) + i\gamma\sigma\sin\xi}$
8	With skew derivative	$O(\tau^2, h^2)$	$\left\{\left(1 + \dfrac{\sigma}{2}\right)^2 - \sigma\left(1 + \dfrac{\sigma}{2}\right)\cos\xi + \dfrac{\sigma^2}{4}\cos\xi - i\left[\sigma\sin\xi\left(1 + \dfrac{\sigma}{2}\right) - \dfrac{\sigma^2}{4}\sin^2\xi\right]\right\}$ $\overline{/\left[1 + \sigma + \dfrac{\sigma^2}{2} - \sigma\left(1 + \dfrac{\sigma}{2}\right)\cos\xi\right]}$
9	Predictor-corrector with central differences in the predictor		The same as for an implicit scheme with central approximation

Table 6 (continued)

No.	Scheme	Order of approximation	ϱ
10	Predictor-corrector with one-sided difference in the predictor	$O(\tau, h^2),\ \gamma \neq \frac{1}{2}$ $O(\tau^2, h^2),\ \gamma = \frac{1}{2}$	$\dfrac{1 + 2\gamma|\sigma|\sin^2\left(\frac{\xi}{2}\right) - i(1-\gamma)\sigma\sin\xi}{1 + 2\gamma|\sigma|\sin^2\left(\frac{\xi}{2}\right) + i\gamma\sigma\sin\xi}$
11	Three-parameter family $\mathscr{L}^{\alpha}_{\beta,\varepsilon}$		The same as for the Lax-Wendroff-scheme
12	Kolgan	$O(\tau, h^2)$	
13	"leap-frog"	$O(\tau^2, h^2)$	$-i\sigma\sin\xi \pm \sqrt{1 - \sigma^2\sin^2\xi}$
14	Iterated schemes	$O(\tau)$	$1 + (1+\eta)\displaystyle\sum_{r=1}^{s+1}(D + iB)^r,\ D = -2\sigma^2\theta\sin^2\frac{\xi}{2},\ B = -\sigma\theta\sin\xi,\ \eta = \dfrac{1-\theta}{\theta}$
15	"Shasta"	$O(\tau^2, h^2)$	$\left[1 + \frac{1}{2}\sin^2\left(\frac{\xi}{2}\right)\right]\left[1 - 2\left(\frac{1}{4} + \sigma^2\right)\sin^2\left(\frac{\xi}{2}\right) - i\sigma\sin\xi\right]$
16	Four-parameter family $\mathscr{L}^{\alpha,\omega}_{\beta,\varepsilon}$	$O\left(\tau^3, \dfrac{h^4}{\tau}\right)$	$1 - \frac{1}{4}\sigma^2 - \frac{\omega}{4} + \left(\frac{\sigma^2}{4} - \frac{\omega}{12}\right)\cos(2\xi) + \frac{1}{3}\omega\cos\xi + i\left[\frac{1}{6}\sigma(1-\sigma^2)\sin(2\xi) + \frac{1}{3}\sigma(-4 + \sigma^2)\sin\xi\right]$
17	Balakin's scheme of third order of approximation	$O\left(\tau^3, \dfrac{h^4}{\tau}\right)$	$\frac{1}{2}\left(\frac{9}{4} - \sigma^2\right)\cos\left(\frac{\xi}{2}\right) - \frac{1}{2}\left(\frac{1}{4} - \sigma^2\right)\cos\left(\frac{3\xi}{2}\right) + i\left[-\sigma\left(\frac{9}{4} - \sigma^2\right)\sin\left(\frac{\xi}{2}\right) + \frac{1}{3}\sigma\left(\frac{1}{4} - \sigma^2\right)\sin\left(\frac{3\xi}{2}\right)\right]$
18	Balakin's scheme of fourth order of approximation	$O(\tau^4, h^4, \tau^2 h^2)$	$1 + \sigma^2\left[-\frac{5}{4} + \frac{\sigma^2}{4} + \frac{1}{12}(\sigma^2 - 1)\cos(2\xi) + \frac{1}{3}(4 - \sigma^2)\cos\xi\right] - i\frac{\sigma}{6}[(\sigma^2 - 1)\sin(2\xi) + 2(4 - \sigma^2)\sin\xi]$
19	Tusheva	$O(\tau^4, h^4)$	$\dfrac{2i\sigma\sin\xi \pm \sqrt{-3\sigma^2\sin^2\xi + 4\cos\xi + \cos^2\xi + 4}}{(-\cos\xi - 2 + i\sigma\sin\xi)}$

No.	Stability condition	χ_h	$\Delta\Phi_h$	Monotonicity
1	$|\sigma| \leq 1$	$\frac{1}{2}(1-\sigma^2)\zeta^2 + O(\zeta^4)$	$\frac{1}{3}(1-\sigma^2)\sigma\zeta^3 + O(\zeta^5)$	yes
2	$|\sigma| \leq 1$	$\frac{1}{2}|\sigma|(1-|\sigma|)\zeta^2 + O(\zeta^4)$	$\frac{1}{6}(1-2\sigma)(1-\sigma)\sigma\zeta^3 + O(\zeta^5)$	yes
3	$\sigma^2 \leq \omega|\sigma| \leq 1$	$\frac{1}{2}|\sigma|(\omega-|\sigma|)\zeta^2 + O(\zeta^4)$	$\frac{1}{6}\sigma(1+2\sigma^2 - 3|\sigma|\omega)\zeta^3 + O(\zeta^5)$	for $\omega \geq 1$
4	$|\sigma| \leq (\sqrt{\alpha_0^2 + 4a^2} - \alpha_0)/2|a|$	$\frac{1}{4}\alpha\zeta^2 + O(\zeta^4)$	$\frac{1}{6}\sigma(1-\sigma^2 - \alpha)\zeta^3 + O(\zeta^5)$	for $|\sigma| \leq \sigma^2 + \alpha \leq 1$
5	$|\sigma| \leq 1$	$\frac{1}{8}\sigma^2(1-\sigma^2)\zeta^4 + O(\zeta^6)$	$\frac{1}{6}\sigma(1-\sigma^2)\zeta^3 + O(\zeta^5)$	no
6	For $\gamma \geq \frac{1}{2}$ absolutely stable, for $\alpha < \frac{1}{2}$ unstable	$\frac{2\gamma-1}{2}\sigma^2\zeta^2 + O(\zeta^4)$, $\gamma \neq \frac{1}{2}$ 0, $\gamma = \frac{1}{2}$	$\frac{1}{6}\sigma[1 + [2 - 6\gamma(1-\gamma)]\sigma^2]\zeta^3 + O(\zeta^5)$	no
7	For $\gamma \geq \frac{1}{2}$ absolutely stable, for $\gamma < \frac{1}{2}$: $\sigma \leq 1/(1-2\gamma)$	$\frac{|\sigma|}{2}\zeta^2 - \frac{1-2\gamma}{2}\sigma^2\zeta^2 + O(\zeta^4)$	$\left[\frac{\sigma^2}{3} - \gamma(1-\gamma)\sigma^2 - \text{sign}\{a\}\frac{1-2\gamma}{2}\sigma + \frac{1}{6}\right]\cdot\sigma\zeta^3 + O(\zeta^5)$	for $\gamma = 1$
8	$-1 \leq \sigma \leq 2$	0	$\left(\frac{1}{6} - \frac{|\sigma|}{4} + \frac{\sigma^2}{12}\right)\sigma\zeta^3 + O(\zeta^5)$	no
9	The same as for an implicit scheme with central approximation			no
10	For $\gamma \geq \frac{1}{2}$ absolutely stable, for $\gamma < \frac{1}{2}$ unstable	$\frac{2\gamma-1}{2}\sigma^2\zeta^2 + O(\zeta^4)$	$\left\{\left[\frac{1}{3} - \gamma(1-\gamma)\right]\sigma^2 + \text{sign}\{a\}\frac{\gamma}{2}\sigma + \frac{1}{6}\right\}\cdot\sigma\zeta^3 + O(\zeta^5)$	no
11	The same as for the Lax-Wendroff-scheme			no
12	$|\sigma| \leq \frac{1}{2}$			yes
13	$|\sigma| \leq 1$	0	$\frac{1}{6}\sigma(1-\sigma^2)\zeta^3 + O(\zeta^5)$	no

Table 6 (continued)

No.	Stability condition	χ_h	$\Delta\Phi_h$	Monotonicity
14	For $\Theta < 0$ unstable, for $\frac{1}{2} \leq \Theta < 1$, $|\sigma| \leq z(\Theta,s)/\sqrt{2\Theta}$, $0.995 \leq z \leq 1$; for $0 < \Theta < \frac{1}{2}$ and even s Θ_0 exists such that for $\Theta_0 \leq \Theta \leq 1/(2\Theta)$, $|\sigma| \leq 1/(2\Theta)$, in the remaining interval $|\sigma| \leq z(\Theta,s)$, $1 \leq z(\Theta,s) \leq 1/\sqrt{2\Theta_0}$. If s is an odd number $\sigma \leq z(\Theta,s)$, $1 \leq z(\Theta,s) \leq 1/\sqrt{2\Theta}$.	s is arbitrary $\sigma^2\Theta\xi^2 + O(\xi^4)$	$s = 1$ $\frac{1}{6}\sigma(1-\sigma^2)\xi^3 + O(\xi^5)$	no
15	$|\sigma| \leq \sqrt{\dfrac{7}{12}}$	$\left[\dfrac{1}{2} + 2\sigma^2\left(1-\sigma^2\right)\right]\dfrac{\xi^4}{32} + O(\xi^6)$	$\dfrac{1}{6}\left(\dfrac{1}{4} - \sigma^2\right)\sigma\xi^3 + O(\xi^5)$	in the 1st step for $|\sigma| \leq \frac{1}{2}$
16	$4\sigma^2 - \sigma^4 \leq \omega \leq 3$; $0 \leq \sigma \leq 1$	$-\dfrac{1}{24}(4\sigma^2 - \omega - \sigma^4)\xi^4 + O(\xi^6)$	$\left(\dfrac{1}{30} + \dfrac{\sigma^2}{8} - \dfrac{\omega}{24} - \dfrac{\sigma^4}{30}\right)\xi^5 + O(\xi^7)$	no
17	$|\sigma| \leq \dfrac{1}{2}$	$-\left(\dfrac{5}{48}\sigma^2 - \dfrac{\sigma^4}{24} - \dfrac{3}{128}\right)\xi^4 + O(\xi^6)$	$\dfrac{\sigma}{32}\left(\dfrac{3}{5} + \dfrac{8}{3}\sigma^2 - \dfrac{16}{15}\sigma^4\right)\xi^5 + O(\xi^7)$	no
18	$|\sigma| \leq 1$	$\dfrac{1}{144}\sigma^2(1-\sigma^2)(4-\sigma^2)\xi^6 + O(\xi^8)$	$-\dfrac{\sigma}{120}(1-\sigma^2)(4-\sigma^2)\xi^5 + O(\xi^7)$	no
19	$|\sigma| \leq 1$	0	$\dfrac{\sigma}{180}(1-\sigma^4)\xi^5 + O(\xi^7)$	no

Table 7

No.	Scheme	First differential representation for the equation (1), $Lu = u_t + a u_x = 0$
1	Lax	$Lu = \dfrac{h^2}{2\tau}(1 - \kappa^2 a^2)u_{xx}$
2	Godunov	$Lu = \dfrac{h}{2}(a\,\text{sign}\{a\} - \kappa a^2)u_{xx}$
3	Rusanov	$Lu = \dfrac{h}{2}(\omega a\,\text{sign}\{a\} - \kappa a^2)u_{xx}$
4	Invariant	$Lu = \dfrac{h^2}{2\tau}\alpha u_{xx}$
5	Lax-Wendroff	$Lu = \dfrac{h^3}{6\tau}(\kappa^3 a^3 - \kappa a)u_{xxx}$
6	Implicit with central approximation	$Lu = \begin{cases} \dfrac{\tau}{2}(2\gamma - 1)a^2 u_{xx}, & \gamma \neq \dfrac{1}{2} \\[2mm] -\dfrac{h^2}{6}a(2 + \kappa^2 a^2)u_{xxx}, & \gamma = \dfrac{1}{2} \end{cases}$
7	Implicit with one-sided approximation	$Lu = \dfrac{ha}{2}[(2\gamma - 1)\kappa a + \text{sign}\{a\}]u_{xx}$
8	Predictor-corrector with central differences in the predictor	$Lu = \begin{cases} \dfrac{\tau}{2}(2\gamma - 1)a^2 u_{xx}, & \gamma \neq \dfrac{1}{2} \\[2mm] -\dfrac{h^2}{6}a(\kappa^2 a^2 + 2)u_{xxx}, & \gamma = \dfrac{1}{2} \end{cases}$

Table 7 (continued)

No.	Scheme	First differential representation for the equation (1), $Lu = u_t + a u_x = 0$
9	Predictor-corrector with one-sided differences in the predictor	$$Lu = \begin{cases} \dfrac{\tau}{2}(2\gamma - 1)a^2 u_{xx}, \quad \gamma \neq \dfrac{1}{2} \\[2mm] -\dfrac{\tau h}{4}(\text{sign}\{a\})a^2 u_{xxx} - \dfrac{\tau^2}{12}a^3 u_{xxx} - \dfrac{h^2}{6}a u_{xxx}, \quad \gamma = \dfrac{1}{2} \end{cases}$$
10	Two-parameter family $\mathcal{L}^{\alpha}_{\beta,0}$	$Lu = \dfrac{h^3}{6\tau}(\kappa^3 a^3 - \kappa a) u_{xxx}$
11	Two-parameter family $\mathcal{L}^{\alpha}_{0,\varepsilon}$	$Lu = \dfrac{h^3}{6\tau}(\kappa^3 a^3 - \kappa a) u_{xxx}$
12	"leap-frog"	$Lu = \dfrac{h^3}{6}(\kappa^3 a^3 - \kappa a) u_{xxx}$
13	Iterated schemes	$Lu = \Theta \tau a^2 u_{xx}$
14	Two-parameter family $\mathcal{L}^{\alpha,\omega}_{1/2,0}$	$Lu = \dfrac{1}{6}\tau h^2 a^2 \left(1 - \dfrac{\kappa^2 a^2}{4}\right) u_{xxxx} - \omega \dfrac{h^4}{24\tau} u_{xxxx}$
15	Three-parameter family $\mathcal{L}^{\alpha,\omega}_{0,\varepsilon}$	$Lu = \dfrac{1}{6}\tau h^2 a^2 \left(1 - \dfrac{\kappa^2 a^2}{4}\right) u_{xxxx} - \omega \dfrac{h^4}{24\tau} u_{xxxx}$
16	Balakin's scheme of third order of approximation	$Lu = \dfrac{5}{48}\tau h^2 a^2 u_{xxxx} - \dfrac{\tau^3}{24}a^4 u_{xxxx} - \dfrac{3h^4}{128\tau} u_{xxxx}$
17	Balakin's scheme of fourth order of approximation	$Lu = -\dfrac{h^5}{120\tau}\kappa a(1 - \kappa^2 a^2)(4 - \kappa^2 a^2) u_{xxxxx}$
18	Tusheva	$Lu = \dfrac{h^5}{180\tau}\kappa a(1 - \kappa^4 a^4) u_{xxxxx}$

No.	First differential representation for the equation $\quad Lu = u_t + f_x = 0, \quad f = \dfrac{u^2}{2}$	Conservativity	Invariance
1	$Lu = \dfrac{h^2}{2\tau}[(1 - \kappa^2 u^2)u_{\bar x}]_{\hat x}$	yes	no
2	$Lu = \dfrac{h}{2}[(u\,\operatorname{sign}\{u\} - \kappa u^2)u_{\bar x}]_{\hat x}$	yes	no
3	$Lu = \dfrac{h}{2}[(\omega\,\operatorname{sign}\{u\} - \kappa u)u\,u_{\bar x}]_{\hat x}$	yes	no
4	$Lu = \dfrac{h^2}{2\tau}(\alpha u_x)_x, \quad \partial\alpha/\partial u = 0$	yes	yes
5	$Lu = \dfrac{\tau^2}{6}(u^3 u_x)_{xx} - \dfrac{h^2}{12}(u^2)_{xxx}$	yes	no
6	$Lu = \begin{cases} \dfrac{\tau}{2}(2\gamma - 1)(u^2 u_x)_x, & \gamma \neq \dfrac{1}{2} \\[2mm] -\dfrac{\tau^2}{12}(u^3 u_x)_{xx} - \dfrac{h^2}{12}(u^2)_{xxx} \end{cases}$	yes	no
7	$Lu = \dfrac{\tau}{2}(2\gamma - 1)(u^2 u_x)_x + \dfrac{h}{4}(\operatorname{sign}\{u\})(u^2)_{xx}$	yes	no
8	$Lu = \begin{cases} \dfrac{\tau}{2}(2\gamma - 1)(u^2 u_x)_x, & \gamma \neq \dfrac{1}{2} \\[2mm] \dfrac{\tau^2}{6}(u^3 u_x)_{xx} - \dfrac{\tau^2}{8}[u^2((u^2)_{xx} + (u_x)^2)]_x - \dfrac{h^2}{12}(u^2)_{xxx}, & \gamma = \dfrac{1}{2} \end{cases}$	yes	no
9	$Lu = \begin{cases} \dfrac{\tau}{2}(2\gamma - 1)(u^2 u_x)_x, & \gamma \neq \dfrac{1}{2} \\[2mm] -\dfrac{\tau h}{4}(\operatorname{sign}\{u\})(u^2 u_{xx})_x + \dfrac{\tau^2}{6}(u^3 u_x)_x - \dfrac{\tau^2}{8}]u^2((u^2)_{xx} + (u_x)^2)]_x - \dfrac{h^2}{12}(u^2)_{xxx}, & \gamma = \dfrac{1}{2} \end{cases}$	yes	no

Table 7 (continued)

No.	First differential representation for the equation $\quad Lu = u_t + f_x = 0, \quad f = \dfrac{u^2}{2}$	Conservativity	Invariance
10	$Lu = \dfrac{\tau^2}{6}(u^3 u_x)_{xx} - \dfrac{\tau^2}{4}\alpha(u^2(u_x)^2)_x - \dfrac{\tau h(1-2\beta)}{4}(u(u_x)^2)_x - \dfrac{h^2\beta(\beta-1)}{4\alpha}((u_x)^2)_x - \dfrac{h^2}{12}(u^2)_{xxx}$	yes	no
11	$Lu = \dfrac{\tau^2}{6}(u^3 u_x)_{xx} - \dfrac{\alpha\tau^2}{4}(u^2(u_x)^2)_x - \dfrac{h^2}{12}(u^2)_{xxx} - \dfrac{\tau h}{4}(1-2\varepsilon)(u(u_x)^2)_x$	yes	no
12	$Lu = \dfrac{\tau^2}{6}(u^3 u_x)_{xx} - \dfrac{h^2}{12}(u^2)_{xxx}$	yes	no
13	$Lu = \tau\,\Theta[2u(u_x)^2 + u^2 u_{xx}]$	yes	no
14	$Lu = \dfrac{\tau h^2}{24\alpha}(4\alpha u_x^3 + (18\alpha - 1)uu_x u_{xx} + 4\alpha u^2 u_{xxx})_x + \dfrac{\tau^3}{18\alpha}[(3+2\alpha)u^3 u_x u_{xx} + (3+4\alpha)u^2(u_x)^3]_x$ $-\dfrac{\tau^3}{24}(u^4 u_x)_{xxx} - \dfrac{\omega h^4}{24\tau}u_{xxxx}$	yes	no
15	$Lu = \dfrac{\tau^2 h}{6}\left(\varepsilon - \dfrac{1}{2}\right)[(u^3 u_{xx} + u^2(u_x)^2)_x] + u(u^2 u_x)_{xx}]_x$ $+\dfrac{\tau^3\alpha}{6}[u^3 u_x u_{xx} + u^2(u_x)^3]_x + \dfrac{\tau^3}{9}[uu_x(u^2 u_x)_x]_x + \dfrac{\tau h^2}{12}[u^2 u_{xxx} + 3uu_x u_{xx} + (u^2 u_x)_{xx}]_x$ $-\dfrac{\tau^3}{24}(u^4 u_x)_{xxx} - \dfrac{\omega h^4}{24\tau}u_{xxxx}$	yes	no
16	$Lu = \dfrac{\tau h^2}{48}(12uu_x u_{xx} + 5u^2 u_{xxx} + 2(u_x)^3)_x + \dfrac{\tau^3}{18}(4u^2(u_x)^3 + u^3 u_x u_{xx} - u^2(u_x)^3)_x - \dfrac{3}{128}\dfrac{h^4}{\tau}u_{xxxx} - \dfrac{\tau^3}{24}(u^4 u_x)_{xxx}$	yes	no
17	$Lu = \dfrac{h^2}{12}u_x u_{xx} - \dfrac{25}{72}\tau^2[u^2(u_x)^2]_x - \dfrac{\tau^2}{6}[u(u^2 u_x)_x]_x + \dfrac{\tau^2}{6}(u^3 u_x)_{xx}$	yes	no
18	$Lu = \dfrac{h^4}{180}[-120\kappa^4 u(u_x)^5 - 300\kappa^3 u^3 u_x(u_{xx})^2 - 200\kappa^4 u^3(u_x)^2 u_{xxx} + u(1-\kappa^4 u^4)u_{xxxxx}$ $-600\kappa^4 u^2(u_x)^3 u_{xx} + 10(1-5\kappa^4 u^4)u_{xx}u_{xxx} + 5(1-5\kappa^4 u^4)u_x u_{xxxx}]$	yes	no

7) In Table 5 the stars of all difference schemes are shown, mentioned above. In Table 6 the classification according to stability, dispersion, dissipation, and monotonicity is given and the order of approximation of these schemes in the case of equation (A 1) is listed. In Table 7 the first differential approximations of the schemes in linear and nonlinear cases are given, and also the characteristics of conservativity and invariance are listed. The property of invariance in the considered case means invariance of the scheme (in the sense of the first differential approximation) in the linear case with respect to shifting along t- and x-direction and to a Galilei-transformation and a dilatation.

A.3 Difference Schemes for the Equations of One-dimensional Gas Dynamics

1) The equations of gas dynamics will be considered in the divergence form

$$w_t + f_x = 0, \tag{A21}$$

where

$$w := \begin{pmatrix} \varrho u \\ \varrho \\ \varrho E \end{pmatrix}; \quad f := \begin{pmatrix} \varrho u^2 + p \\ \varrho u \\ \varrho u (E + p/\varrho) \end{pmatrix};$$

in the case of Eulerian coordinates and

$$w := \begin{pmatrix} u \\ v \\ E \end{pmatrix}; \quad f := \begin{pmatrix} p \\ -u \\ up \end{pmatrix};$$

in the case of Lagrangean coordinates. Here u is the velocity, ϱ the density, p the pressure, $E := \varepsilon + \frac{1}{2} u^2$, ε is the specific internal energy, $p = p(\varepsilon, \varrho)$, $v = \varrho^{-1}$ is the specific volume. By c we denote the sound velocity, $A = df/dw$.

2) First we describe a class of two-layer three-point, homogeneous difference schemes with respect to the space variable with weights:

$$\frac{\Delta_0 w_i^n}{\tau} + \frac{\Delta_1 + \Delta_{-1}}{2h} f_i^\alpha = \frac{h}{2\tau} \left[\Omega_{i+\frac{1}{2}}^n \frac{\Delta_1}{h} - \Omega_{i-\frac{1}{2}}^n \frac{\Delta_{-1}}{h} \right] \phi_i^\beta. \tag{A22}$$

Such schemes are used as in Lagrangean as in Eulerian coordinates. Here:

$$0 \le \alpha; \quad \beta \le 1;$$
$$f^\alpha := \alpha f^{n+1} + (1-\alpha) f^n,$$
$$\phi^\beta := \beta \phi^{n+1} + (1-\beta) \phi^n,$$
$$\Omega = \Omega(w) \text{ is a } (3 \times 3)\text{-matrix};$$
$$\Omega_{i+\frac{1}{2}}^n = \Omega(\mu w_i^n).$$

The parameters of special difference schemes are listed in Table 8. For invariant difference schemes the matrix Ω should satisfy conditions of invariance of

Table 8

Scheme	Order of approximation	α	β	Ω	ϕ_i	Coordinates	Star
Lax	$O\left(\tau, h^2, \dfrac{h^2}{\tau}\right)$	0	0	I	w_i	Lagrangean Eulerian	(triangle stencil: $n{+}1$, n; $i{-}1$, i, $i{+}1$)
Rusanov	$O(\tau, h)$	0	0	$\kappa\omega\varrho c\,I$ $\kappa\omega(|u|+c)\,I$ $\omega = \text{const}$	w_i	Lagrangean Eulerian	(triangle stencil: $n{+}1$, n; $i{-}1$, i, $i{+}1$)
Lax-Wendroff	$O(\tau^2, h^2)$	0	0	$\kappa^2 A$	f_i	Lagrangean Eulerian	(triangle stencil: $n{+}1$, n; $i{-}1$, i, $i{+}1$)
Invariant, explicit	$O(\tau, h)$	0	0	Ω satisfies the conditions of an invariant scheme	w_i	Lagrangean Eulerian	(triangle stencil: $n{+}1$, n; $i{-}1$, i, $i{+}1$)
Invariant, implicit	$O(\tau, h)$	α	β	Ω satisfies the conditions of an invariant scheme	w_i	Lagrangean Eulerian	(box stencil: $n{+}1$, n; $i{-}1$, i, $i{+}1$)
Implicit with weights and central difference	$O(\tau, h^2),\ \alpha \neq \tfrac{1}{2}$ $O(\tau^2, h^2),\ \alpha = \tfrac{1}{2}$	α	0	0	w_i	Lagrangean Eulerian	(box stencil: $n{+}1$, n; $i{-}1$, i, $i{+}1$)

the first differential approximation which are given in this book and also in the publication [78]. For the realization of implicit schemes usually one uses the method of iteration according to nonlinearity with a following sweep.

3) Especially the family of explicit two-step schemes of second order of approximation $\mathscr{L}^{\alpha}_{\beta,\varepsilon}$ is very popular, (α-arbitrary parameter; $\beta = 0$ or $\varepsilon = 0$):

$$\frac{\tilde{w}_i - w_i^n}{\tau} + \alpha \frac{\Delta_1 f_i^n}{h} = \frac{\beta h}{\tau} \frac{\Delta_1}{h} w_i^n + \alpha h \varepsilon \frac{\Delta_1 \Delta_{-1}}{h^2} f_i^n,$$

$$\frac{\Delta_0 w_i^n}{\tau} + \frac{\Delta_1 + \Delta_{-1}}{2h} f_i^n = \frac{1}{2\alpha} \frac{\Delta_{-1}}{h} f_i^n - \frac{1}{2\alpha} \frac{\Delta_{-1}}{h} \tilde{f}_i + \frac{\varepsilon + \beta}{2\alpha} h \frac{\Delta_1 \Delta_{-1}}{h^2} f_i^n$$

$$- \frac{\varepsilon}{2\alpha} h \frac{\Delta_1 \Delta_{-1}}{h^2} \tilde{f}_i.$$

(A23)

The parameters for well-known difference schemes from this family are listed in Table 8.

The schemes of family (A23) can be considered as schemes of predictor-corrector type. To get absolute stability of the scheme sometimes as a predictor a linear implicit scheme is taken [8], [62]:

$$\frac{w_i^* - w_i^n}{\tau^*} + A_i^n \frac{\Delta_1 + \Delta_{-1}}{2h} w_i^* = \frac{h}{2} G_i^n \frac{\Delta_1 \Delta_{-1}}{h^2} w_i^*,$$

(A24)

where $A_i^n = A(w_i^n)$; $G_i^n = G(w_i^n)$ are (3×3)-matrices $\tau^* = \gamma\tau$, $0 \leq \gamma \leq 1$. Especially one can choose $G \equiv 0$ or $G = A - u(1 - \text{sign}\{u\})I$, where I is the identity matrix. The corrector usually has the form

$$\frac{\Delta_0 w_i^n}{\tau} + \frac{\Delta_1 + \Delta_{-1}}{2h} f_i^* = 0; \quad f_i^* = f(w_i^*).$$

(A25)

In [62] a class of invariant difference schemes of predictor-corrector type for Eulerian coordinates is constructed, where the scheme (A24) is taken as a predictor with $\tau^* = \tau/2$, but as corrector step the scheme

$$\frac{\Delta_0 w_i^n}{\tau} + \frac{\Delta_1 + \Delta_{-1}}{2h} f_i^* = \frac{h}{2\tau} \left[\Omega^*_{i+\frac{1}{2}} \frac{\Delta_1}{h} - \Omega^*_{i-\frac{1}{2}} \frac{\Delta_{-1}}{h} \right] w_i^*$$

(A26)

is used where the matrix Ω is chosen such that the first differential approximation of the scheme (A24), (A26) is invariant.

To realize the scheme (A24) a sweep is necessary which makes the algorithm complicated. In the paper of Yanenko and Yaushev [229] a scheme is proposed which is free of such disadvantage. As a predictor the authors propose a scheme for the inner grid points on the basis of invariants. Depending on the family of flows either an explicit scheme (where the Courant-criterion is not violated) or an implicit scheme is used, which draws together the regions of dependence for the difference and the differential equation. As a corrector a normal divergence scheme with central differences is chosen.

4) Often the difference schemes for the equations of gas dynamics in Lagrangean coordinates are considered in the form:

$$\frac{\Delta_0 u^n_{i+\frac{1}{2}}}{\tau} + \frac{\Delta_1 p^*_i}{h} = 0,$$

$$\frac{\Delta_0 v^n_{i+\frac{1}{2}}}{\tau} - \frac{\Delta_1 u^*_i}{h} = 0, \qquad (A27)$$

$$\frac{\Delta_0 E^n_{i+\frac{1}{2}}}{\tau} + \frac{\Delta_1}{h} p^*_i u^*_i = 0.$$

This is the case in the Godunov-scheme [3] and in the schemes of Kuropatenko [230]. The kind of defining u^* and p^* leads to a specific scheme. As in the Godunov-scheme these quantities are found from the solution of the problem of resolution of a discontinuity, in the Kuropatenko-scheme they are found taking into account the Hugoniot-condition.

5) The "leap-frog" scheme of second order of approximation is a three-level scheme:

$$\frac{w^{n+1}_i - w^{n-1}_i}{2\tau} + \frac{\Delta_1 + \Delta_{-1}}{2h} f^n_i = 0. \qquad (A28)$$

6) We consider difference schemes for the equations of gas dynamics in the non-divergence form

$$\tilde{w}_t + \tilde{A}\tilde{w}_x = 0; \quad \tilde{A} = QAQ^{-1}, \qquad (A29)$$

where

$$\tilde{w} := \begin{pmatrix} \varrho u \\ \varrho \\ \varrho \varepsilon \end{pmatrix}; \quad Q := \begin{pmatrix} 1 & 0 & 0 \\ 0 & 1 & 0 \\ -u & u^2/2 & 1 \end{pmatrix}; \quad \det\{Q\} = 1,$$

in Eulerian coordinates, and

$$\tilde{w} := \begin{pmatrix} u \\ v \\ \varepsilon \end{pmatrix}; \quad Q := \begin{pmatrix} 1 & 0 & 0 \\ 0 & 1 & 0 \\ -u & 0 & 1 \end{pmatrix}; \quad \det\{Q\} = 1,$$

in Lagrangean coordinates.

We will discuss in the following only the case of Lagrangean coordinates, because in the Eulerian coordinates good schemes are not known.

The von Neumann-Richtmyer-scheme [8], [231] has the form

$$\frac{u^{n+\frac{1}{2}}_i - u^{n-\frac{1}{2}}_i}{\tau} + \frac{\delta}{h} \bar{p}^n_i = 0,$$

$$\frac{\Delta_0 v^n_{i+\frac{1}{2}}}{\tau} - \frac{\Delta_1}{h} u^{n+\frac{1}{2}}_i = 0, \qquad (A30)$$

$$\frac{\Delta_0 \varepsilon^n_{i+\frac{1}{2}}}{\tau} + \frac{\bar{p}^{n+1}_{i+\frac{1}{2}} + \bar{p}^n_{i+\frac{1}{2}}}{2} \cdot \frac{\Delta_0 v^n_{i+\frac{1}{2}}}{\tau} = 0.$$

Table 9

"Cross-type" schemes	Authors who suggested a "viscous pressure"	Form of the "viscous pressure"
(A 30)	von Neumann, J., Richtmyer, R. D. [231]	$-\mu_0 \varrho h^2 |u_x| u_x$, $\quad \mu_0 = $ const
(A 30)	Latter, R. [8], [234]	$\begin{cases} 0, & u_x \geqq 0 \\ -\mu_0 \varrho h^2 |u_x| u_x, & u_x < 0, \end{cases}$ $\quad \mu_0 = $ const
(A 30)	Samarskii, A. A., Arsenin, V. Ya. [232]	$-0.5\, \mu_0 \varrho |u_x|^\nu (u_x - \lambda |u_x|)$, $\nu = $ const, $\lambda = $ const.
(A 30)	Kuropatenko, V. F. [230]	$\begin{cases} 0 & , u_x \geqq 0 \\ (1 - D\tau/h)\left[\dfrac{\gamma+1}{4v} h^2 u_x^2 + \sqrt{\left(\dfrac{\gamma+1}{4v} h^2 u_x^2\right)^2 + \dfrac{\gamma P}{v} h^2 u_x^2} \right], & u_x < 0 \end{cases}$ D is the velocity of the shock wave, γ is the exponent of a polytropic gas
(A 31)	Kuropatenko, V. F. [230]	$\begin{cases} 0 & , u_x \geqq 0 \\ D(1 - D\tau/h)h|u_x|, & u_x < 0 \end{cases}$

Table 10

Scheme	Order of approximation	Coordinates	Star
1	2	3	4
Class $\mathscr{L}^{\alpha}_{\beta,\varepsilon}$	$O(\tau^2, h^2)$	Lagrangean, Eulerian	(stencil: $n{+}1$ at i; n at $i{-}1$, i, $i{+}1$ — triangle)
Predictor-corrector	$O(\tau, h^2)$	Lagrangean, Eulerian	(stencil: $n{+}1$ at i; $n{+}\gamma$ at $i{-}1$, $i{+}1$; n at i — diamond)
Predictor-corrector	$O(\tau^2, h)$	Eulerian	(stencil: $n{+}1$ at i; $n{+}\gamma$ at $i{-}1$, $i{+}1$; n at i — diamond)
"leap-frog"	$O(\tau^2, h^2)$	Lagrangean, Eulerian	(stencil: $n{+}1$ at i; n at $i{-}1$, $i{+}1$; $n{-}1$ at i — diamond)
von Neumann-Richtmyer	$O(\tau^2, h^2)$	Lagrangean	(stencil, Equation of motion: $n{+}\frac12$ at i (u); n at $i{-}1$, $i{+}1$ (ϑ,p); $n{-}\frac12$ at $i{-}1$, i, $i{+}1$ (u))
			(stencil, Continuity equation: $n{+}1$ (ϑ); $n{+}\frac12$ at i, $i{+}1$ (u); n (ϑ); columns i, $i{+}\frac12$, $i{+}1$)
			(stencil, Energy equation: $n{+}1$ (p,ϑ,ε); $n{+}\frac12$ at i, $i{+}1$ (u); n (p,ϑ,ε); $n{-}\frac12$ at i, $i{+}1$ (u))
Kuropatenko (scheme (A 30))	$O(\tau^2, h)$	Lagrangean	(stencil: $n{+}\frac32$ at i (u); $n{+}1$ at $i{-}1$, $i{+}1$ (p,ϑ); $n{+}\frac12$ at $i{-}1$, i, $i{+}1$ (u))
			Equation of motion; for the other equations the same star is used as in the Neumann-Richtmyer scheme

Table 10 (continued)

Scheme	Order of approximation	Coordinates	Star	
1	2	3	4	
Popov-Samarskii	$O(\tau, h^2)$	Lagrangean		Equation of motion
				Continuity and energy equation

Here

$$\bar{p}_i^n = p_i^n + \omega_i^n;$$

$$\omega_{i+\frac{1}{2}}^n = -\mu_0 h^2 \varrho_{i+\frac{1}{2}}^n \frac{|\Delta_1 u_i^{n-\frac{1}{2}}|}{h} \cdot \frac{\Delta_1 u_i^{n-\frac{1}{2}}}{h},$$

$$\varepsilon_{i+\frac{1}{2}}^n = \varepsilon(p_{i+\frac{1}{2}}^n, v_{i+\frac{1}{2}}^n);$$

$$\mu_0 = \text{const.}$$

This was the first scheme from the family of difference scheme of the cross-type. Later other forms of the "viscous pressure" term ω were proposed; for specific schemes the formulas for ω are listed in differential form in Table 9 [8], [231, 232] where another Kuropatenko-scheme [230] of the cross-type is given, which has some features in common with the scheme (A30):

$$\frac{u_i^{n+\frac{3}{2}} - u_i^{n+\frac{1}{2}}}{\tau} + \frac{\delta}{h} \bar{p}_i^{n+1} = 0,$$

$$\frac{\Delta_0 v_{i+\frac{1}{2}}^n}{\tau} - \frac{\Delta_1}{h} u_i^{n+\frac{1}{2}} = 0,$$

$$\frac{\Delta_0 \varepsilon_{i+\frac{1}{2}}^n}{\tau} + \frac{\bar{p}_{i+\frac{1}{2}}^{n+1} + \bar{p}_{i+\frac{1}{2}}^n}{2} \frac{\Delta_0}{\tau} v_{i+\frac{1}{2}}^n = 0.$$

(A31)

Here

$$\bar{p}_{i+\frac{1}{2}}^{n+1} = p_{i+\frac{1}{2}}^{n+1} + \omega_{i+\frac{1}{2}}^{n+1}.$$

Table 11

Scheme	Π-form of the first differential representation	Eulerian coordinates			Lagrangean coordinates		
		K	M	$\mathscr{I}$	K	M	$\mathscr{I}$
1	2	3			4		
Lax	$w_t + f_x = \dfrac{h^2}{2\tau}[(I - \kappa^2 A^2)w_x]_x$	no	no		no	no	yes
Rusanov	$\sigma = \begin{cases} \varrho c & \text{(Lagrangean coordinates)} \\ |u| + c & \text{(Eulerian coordinates)} \end{cases}$ $\;$ $w_t + f_x = \dfrac{h}{2}[(\omega\sigma I - \kappa A^2)w_x]_x$	no	no		no	no	yes
Invariant with weights	$w_t + f_x = \dfrac{h^2}{2\tau}[(\Omega - \kappa^2(1 - 2\alpha)A^2)w_x]_x$	yes	yes	yes	yes	yes	yes
		For a special choice of the matrix Ω					
Lax-Wendroff	$w_t + f_x = \dfrac{\tau^2}{6}[f_{ww}f_x^2 + A(Af_x)_x]_x - \dfrac{h^2}{6}f_{xxx}$	no	no	no	no	no	yes
Lerat-Peyret $\mathscr{L}^\alpha_{\beta,0}$	$w_t + f_x = \dfrac{\tau^2}{2}\left[\dfrac{2-\alpha}{2}(f_{ww}f_x^2) + A(Af_x)_x\right]_x$ $- \dfrac{h^2}{6}f_{xxx} - \dfrac{h^2\beta(\beta-1)}{4\alpha}(f_{ww}w_x^2)_x - \dfrac{\tau h(1-2\beta)}{4}(f_{ww}w_xf_x)_x$	no	no	no	no	no	yes
Warming, Kutler, Lomax $\mathscr{L}^\alpha_{0,\varepsilon}$	$w_t + f_x = \dfrac{\tau^2}{2}\left[\dfrac{2-\alpha}{2}(f_{ww}f_x^2) + A(Af_x)_x\right]_x$ $- \dfrac{h^2}{6}f_{xxx} - \dfrac{\tau h}{4}(1 - 2\varepsilon)(f_{ww}w_xf_x)_x$	no	no	no	no	no	yes

Scheme							
Implicit with weights	$w_t + f_x = \begin{cases} -\dfrac{\tau}{2}(1-2\alpha)(A^2 w_x)_x, & \alpha \neq \tfrac{1}{2} \\[2mm] -\dfrac{\tau^2}{12}\left[f_{ww}f_x^2 + A(Af_x)_x\right]_x - \dfrac{h^2}{6}f_{xxx}, & \alpha = \tfrac{1}{2} \end{cases}$	no	no	no	yes	no	yes
"leap-frog"	$w_t + f_x = \dfrac{\tau^2}{6}[f_{ww}f_x^2 + A(Af_x)_{\pm x}]_{\mp x} - \dfrac{h^2}{6}f_{xxx}$	no	no	no	no	no	yes
Predictor-corrector	$w_t + f_x = \begin{cases} -\dfrac{\tau}{2}(1-2\gamma)(A^2 w_{\mp x})_{\pm x}, & \gamma \neq \tfrac{1}{2} \\[2mm] \dfrac{\tau^2}{24}\left[f_{ww}f_x^2 + 4A(Af_x)_x - 6A^2 f_{xx}\right]_x \\ \qquad - \dfrac{\tau h}{4}(AGw_{xx})_x - \dfrac{h^2}{2}f_{xxx}, & \gamma = \tfrac{1}{2} \end{cases}$	no	no	no	yes	no	yes
Invariant scheme of predictor-corrector type in Eulerian coordinates	$w_t + f_x = \dfrac{h^2}{2\tau}(\Omega w_x)_x$	yes	yes (for a special choice of the matrix Ω)	yes	no	no	yes for $G=0$

In the publications of Samarskii and Popov [13], [233] a four-parameter family of schemes was considered:

$$\frac{\Delta_0 u_i^n}{\tau} + \frac{\delta}{h} p_i^{\alpha_1} = 0,$$

$$\frac{\Delta_0 v_{i+\frac{1}{2}}^n}{\tau} - \frac{\Delta_1}{h} u_i^{\alpha_2} = 0, \qquad (A32)$$

$$\frac{\Delta_0 \varepsilon_{i+\frac{1}{2}}^n}{\tau} + p_{i+\frac{1}{2}}^{\alpha_3} \frac{\Delta_1 u_i^{\alpha_4}}{h} = 0,$$

where

$$\phi^{\alpha_j} := \alpha_j \phi^{n+1} + (1 - \alpha_j) \phi^n,$$

$$0 \le \alpha_j \le 1; \quad j = 1, 2, 3, 4$$

$$\phi = u, v, p, \varepsilon.$$

For $\alpha_1 = \alpha_3 = \alpha$, $\alpha_2 = \alpha_4 = 0.5$ the family (A32) is a one-parameter family of totally conservative schemes.

7) In Table 10 the stars of all difference schemes mentioned in 3)–5) are listed [8], [62]. In the Tables 11 and 12 the dissipative properties of the described difference schemes and the first differential approximation are shown in parabolic form. Table 11 contains a summary of difference schemes as in Eulerian

Table 12

Scheme	Π-form of the first differential representation	Features		
		K	M	$\mathscr{I}$
von Neumann-Richtmyer	$u_t + \bar{p}_x = (z_1 p_{xx} + z_2 u_x^2)_x,$	no	no	yes
	$v_t - u_x = -(z_1 u_x)_{xx},$			
	$E_t + (u\bar{p})_x = [z_1 u p_{xx} + p(z_1 u_x)_x + 2 z_1 u_x p_x + z_2 u u_x^2]_x$			
	or			
	$\varepsilon_t + \bar{p} u_x = z_1 u_x p_{xx} + z_2 u_x^3 + [p(z_1 u_x)_x + 2 z_1 u_x p_x]_x,$			
	$z_1 = \frac{\tau^2 \varrho^2 c^2}{24} - \frac{h^2}{2}, \quad z_2 = (\varrho^2 c^2 p_\varepsilon + p^2 p_{\varepsilon\varepsilon} - 2 p p_{\varepsilon v} + p_{vv}),$			
	$\bar{p} = p + \omega.$			
Popov and Samarskii (fully conservative)	$u_t + p_x = -\left(\frac{\tau}{2} - \alpha \tau\right)(\varrho^2 c^2 u_x)_x,$	yes	yes	yes
	$v_t - u_x = 0,$			
	$E_t + (up)_x = -\left(\frac{\tau}{2} - \alpha \tau\right)(\varrho^2 c^2 u u_x)_x,$			
	$\varepsilon_t + p u_x = -\left(\frac{\tau}{2} - \alpha \tau\right)\varrho^2 c^2 u_x^2.$			

as in Lagrangean coordinates. In Table 12 schemes are described for the equations of gas dynamics in non-divergence form in Lagrangean coordinates. Both the tables contain information whether the schemes belong to class K, M, $\mathscr{I}$ (according to the definitions given in this book and in [78]).

8) Schemes of higher order of accuracy for the system of one-dimensional gas dynamics (class $\mathscr{L}_{\beta,\,\varepsilon}^{\alpha;\,\omega}$ Balakin's third-order scheme and Tusheva's scheme) are described not only for the case of Lagrangean coordinates but also for Eulerian coordinates. For their description in the equation (A14), (A15), (A19) it is necessary to change the scalar functions u and f into vector functions w and f, which have a corresponding form in the coordinate system. The Balakin's scheme (A17) as already mentioned has an accuracy of second order in the nonlinear case. The stars of all schemes are the same as it was shown in Table 5. All schemes are of divergence form.

For the realization of schemes of family $\mathscr{L}_{\beta,\,\varepsilon}^{\alpha;\,\omega}$ and of Balakin's scheme of third order it is necessary to introduce two intermediate layers. Tusheva's scheme needs the application of iterations.

References*

[1] Belotserkovskii, O. M., Golovachev, Yu. P., Grudnitskii, V. G., Davydov, Yu. M., Dyshin, V. K., Lun'kin, Yu. P., Magomedov, K. M., Molodkov, V. K., Popov, F. D., Tolstykh, A. I., Fomin, V. N., Kholodov, A. S. *Chislennoe issledovanie sovremennykh zadach gazovoi dinamiki* (Numerical Investigation of Modern Problems in Gas Dynamics) (Nauka, Moscow 1974)
[2] Godunov, S. K. *Uravneniya matematicheskoi fiziki* (Equations of Mathematical Physics) (Nauka, Moscow 1971)
[3] Godunov, S. K., Zabrodin, A. V., Ivanov, M. Ya., Kraiko, A. N., Prokopov, G. P. *Chislennoe reshenie mnogomernykh zadach gazovoi dinamiki* (Numerical Solution of Multidimensional Problems of Gas Dynamics) (Nauka, Moscow 1976)
[4] Godunov, S. K., Ryabenkii, V. S. *Raznostnye skhemy* (Difference Schemes) (Nauka, Moscow 1973)
[5] Marchuk, G. I. *Metody vychislitelnoi matematiki* (Methods of Computational Mathematics) (Nauka, Novosibirsk 1973)
[6] Marchuk, G. I. *Metody vychislitelnoi matematiki* (Methods of Computational Mathematics) (Nauka, Moscow 1977)
[7] Richtmyer, R. D., Morton, K. W. *Raznostnye metody resheniya kraevykh zadach* [Translated from English: *Difference Methods for Initial-value Problems*, 2nd ed. (Wiley-Interscience, New York 1967)] (Mir, Moscow 1972)
[8] Rozhdestvenskii, B. L., Yanenko, N. N. *Sistemy kvazilinienykh uravnenii* (Systems of Quasilinear Equations) (Nauka, Moscow 1978)
[9] Samarskii, A. A. *Vvedenie v teoriyu raznostnykh skhem* (Introduction to the Theory of Difference Schemes) (Nauka, Moscow 1971)
[10] Samarskii, A. A. *Theoriya raznostnykh skhem* (Theory of Difference Schemes) (Nauka, Moscow 1977)
[11] Samarskii, A. A., Andreev, V. B. *Raznostnye metody dlya ellipticheskikh uravnenii* (Difference Methods for Elliptic Equations) (Nauka, Moscow 1976)
[12] Samarskii, A. A., Gulin, A. V. *Ustoichivost' raznostnykh skhem* (Stability of Difference Schemes) (Nauka, Moscow 1973)
[13] Samarskii, A. A., Popov, Yu. P. *Raznostnye skhemy gazovoi dinamiki* (Difference Schemes of Gas Dynamics) (Nauka, Moscow 1975)
[14] Yanenko, N. N. *Metod drobnykh shagov resheniya mnogomernykh zadach matematicheskoi fiziki* [English transl.: The Method of Fractional Steps. The Solution of Problems of Mathematical Physics in Several Variables (Springer, Berlin, Heidelberg, New York 1971)] [German transl.: Die Zwischenschrittmethode zur Lösung mehrdimensionaler Probleme der mathematischen Physik, Lecture Notes in Mathematics, Vol. 91 (Springer, Berlin, Heidelberg, New York 1969)] [French transl.: Méthode à pas fractionnaires (Librairie Armand Colin, Paris 1968)] (Nauka, Novosibirsk 1967)

---

* Please note that Soviet publications are in Russian. The titles have been translated for the reader's convenience .

[15] Yanenko, N. N. *Vvedenie v raznostnye metody matematicheskoi fiziki*, chasti 1, 2 (Introduction to Difference Methods of Mathematical Physics, Part 1, 2) (Novosib. Gos.
Univ., Novosibirsk 1968)

[16] Anuchina, N. N. O metodakh rascheta techenii szhimaemoi zhidkosti s bol'shimi
deformatsiyami. (On methods for the calculation of viscous fluid flows with large
deformations) Chislennye Metody Mekh. Sploshnoi Sredy 1 (4), 3–84 (1970)

[17] Anuchina, N. N., Petrenko, V. E., Shokin, Yu. I., Yanenko, N. N. On numerical methods of solving gasdynamic problems with large deformations. Fluid Dyn. Trans. 5,
9–32 (1971)

[18] Bakhrakh, S. M., et al. Zakhlopyvanie shcheli pri normal'nom padenii na stenku
udarnoi volny. (Closing of a gap under the effect of an incident shock wave normal to
the wall) Chislennye Metody Mekh. Sploshnoi Sredy 9 (5), 5–12 (1978)

[19] Belotserkovskii, O. M., Gushchin, V. A., Shchennikov, V. V. Metod rasshchepleniya v
primenenii k resheniyu zadach dinamiki vyazkoi zhidkosti. (Splitting method applied
to the solution of dynamic problems of viscous fluids) Zh. Vychisl. Mat. Mat. Fiz.
15 (1), 197–207 (1975)

[20] Belotserkovskii, O. M., Davydov, Yu. M. *Nestatsionarny metod "krupnykh chastits"
dlya resheniya zadach vneshnei aerodinamiki* (Time-Dependent Method of "Large Particles" for the Solution of Problems of Outer Aerodynamics) (Vychicl. Tsentr Akad.
Nauk SSSR, Moscow, 1970)

[21] Belotserkovskii, O. M., Davydov, Yu. M. "Issledovanie skhem metoda "krupnykh
chastits" s pomoshch'yu differentsialnykh priblizhenii", (Investigation of Schemes of
the Method of "Large Particles" Using the Differential Approximations) in *Probl. prikl.
mat. i. mekh.* (Nauka, Moscow 1971), pp. 145–155

[22] Belotserkovskii, O. M., Davydov, Yu. M. "Novy chislennyi metod dlya issledovaniya
slozhnykh zadach gazovoi dinamiki", (A new Computational Method for the Investigation of Complicated Problems of Gas Dynamics") in *Trudy simpoz. po mekh.
sploshnoi sredy i rodstvennym problemam analiza* (Metsniereba, Tbilisi 1974) pp. 73–85

[23] Belotserkovskii, O. M., Davydov, Yu. M. Method "krupnykh chastits" dlya zadach
gazovoi dinamiki. ("Large particle" method for problems of Gas Dynamics). Chislennye Metody Mekh. Sploshnoi Sredy 1 (3), 3–23 (1970)

[24] Vorozhtsov, E. V., Fomin, V. M., Yanenko, N. N. Differentsialnye analizatory udarnykh voln. Prilozhenie teorii. (Differential analyzers of shock waves) Chislennye Metody Mekh. Sploshnoi Sredy 7 (6), 8–23 (1976)

[25] Voinovich, P. A., Zhmakin, A. I., Popov, F. D., Fursenko, A. A. *O raschete razryvnykh
techenii gaza* (The Calculation of the Flow of Gas Explosions) (Fiz. Tekh. Inst. im.
A. F. Ioffe, Leningrad 1977)

[26] Davydov, Yu. M. Ob odnom metode issledovaniya ustoichivosti nelineinykh raznostnykh uravnenii. (A method for the investigation of the stability of nonlinear difference
equations) Tr. Mosk. Fiz. Tekh. Inst. Ser. Aerofiz. Prikl. Mat., 79–92 (1971)

[27] Davydov, Yu. M. "Issledovanie nelineinykh kolebanii, voznikayushchikh pri reshenii
evolyutsionnykh raznostnykh skhem", ("Investigation of Nonlinear Oscillations
Which Arise in the Solution of Evolution Difference Schemes") in *Probl. nelineinykh
kolebanii mekhanicheskykh sistem.* (Naukova Dumka, Kiev 1974)

[28] Davydov, Yu. M. "Chislennoe issledovanie nekotorykh perekhodnykh yavlenii," in
Asimptoticheskie metody v teorii sistem ("Numerical Investigation of Some Transition
Phenomena") in: Asymptotic Methods in the Theory of Systems (Irkutskii Gos. Uni-
versitet, Irkutsk 1974) pp. 91–101

[29] Davydov, Yu. M. "Chislennoe issledovanie gidrodinamicheski neustoichivykh techenii
metodom krupnykh chastits," ("Numerical Investigation of Hydrodynamic Unstable
Flows by the Method of "Large Particles"") in *Trud. Vses. shkoly-seminara "Nelineinye
zadachi teorii gidrodinamicheskoi ystoichivosti"* (MGU, Moscow 1978)

[30] Davydov, Yu. M. "Chislennoe eksperimentirovanie metodom "krupnykh chastits"
(teoreticheskie osnovy chislennogo eksperimenta i ego realizatsii)", ("Numerical Experiments by the "Large Particle" Method (Theoretical Foundation of Numerical Expe-

riments and Their Realization)") in *Pryamoe chislennoe modelirovanie techenii gaza.* (Vychisl. Tsentr. Akad. Nauk SSSR, Moscow 1978)

[31] Davydov, Yu. M. Dokl. Akad. Nauk SSSR Ser. Mat. Fiz. **244** (6), 1298–1302 (1979) [English transl.: An investigation of the stability of difference schemes on the boundaries of the computational domain by the method of differential approximation. Sov. Math. Dokl. **20** (1), 210–214 (1979)]

[32] Davydov, Yu. M. Struktura approximatsionnoi vyazkosti (The structure of the artificial viscosity) Dokl. Akad. Nauk SSSR Ser. Mat. Fiz. **245** (4), 812–816 (1979)

[33] Davydov, Yu. M. "Primenenie metoda differentsialnykh priblizhenii dlya issledovaniya ustoichivosti raznostnykh skhem gazovoi dinamiki", ("Application of the Method of Differential Approximations for the Investigation of the Stability of Difference Schemes in Gas Dynamics") in *Doklady VI Mezhdynarodnoi konfer. po chisl. metodam v gidrodinamike, Tom 2,* (Nauka, Moscow, 1978) pp. 95–100

[34] Davydov, Yu. M., Skotnikov, V. P. *Issledovanie drobnykh yacheek v metode "krupnykh chastits",* (The Investigation of Small Boxes in the Method of "Large Particles") (Vychisl. Tsentr Akad. Nauk SSSR, Moscow 1978)

[35] Davydov, Yu. M., Skotnikov, V. P. *Differentsial'nye priblizheniya raznostnykh skhem* (Differential Approximation of Difference Schemes) (Vychisl. Tsentr Akad. Nauk SSSR, Moscow 1978)

[36] Davydov, Yu. M., Skotnikov, V. P. *Metod "krupnykh chastits": voprosy approksimatsii, skhemnoi* vyazkosti i ustoichivosti (The Method of "Large Particles": The Questions of Approximation, Viscosity and Stability of the Scheme) (Vychisl. Tsentr Akad. Nauk SSSR, Moscow 1978)

[37] Davydov, Yu. M., Shidlovskaya, L. V. "Provedenie chislennykh eksperimentov po issledovaniyu fizicheskikh protsessov v blizhnem kosmose s pomoshch'yu metoda krupnykh chastits," ("Carrying out Numerical Experiments for the Investigation of Physical Processes in the Near Cosmic Space Using the Method of "Large Particles"") in *Matem. modeli blizhnego kosmosa* (Krasnoyarsk, 1977) pp. 67–87

[38] Zhukov, A. I. O skhodimosti resheniya raznostnogo uravneniya k resheniyu differentsialnogo uravneniya. (Convergence of the solution of a difference equation to the solution of the differential equation) Dokl. Akad. Nauk SSSR Ser. Mat. Fiz. **117** (2), 174–176 (1957)

[39] Zhukov, A. I. Predel'naya teorema dlya raznostnykh operatorov. (Theorem on the limit of difference operators) Usp. Mat. Nauk **14** (5), (1959)

[40] Zhukov, A. I. K voprosu o skhodimosti raznostnykh metodov resheniya zadachi Koshi. (On the problem of convergence of difference methods for the solution of the Cauchy problem.) Vychisl. Prikl. Mat. **6**, 34–62 (1960)

[41] Ivandaev, A. I. Ob odnom sposobe vvedeniya "psevdovyazkosti", i ego primenenie k utochneniyu raznostnykh reshenii uravnenii gidrodinamiki. (On a method to introduce the "pseudoviscosity" and its application to the accuracy of difference solutions of hydrodynamic equations.) Zh. Vychisl. Mat. Mat. Fiz. **15** (2) 523–527 (1975)

[42] Kochergin, V. P., Sherbakov, A. V. O raznostnykh skhemakh vtorogo poryadka approksimatsii dlya ellipticheskogo uravneniya s malym parametrom pri starshykh proizvodnykh. (On difference schemes of second order of approximation for elliptic equations with a small parameter in the highest derivatives.) Chislennye Metody Mekh. Sploshnoi Sredy **5** (1), 88–97 (1974)

[43] Kuznetsov, N. N. *Raznostnye skhemy v prostranstvakh rastushchikh funktsii* (Difference schemes in spaces of increasing functions) (Vychisl. Tsentr Mosk. Gos. Univ., Moscow 1969)

[44] Kuznetsov, N. N. Asimptoticheskii analiz pogreshnosti konechnoraznostnykh metodov resheniya giperbolicheskikh uravnenii. I. (Asymptotic error analysis of finite-difference methods for the solution of hyperbolic equations, I.) Chislennye Metody Mekh. Sploshnoi Sredy **1** (2), 42–64 (1970)

[45] Kuznetsov, N. N. Dokl. Akad. Nauk SSR Ser. Mat. Fiz. **200** (5), 1026–1029 (1971) [English transl.: Weak stability and asymptotics for solutions of finite-difference approximations of differential equations. Sov. Math. Dokl. **12** (5), 1529–1533 (1971)]

[46] Kuznetsov, N. N. Asimptotika reshenii konechnoraznostnoi zadachi Koshi. (Asymptotic solution of the finite difference Cauchy problem.) Zh. Vychisl. Mat. Mat. Fiz. **12** (2), 334–351 (1972)

[47] Kukudzhanov, V. I. *Chislennoe reshenie neodnomernykh zadach rasprostraneniya voln napryazhenii v tverdykh telakh* (Numerical Solution of Multi-Dimensional Problems of the Propagation of Pressure Waves in Rigid Bodies) (Vychisl. Tsentr Akad. Nauk SSSR, Moscow 1976)

[48] Kuropatenko, V. F. O tochnosti vychisleniya entropii v raznostnykh skhemakh dlya uravnenii gazovoi dinamiki. (On the accuracy of entropy calculation in difference schemes for the equations of gas dynamics) Chislennye Metody Mekh. Sploshnoi Sredy **9** (7), 45–59 (1978)

[49] Levi, B. I., Zaidel', Ya. M., Saikin, V. M. O metode ponizheniya orientatsionnoi pogreshnosti pri chislennom modelirovanii dvukhfaznoi fil'tratsii. (On a method to lower the orientation error in the numerical modelling of two-phase filtration.) Chislennye Metody Mekh. Sploshnoi Sredy **9** (6), 105–114 (1978)

[50] Marchuk, An. G. Odin klass invariantnykh raznostnykh skhem dlya rascheta vzaimodeistviya nestationarnykh udarnykh voln s prepyatstviyami. (A class of invariant difference schemes for the calculation of the interaction of nonstationary shock waves with obstacles.) Chislennye Metody Mekh. Sploshnoi Sredy **8** (2), 73–81 (1977)

[51] Meskal'kov, M. N. "Issledovanie dispersionnykh svoistv raznostnykh skhem dlya uravneniya perenosa", ("Investigation of dispersive features of difference schemes for the transport equations") in *Chislenny analiz* (Kiev, 1978) pp. 75–86

[52] Paasonen, V. I., Shokin, Yu. I., Yanenko, N. N. "On the Theory of Difference Schemes for Gas Dynamics Equations", in Lecture Notes in Physics, Vol. 35, ed. by R. D. Richtmyer (Springer, Berlin, Heidelberg, New York 1975) pp. 293–303

[53] Paasonen, V. I. Dissipativnye neyavnye skhemy s psevdovyazkost'yu vysshikh poryadkov dlya giperbolicheskikh sistem uravnenii. (Dissipative implicit schemes with pseudoviscosity of high orders of approximation for hyperbolic systems of equations) Chislennye Metody Mekh. Sploshnoi Sredy **4** (4), 44–57 (1973)

[54] Petrenko, V. E., Vorozhtsov, E. V. Primenenie chastits – sloev pri raschetakh po metodu chastits v yacheikakh. (Application of particle layers in the calculation by the particle-in-cell method) Chislennye Metody Mekh. Sploshnoi Sredy **4** (2), 132–141 (1973)

[55] Petrenko, V. E., Sapozhnikov, G. A. Ob usilenii ustoichivosti metoda chastits v yacheikakh dlya techenii vyazkoi zhidkosti. (On the increasing of stability of the particle-in-cell method for viscous fluid flows) Chislennye Metody Mekh. Sploshnoi Sredy **7** (4), 130–148 (1976)

[56] Popov, Yu. P., Samarskii, A. A. O metodakh chislennogo resheniya odnomernykh nestationarnykh zadach gazovoi dinamiki. (On methods of numerical solution of one-dimensional unstationary problems of gas dynamics) Zh. Vychisl. Mat. Mat. Fiz. **16** (6), 1503–1518 (1976)

[57] Tusheva, L. A. Ob odnoi neyavnoi skheme 4-go poryadka approksimatsii dlya sistemy uravnenii gazovoi dinamiki. (On an implicit scheme of fourth order of approximation for the system of equations of gas dynamics) Chislennye Metody Mekh. Sploshnoi Sredy **8** (5), 120–131 (1977)

[58] Tusheva, L. A. "K analizu iteratsionnykh raznostnykh skhem dlya giperbolicheskikh sistem uravnenii", ("On the Analysis of Iterative Difference Schemes for Hyperbolic Systems of Equation") (Inst. Teor. Prikl. Mekh. Sib. Otd. Akad. Nauk in *Chislenny analiz* SSSR, Novosibirsk 1978) pp. 83–93

[59] Tusheva, L. A., Shokin, Yu. I., Yanenko, N. N. "Ob odnom metode postroeniya skhem povyshennogo poryadka approksimatsii," ("On a Method of Construction of Schemes

with a Higher Order of Approximation") in: *Izbran. probl. prik. mekh.* (VINITI, Moscow 1974) pp. 681–689

[60] Tusheva, L. A., Shokin, Yu. I., Yanenko, N. N. "O postroenii raznostnykh skhem povyshennogo poryadka approksimatsii na osnove differentialnykh sledstvii", ("On the Construction of Difference Schemes of Higher Order of Approximation on the Basis of Differential Investigations") in *Nekotorye problemy vychislitel'noi i prikladnoi matematiki*, (Nauka, Novosibirsk 1975) pp. 184–191

[61] Fedotova, Z. I. Issledovanie approksimatsionnoi vyazkosti raznostnykh skhem dlya dvumernykh uravnenii gazovoi dinamiki. (Investigation of Artificial Viscosity of difference schemes for the twodimensional equations of gas dynamics) Chislennye Metody Mekh. Sploshnoi Sredy **6** (5), 112–126 (1975)

[62] Fedotova, Z. I. "Invariantnye raznostnye skhemy tipa prediktor-korrektor dlya odnomernykh uravnenii gazovoi dinamiki v Eilerovykh koordinatakh", ("Invariant Difference Schemes of Predictor-Corrector Type for One-Dimensional Equations of Gas Dynamics in Eulerian Coordinates") in *Trud. V Vsesoynznogo seminara po chisl. met. mekh. vyazkoi zhidkosti*, chast' II, (Novosibirsk, 1975) pp. 160–176

[63] Fedotova, Z. I. O primenenii invariantnoi raznostnoi skhemy k raschetu kolebanii zhidkosti v basseine. (Application of invariant difference schemes to the calculation of the oscillation of fluid in a basin) Chislennye Metody Mekh. Sploshnoi Sredy **9** (3), 137–146 (1978)

[64] Harlow, F. H., Evans, M. W. The particle-in-cell method for hydrodynamic calculations. Los Alamos Scient. Lab., Rept No. La-2139 (1957)

[65] Kholodov, A. S. O postroenii raznostnykh skhem s polozhitel'noi approksimatsiei dlya uravnenii giperbolicheskogo tipa. (Construction of difference schemes with total approxiamtion for equations of hyperbolic type.) Zh. Vychisl. Mat. Mat. Fiz. **18** (6), 1476–1492 (1978)

[66] Khonichev, V. I., Yakovlev, V. I. Modifikatsia metoda "krupnykh chastits" dlya issledovaniya otrazheniya difragirovannoi udarnoi volny ot tverdoi stenki, (Modification of the "large particle" method for the investigation of the reflection of a diffracted shock wave from a solid wall) Chislennye Metody Mekh. Sploshnoi Sredy **8** (6), 120–131 (1977)

[67] Shashkin, A. P. "O postroenii monotonnoi skhemy vtorogo poryadka approksimatsii", ("On the Construction of a Monotone Scheme of Second Order of Approximation") (Inst. Teor. Prikl. Mekh. Sib. Otd. Akad. Nauk in *Chislenny analiz* SSSR, Novosibirsk 1978) pp. 111–118

[68] Shokin, Yu. I. "O svyazi korrektnosti pervykh differentsial'nykh priblizhenii i ustoichivosti nekotorykh raznostnykh skhem", ("On the Connection Between the Correctness of the First Differential Approximation and the Stability of Some Difference Schemes") in *Trudy Vses. seminara po chisl. metodam mekh. vyazk. zhidkosti* (Nauka, Novosibirsk 1969) pp. 262–268

[69] Shokin, Yu. I. "Ob approksimatsionnoi vyazkosti neyavnykh raznostnykh skhem", ("On the Artificial Viscosity of Implicit Difference Schemes") in *Trudy Vses. seminara po vychl. metodam mekh. vyazkoi zhidkosti* (Nauka, Novosibirsk 1969) pp. 256–261

[70] Shokin, Yu. I. Ob asimptoticheskom povedenii reshenii raznostnykh skhem. (On the asymptotic behavior of the solutions of difference schemes.) Izv. Sib. Otd. Akad. Nauk SSSR, Ser. Tekh. Nauk **1** (3), 65–68 (1969)

[71] Shokin, Yu. I. Nekotorye voprosy teorii raznostnykh skhem dlya giperbolicheskikh sistem uravnenii. (Some questions to the theroy of difference schemes for hyperbolic systems of equations) Dissertatsiya kand. fiz.-mat. Nauk, Novosibirsk (1969)

[72] Shokin, Yu. I. O cvyazi ustoichivosti raznostnykh skhem i parabolichnosti ikh pervykh differentsial'nykh priblizhenii dlya giperbolicheskoi sistem uravnenii. (The connection between the stability of difference schemes and the parabolicity of their first differential approximation for hyperbolic systems of equations) Izv. Sib. Otd. Akad. Nauk SSSR, Ser. Tekh. Nauk **2** (8), 81–85 (1970)

[73] Shokin, Yu. I. O metode pervogo differentsial'nogo priblizheniya v teorii raznostnykh skhem dlya giperbolicheskykh sistem uravnenii. (The method of the first differential approximation in the therory of difference schemes for hyperbolic equations.) Tr. Mat. Inst. Akad. Nauk SSSR **122**, 66–84 (1973)

[74] Shokin, Yu. I. Neobkhodimoe i dostatochnoe uslovie invariantnosti raznostnykh skhem v terminakh pervogo differentsial'nogo priblizheniya. (A necessary and sufficient condition for the invariance of difference schemes in terms of the first differential approximation) Chislennye Metody Mekh. Sploshnoi Sredy **5** (5), 120–122 (1974)

[75] Shokin, Yu. I. *O primenimosti metoda differentsial'nogo priblizheniya pri issledovanii effekta nelineinykh preobrazovanii v raschetakh slabykh reshenii.* (On the Applicability of the Method of the First Differential Approximation in the Investigation of the Effect of Nonlinear Transformation in the Calculation of Weak Solutions) (Vychisl. Tsentr Sib. Otd. Akad. Nauk SSSR, Novosibirsk, 1974)

[76] Shokin, Yu. I. K analizu dissipatsii i dispersii raznostnykh skhem. (The analysis of dissipation and dispersion of difference schemes) Chislennye Metody Mekh. Sploshnoi Sredy **7** (7), 131–141 (1976)

[77] Shokin, Yu. I. "Analysis of the Properties of Approximation Viscosity of Difference Schemes by Means of the Method of Differential Approximation", in Lecture Notes in Physics, Vol. 59, ed. by. A I. van de Vooren, P. J. Zandbergen (Springer, Berlin, Heidelberg, New York 1976) pp. 410–414

[78] Shokin, Yu. I. *Chislennye metody gazovoi dinamiki. Invariantnye raznostnye skhemy* (Numerical Methods of Gas Dynamics. Invariant Difference Schemes) (NGU, Novosibirsk 1977)

[79] Shokin, Yu. I. "K analizu svoistv raznostnykh skhem metodom differentsialnogo priblizheniya", ("Analysis of Characteristics of Difference Schemes by the Method of the First Differential Approximation) in *Vychisl. met. v mat. fizike, geofizike i optimal'nom upravlenii*, (Nauka, Novosibirsk 1978) pp. 138–145

[80] Shokin, Yu. I. *Metod differentsial'nogo priblizheniya* (The Method of Differential Approximation) (Nauka, Novosibirsk 1979)

[81] Shokin, Yu. I., Talyshev, A. A. Ob ekvivalentnykh raznostnykh skhemakh. (Equivalent difference schemes) Chislennye Metody Mekh. Sploshnoi Sredy **6** (2), 120–125 (1975)

[82] Shokin, Yu. I., Tusheva, L. A. O dissipativnykh raznostnykh skhemakh dlya giperbolicheskikh sistem uravnenii. (Dissipative difference schemes for hyperbolic systems of equations.) Chislennye Metody Mekh. Sploshnoi Sredy **2** (1), 91–98 (1971)

[83] Shokin, Yu. I., Urusov, A. I. "Ob invariantnykh raznostnykh skhemakh rasshchepleniya", ("Invariant Difference Schemes with Splitting") in *Trudy chetvertogo vsesoyuznogo seminara po chisln. met. mekh. vyazkoi zhidkosti* (Novosibirsk, 1973) pp. 192–209

[84] Shokin, Yu. I., Urusov, A. I. [English transl.: Difference schemes in spaces of generalized functions and their differential representations. Sov. Math. Dokl. **20** (5), 1085–1088 (1979)] Dokl. Akad. Nauk. SSSR Ser. Mat. Fiz. **248** (4), 810–813 (1979)

[85] Shokin, Yu. I., Urusov, A. I. Differentsialnye predstavleniya raznostnykh skhem v prostranstvakh obobshchennykh funktsii. (Differential representations of difference schemes in spaces of generalized functions) Chislennye Metody Mekh. Sploshnoi Sredy **10** (4), 125–147 (1979)

[86] Shokin, Yu. I., Fedotova, Z. I. Ob odnom klasse invariantnykh raznostnykh skhem. (A class of invariant difference schemes) Chislennye Metody Mekh. Sploshnoi Sredy **3** (5), 85–94 (1972)

[87] Shokin, Yu. I., Fedotova, Z. I. [English transl.: Invariant difference schemes with a polynomial viscosity matrix. Sov. Math. Dokl. **16** (3), 601–604 (1975)] Dokl. Akad. Nauk SSSR Ser. Mat. Fiz. **222** (1), 51–53 (1975)

[88] Shokin, Yu. I., Fedotova, Z. I. "Issledovanie approksimatsionnoi vyazkosti raznostnykh skhem dlya odnomernykh uravnenii gazovoi dinamiki v Eilerovykh koordina-

takh", ("Investigation of Artificial Viscosity of Difference Schemes for the One-Dimensional Equations of Gas Dynamics in Eulerian Coordinates") in *Chisl. met. resheniya zadach filtratsii neszhim. zhidkosti* (Vychisl. Tsentr Sib. Otd. Akad. Nauk SSSR, Novosibirsk 1975) pp. 297–314

[89] Shokin, Yu. I., Fedotova, Z. I., Marchuk, An. G. [English transl.: On the connection between the conservative property of difference approximations. Sov. Math. Dokl. **19** (5), 1104–1108 (1978)] Dokl. Akad. Nauk SSSR Ser. Mat. Fiz. **242** (2), 290–293 (1978)

[90] Shokin, Yu. I., Yanenko, N. N. O svyazi korrektnosti pervykh differentsialnykh priblizhenii i ustoichivosti raznostnykh skhem dlya giperbolicheskikh sistem uravnenii. (On the connection of the correctness of the first differential approximation and the stability of difference schemes for hyperbolic systems of equations) Mat. Zametki **4** (5), 493–502 (1968)

[91] Shokin, Yu. I., Yanenko, N. N. "Gazovoi dinamiki chislennye metody", ("Numerical Methods of Gas Dynamics") in *Matematicheskaya entsiklopediya*, T. 1. M., 1977, pp. 835–839

[92] Shokin, Yu. I., Yanenko, N. N. "Giperbolicheskogo tipa uravnenie chislennye metody resheniya", ("Numerical Methods for the Solution of Hyperbolic Equations") in *Matematicheskaya entsiklopediya*, T. 1. M., 1977, pp. 993–998

[93] Yanenko, N. N. "Chislennoe reshenie zadach mekhaniki zhidkosti", ("Numerical Solution of Problems of Fluid Mechanics") in *Trudy tret' ego vsesoyuzn. seminara po modelyam mekhaniki sploshnoi sredy*, (Vychisl. Tsentr Sib. Otd. Akad. Nauk SSSR, Novosibirsk 1976) pp. 177–199

[94] Yanenko, N. N., Anuchina, N. N., Petrenko, V. E., Shokin, Yu. I. O metodakh rascheta zadach gazovoi dinamiki s bol'shimi deformatsiyami. (On methods for the calculation of problems of gas dynamics with large deformations) Chislennye Metody Mekh. Sploshnoi Sredy **1** (1), 40–62 (1970)

[95] Yanenko, N. N., Anuchina, N. N., Petrenko, V. E., Shokin, Yu. I. On numerical methods of solving gas dynamic problems with large deformations. Fluid Dyn. Trans. **5** (1), 9–32 (1971)

[96] Yanenko, N. N., Anuchina, N. N., Petrenko, V. E., Shokin, Yu. I. "O metode rascheta techenii szhimaemoi zhidkosti s bol'shimi deformatsiyami i approksimatsionnoi vyazkosti", ("A Method of the Calculation of Viscous Flows with Large Deformations and Artificial Viscosity") in *Trudy sektsii po chisl. metodam v gazodinamike vtorovo mezhdunar. kollokviuma po gazodinamike vzryva i reagir. sistem* (Vychisl. Tsentr Akad. Nauk SSSR, Moscow, 1971) pp. 159–187

[97] Yanenko, N. N., Vorozhtsov, E. V., Fomin, V. M. [English transl.: Differential analyzers of shock waves. Sov. Math. Dokl. **17** (2), 358–362 (1976)] Dokl. SSSR Ser. Mat. Fiz. **227** (1), 50–53 (1976)

[98] Yanenko, N. N., Kovenya, V. M., Liseikin, V. D., Fomin, V. M., Vorozhtsov, E. V. "O nekotorykh metodakh chislennogo modelirovaniya techenii slozhnoi struktury", ("Some Methods of Numerical Modelling of Flows with Complicated Structure") in *Doklady VI mezhdunar. konfer. po chislennym metodam v gidrodinamiki*, (Nauka, Moscow 1978) pp. 210–224

[99] Yanenko, N. N., Shokin, Yu. I. [English transl.: On the artificial viscosity of difference schemes. Sov. Math. Dokl. **9** (5), 1153–1155 (1968)] Dokl. Akad. Nauk SSSR Ser. Mat. Fiz. **182** (2), 280–281 (1968)

[100] Yanenko, N. N., Shokin, Yu. I. O korrektnosti pervykh differentsial'nykh priblizhenii raznostnykh skhem. [English transl.: Correctness of first differential approximations of difference schemes. Soviet Math. Dokl. **9** (5), 1215–1217 (1968)] Dokl. Akad. Nauk SSSR Ser. Mat. Fiz. **182** (4), 776–778 (1968)

[101] Yanenko, N. N., Shokin, Yu. I. "Ob approksimatsionnoi vyazkosti raznostnykh skhem dlya giperbolicheskikh sistem uravnenii", ("Artificial Viscosity of Difference Schemes for Hyperbolic Systems of Equations") in *Trudy vsesoyuznogo seminara po chislennym metodam mekhaniki vyazkoi zhidkosti* (Nauka, Novosibirsk 1969)

[102] Yanenko, N. N., Shokin, Yu. I. "O zagadnieniu lepkosci approksymacyjnej w schema-
tach roznicowich", ("On the Problem of Artificial Viscosity in Difference Schemes")
in *Metody Numericzne w Mechanice Płynów*. (Warszawa, 1969)

[103] Yanenko, N. N., Shokin, Yu. I. O pervom differentsial'nom priblizhenii raznostnykh
skhem dlya giperbolicheskikh sistem uravnenii. (First differential approximation of
difference schemes for hyperbolic systems of equations) Sib. Mat. Z. **10** (5), (1969)

[104] Yanenko, N. N., Shokin, Yu. I. First differential approximation method and approximate viscosity of difference schemes. Phys. Fluids **12** (12), 28–33 (1969)

[105] Yanenko, N. N., Shokin, Yu. I. [English transl.: On a group classification of difference
schemes for the system of equations of gas dynamics. Proc. Steklov Inst. Math. **122**,
87–99 (1973)] in *Nekotorye probl. matem. i mekh.* (Nauka, Leningrad 1970)

[106] Yanenko, N. N., Shokin, Yu. I. "On the Group Classification of Difference Schemes for
Systems of Equations in Gas Dynamics", in Lecture Notes in Physics, Vol. 8, ed by
M. Holt (Springer, Berlin, Heidelberg, New York 1971) pp. 3–17

[107] Yanenko, N. N., Shokin, Yu. I. Gruppovaya klassifikatsiya neyavnykh raznostnykh
skhem dlya sistem uravnenii gazovoi dinamiki. (Group classification of implicit difference schemes for systems of equations of gas dynamics) Chislennye Metody Mekh.
Sploshnoi Sredy **2** (2), 85–92 (1971)

[108] Yanenko, N. N., Shokin, Yu. I. "Primenenie invariantnykh raznostnykh metodov k
zadacham gazovoi dinamiki", ("Application of Invariant Difference Methods to Problems of Gas Dynamics") *Tezisy XIII mezhdunarodnogo kongressa po teoreticheskoi i
prikladnoi mekhaniki* (Nauka, Moscow 1972)

[109] Yanenko, N. N., Shokin, Yu. I. "Schémas numériques invariants de groupe pour les
équations de la dynamique de gas", in Lecture Notes in Physics, Vol. 18, ed. by
H. Cabannes, R. Temam (Springer, Berlin, Heidelberg, New York 1973)
pp. 174–186

[110] Yanenko, N. N., Shokin, Yu. I. O gruppovoi klassifikatsii raznostnykh skhem dlya
sistemy uravnenii gazovoi dinamiki. (Group classification of difference schemes for the
system of equations of gas dynamics) Tr. Mat. Inst. Akad, Nauk **122, 85**–96 (1973)

[111] Yanenko, N. N., Shokin, Yu. I. "K teorii raznostnykh skhem gazovoi dinamiki",
("Theory of Difference Schemes of Gas Dynamics") in *Trudy simpoz. po mekhaniki
sploshnoi sredy i rodstvennym problemam analiza* (Metsniereba, Tbilisi 1974)
pp. 292–306

[112] Yanenko, N. N., Shokin, Yu. I. 'Giperbolicheskogo tipa differentsial'-nykh uravnenii
v chastnykh proizvodnykh sposoby resheniya", ("Hyperbolic Differential Equations
with Partial Derivatives and Their Solution") in *Entsiklopediya kibernetiki* (Kiev, 1975)
pp. 226–229

[113] Yanenko, N. N., Shokin, Yu. I., Urusov, A. I. [English transl.: On difference schemes
in an arbitrary curvilinear coordinate system. Sov. Math. Dokl. **19** (5), 1167–1170
(1978) Dokl. Akad. Nauk SSSR Ser. Mat. Fiz. **242** (3), 552–555 (1978)

[114] Anderson, D. A comparison of numerical solutions to the inviscid equations of fluid
motion. J. Comput. Phys. **15** (1), 1–20 (1974)

[115] Ballhaus, W. F., Lomax, H. "The Numerical Simulation of Low Frequency Unsteady
Transonic Flow Fluids, in Lecture Notes in Physics, Vol. 35 ed. by R. D. Richtmyer
(Springer, Berlin, Heidelberg, New York 1975) pp. 53–63

[116] Beam, R. M., Warming, R. F. An implicit finite-difference algorithm for hyperbolic
systems in conservation-law form. J. Comput. Phys. **22** (1), 87–110 (1976)

[117] Chan, R. K.-C., Street, R. L. A computer study of finite-amplitude water waves. J.
Comput. Phys. **6** (1), 68–94 (1970)

[118] Cheng, S. I. Numerical integration of Navier-Stokes equation. AIAA Pap. No. 70–2
(1970)

[119] Chin, R. C. Y., Hedstrom, G. W. A dispersion analysis for difference schemes: Tables
of generalized Airy functions. Math. Comput. **32** (144), 1163–1170 (1978)

[120] Cloutman, L. D., Fullerton, L. W. Automated heuristic stability analysis for nonlinear
equations. Los Alamos Scientific Laboratory Report No. LA-6885 (1977) pp. 1–52

[121] Cloutman, L. D., Fullerton, L. W. Automated computation of modified equations. J. Comput. Phys. **29** (1), 141–144 (1978)

[122] Daly, B. J. The stability properties of a coupled pair of nonlinear partial differential equations. Math. Comput. **17** (84), 346–360 (1963)

[123] Daly, B. J. A numerical study of turbulence transitions in convective flow. J. Fluid Mech. **64** (1), 129–165 (1974)

[124] Daly, B. J., Pracht, W. F. Numerical study of density-current surges. Phys. Fluids **11** (1), 15–30 (1968)

[125] Fomin, V. M., Vorozhtsov, E. V., Yanenko, N. N. Differential analysers of shock waves: Theory. Comput Fluids **4** (2), 171–183 (1976)

[126] Fromm, J. E. Practical investigation of convective difference approximations of reduced dispersion. Phys Fluids **12** (12), 3–12 (1969)

[127] Genttry, R. A., Martin, R. E., Daly, B. J. An Eulerian differencing method for unsteady compressible flow problems. J. Comput. Phys. **1** (1), 87–118 (1966)

[128] Harlow, F. H., Amsden, A. A. Numerical calculation of almost incompressible flow. J. Comput. Phys. **3** (1), 80–93 (1968)

[129] Harlow, F. H., Amsden, A. A. Fluid dynamics. An introductory text. Los Alamos Scientific Laboratory Report No. LA-4100 (1970)

[130] Harlow, F. H., Amsden, A. A. A numerical fluid dynamics calculation method for all flow speeds. J. Comput. Phys. **8** (2), 197–213 (1971)

[131] Harlow, F. H., Amsden, A. A. Flow of interpretating material phases. J. Comput. Phys. **18** (4), 440–464 (1975)

[132] Harten, A. A method of artificial compression: 1. Shocks and contact discontinuities. AEC Research and Development Report COO-3077-50, New-York University (1974)

[133] Harten, A. The artificial compression method for computation of shocks and contact discontinuities. I. Single conservation laws. Commun. Pure Appl. Math. **30** (5), 611–638 (1977)

[134] Harten, A. The artificial compression method for computation of shocks and contact discontinuities. III. Self-adjusting hybrid schemes. Math. Comput. **32** (142), 363–389 (1978)

[135] Hedstrom, G. W. Models of the difference schemes for $U_t + U_x = 0$ by partial differential equations. Math. Comput. **25** (139), 969–977 (1975)

[136] Hirt, C. W. Heuristic stability theory for finite-difference equations. J. Comput. Phys. **2** (4), 339–355 (1968)

[137] Hirt, C. W. Computer studies of time-dependent turbulent flows. Los Alamos Scientific Laboratory Report No. LA-DC-9578 (1970)

[138] Hirt, C. W., Amsden, A. A., Cook, J. L. An arbitrary Lagrangean-Eulerian computing method for all flow speeds. J. Comput. Phys. **14** (3), 227–253 (1974)

[139] Hirt, C. W., Cook, J. L. Calculating threedimensional flows around structures and over rough terrain. J. Comput. Phys. **10** (2), 324–340 (1972)

[140] Kutler, P., Lomax, H. "The computation of supersonic flow fields about wing-body combinations by "shock-capturing" finite difference techniques", in Lecture Notes in Physics, Vol. 8, ed. by M. Holt (Springer, Berlin, Heidelberg, New York 1971) pp. 24–29

[141] Lerat, A. "Numerical shock structure and nonlinear corrections for difference schemes in conservation form", in VI. mezhdun. konfer. po chisl. metodam v gidrodinam., T. 1, (Nauka, Moscow, 1978) pp. 178–183

[142] Lerat, A., Peyret, R. Sur l'origine des oscillations apparaissant dans les profils de choc calculés par des méthodes aux différence. C. R. Acad. Sc. Paris, série A, **276** (10), 759–762 (1973)

[143] Lerat, A., Peyret, R. Sur le choix de schémas aux différences du second ordre fournissant des profils de choc sans oscillations. C. R. Acad. Sc. Paris, série A **277**, (9) 363–366 (1973)

[144] Lerat, A., Peyret, R. Noncentered schemes and shock propagation problems. Comput. Fluids **2** (1), 35–52 (1974)

[145] Lerat, A., Peyret, R. Propriétés dispersives et dissipatives d'une classe de schémas aux différences pour les systèmes hyperboliques non linéaires. Rech. Aerosp. **2**, 61–79 (1975)

[146] Lerat, A., Peyret, R. "The Problem of Spurious Oscillations in the Numerical Solution of the Equations of Gas Dynamics", in Lecture Notes in Physics, Vol. 35, ed, by R. D. Richtmyer (Springer, Berlin, Heidelberg, New York 1975) pp. 251–256

[147] Majda, A., Osher, St. A systematic approach for correcting nonlinear instabilites. Numer. Math. **30**, 429–459 (1978)

[148] Mc Guire, G. R., Morris, J. Ll. A class of second-order accurate methods for the solutions of systems of conservation laws. J. Comput. Phys. **11** (4), 531–549 (1973)

[149] Nichols, B. D., Hirt, C. W. Calculating three-dimensional free surface flows in the vicinity of submerged and exposed structures. J. Comput. Phys. **12** (2), 234–246 (1973)

[150] Rivard, W. C., Butler, T. D., Farmer, O. A., O'Rourhe, P. J. A method for increased accuracy in Eulerian fluid dynamics calculations. Los Alamos Scientific Laboratory Report N LA-5426-MS (1973)

[151] Roache, P. J. *Computational Fluid Dynamics* (Hermosa, Albuquerque 1972)

[152] Roache, P. J. On artificial viscosity. J. Comput. Phys. **10** (2), 169–184 (1972)

[153] Roseman, J., Zwas, G. Nonlinear transformations and the numerical treatment of shocks. J. Comput. Phys. **19**, 229–235 (1975)

[154] Srinivas, K., Guriraja, J., Prasad, K. K. On the first order local stability scheme for the numerical solution of time-dependent compressible flows. Comput. Fluids **5** (2), 87–97 (1977)

[155] Tyler, L. D. "Heuristic Analysis of Convective Finite Difference Techniques", in Lecture Notes in Physics, Vol. 8, ed. by M. Holt (Springer, Berlin, Heidelberg, New York 1971) pp. 314–319

[156] Viecelli, J. A. A computing method for incompressible fluid bounded by moving walls. J. Comput. Phys. **8** (1), 119–143 (1971)

[157] Vreugdenhil, C. B. On the effect of artificial viscosity methods in calculating shocks. J. Eng. Math. **3** (4), 285–288 (1969)

[158] Warming, R., Kutler, P., Lomax, H. Second- and third-order noncentered difference schemes for nonlinear hyperbolic equations. AIAA J. **11** (2), 189–195 (1973)

[159] Warming, R., Kutler, P., Lomax, H. Computation of space shuttle flowfields using noncentered finite-difference schemes. AIAA J. **11** (2), 196–204 (1973)

[160] Warming, R. F., Hyett, B. J. The modified equation approach to the stability of finite-difference methods. J. Comput. Phys. **14** (2), 159–179 (1974)

[161] Zwas, G., Roseman, J. The effect of nonlinear transformation on the computation of weak solutions. J. Comput. Phys. **12**, (2), 179–186 (1973)

[162] Gel'fand, I. M., Shilov, G. E. *Nekotorye voprosy teorii differentsial'nykh uravnenii. Obobshchennye funktsii*, Vyp. 3 (Some Questions to the Theory of Differential Equations. Generalized Functions) (Fizmatgiz, Moscow 1958)

[163] Petrovskii, I. G. O probleme Koshi dlya sistem lineinykh uravnenii s chastnymi proizvodnykh v oblasti neanaliticheskikh funktsii. (The Cauchy problem for systems of linear equations with partial derivatives in the field of nonanalytic functions.) Byull. Mosk. Gos. Univ., A, **7**, 1–74 (1938)

[164] Courant, R., Friedrichs, K. O., Lewy, H. O raznostnykh uravneniyakh matematicheskoi fiziki. (Difference schemes of mathematical physics) Usp. Mat. Nauk **8**, 125–160 (1940)

[165] Rudin, U. *Funktsional'ny analiz* (Functional Analysis) (Mir, Moscow 1975)

[166] Gel'fand, I. M., Shilov, G. E. *Prostranstvo osnovnykh i obobshchennykh funktsii. Obobshchennye funktsii*, Vyp. 2 (The Space of Fundamental and Generalized Functions. Generalized Functions) (Fizmatgiz, Moscow 1958)

[167] Harten, A. The method of artificial compression: 1. Shocks and contact discontinuities. AEC Research and Development Report COO-3077-50, New York University (1974)

[168] Harten, A., Hyman, J. M., Lax, P. D. On finite-difference approximations and entropy conditions for shocks. Commun. Pure Appl. Math. **29** (2), 297–322 (1976)

[169] Hahn, S. Stability criteria for difference schemes. Commun. Pure Appl. Math. **11** (1), 243–255 (1958)

[170] Lax, P. D. The scope of the energy method. Bull. Am. Math. Soc. **66** (1), 32–36 (1960)

[171] Anuchina, N. N. Nekotorye raznostnye skhemy dlya giperbolicheskikh sistem. (Some difference schemes for hyperbolic systems) Tr. Mat. Inst. Akad. Nauk SSSR **74**, 5–15 (1966)

[172] Yamagutti, M., Nogi, T. An algebra of pseudo difference schemes and its applications. Publ. RIMS Kyoto Univ., Ser. A, **3** (1), 151–162 (1967)

[173] Lax, P. D. Differential equations, difference equations and matrix theory. Commun Pure Appl. Math. **11** (1), 175–194 (1958)

[174] Kreiss, H.-O. On difference approximations of the dissipative type for hyperbolic differential equations. Commun. Pure Appl. Math. **17** (3), 335–353 (1964)

[175] Parlett, B. Accuracy and dissipations in difference schemes. Commun Pure Appl. Math. **19** (1), 111–123 (1966)

[176] Richtmyer, D. "O nelineinoi ustoichivosti raznostnykh skhem", ("Nonlinear Stability of Difference Schemes") in *Nekotorye voprosy vychislitel'noi i prikladnoi matematiki* (Nauka, Novosibirsk 1966) pp. 54–59

[177] Gantmakher, F. R. *Teoriya matrits* (Matrix Theory) (Nauka, Moscow 1967)

[178] Kreiss, H.-O. Über die Stabilitätsdefinition für Differenzengleichungen, die partielle Differentialgleichungen approximieren. BIT **2**, 153–181 (1962)

[179] Lax, P. D. On the stability of difference approximations to solutions of hyperbolic equations with variable coefficients. Commun Pure Appl. Math. **14** (4), 497–520 (1961)

[180] Grudnitskii, V. G., Prokhorchuk, Yu. A. [English transl.: A method of constructing difference schemes with arbitrary order of approximation for partial differential equations. Sov. Math. Dokl. **18** (3), 832–836 (1977)] Dokl. Akad. Nauk SSSR Ser. Mat. Fiz. **234** (6), 1249–1252 (1977)

[181] Strang, G. Difference methods for mixed boundary problem. Duke Math. J. **27** (2), 221–231 (1960)

[182] Kreiss, H.-O., Lundqvist, E. On difference approximations with wrong boundary values. Math. Comput. **22** (10), 1–12 (1968)

[183] Strang, G. Wiener-Hopf difference equations. J. Math. Mech. **13** (1), 85–96 (1964)

[184] Fromm, J. E. A method for reducing dispersion in convective difference schemes. J. Comput. Phys. **3** (2), 176–189 (1968)

[185] Greig, I. S., Morris, J. Ll. A hopscotch method for the Korteweg-deVries equation. J. Comput. Phys. **20** (1), 64–80 (1976)

[186] Leer, B. Stabilization of difference schemes for the equations of ideal compressible flow by artificial diffusion. J. Comput. Phys. **3** (4), 473–485 (1969)

[187] Turkell, E. Phase error and stability of second-order methods for hyperbolic problems. I. J. Comput. Phys. **15** (2), 226–250 (1974)

[188] Turkell, E., Gottlieb, D. Phase error and stability of second order methods for hyperbolic problems. II. J. Comput. Phys. **15** (2), 251–265 (1974)

[189] Vreugdenhil, C. B. On the effect of artificial viscosity method in calculating shocks. J. Eng. Math. **3** (4), 285–288 (1969)

[190] Roberts, K. V., Weiss, N. O. Convective difference schemes. Math. Comput. **20** (94), 272–299 (1966)

[191] Godunov, S. K. Raznostnyi metod chislennogo rascheta razryvnykh reshenii gidrodinamiki. (A difference method for the numerical calculation of solutions of detonation problems in hydrodynamics) Mat. Sb. **47** (89), 271–306 (1959)

[192] Rusanov, V. V. [English transl.: Difference schemes of the third order of accuracy for continuous computation of discontinuous solutions. Sov. Math. Dokl. **9** (3), 771–774 (1968)] Dokl. Akad. Nauk SSSR Ser. Mat. Fiz. **180** (6), 1303–1305 (1968)

[193] Wesseling, P. On the construction of accurate differential equations. J. Eng. Math. **7** (1), 19–31 (1973)

[194] Samarskii, A. A. "O konservativnykh raznostnykh skhemakh", ("Conservative Difference Schemes") in *Problemy prikladnoi matematiki i mekhaniki* (Nauka, Moscow 1971) pp. 129–136

[195] Tikhonov, A. K., Samarskii, A. A. Ob odnorodnykh raznostnykh skhemakh. (Homogeneous difference schemes) Zh. Vychisl. Mat. Mat. Fiz., **1** (1), 5–63 (1961)

[196] Popov, Yu. P., Samarskii, A. A. Polnost'yu konservativnye raznostnye skhemy. (Totally conservative difference schemes) Zh. Vychisl. Mat. Mat. Fiz., **9** (4), 953–958 (1969)

[197] Rusanov, V. V. "Non-linear analysis of the shock profile in difference schemes", in Lecture Notes in Physics Lecture Notes in Physics, Vol. 35, ed. by R. D. Richtmyer (Springer, Berlin, Heidelberg, New York 1975) pp. 270–278

[198] Ovsyannikov, L. V. *Gruppovoi analiz differentsial'-nykh uravnenii* (Group Analysis of Differential Equations) (Nauka, Moscow 1978)

[199] Lax, P. D., Wendroff, B. Systems of conservation laws III. Commun. Pure Appl. Math. **13** (1), 217–238 (1960)

[200] Iskander-Zade, Z. A. K voprosu ob ustoichivosti trivial'nykh reshenii parabolicheskikh sistem uravnenii v chastnykh proizvodnykh. (The question of stability of trivial solutions of parabolic systems of equations with partial derivatives) Zh. Vychisl. Mat. Mat. Fiz. **6** (5), 921–927 (1966)

[201] Kuznetsov, N. N. Asimptotika reshenii konechnoraznostnoi zadachi Cauchy. (Asymptotic solutions of the finite-difference Cauchy problem) Zh. Vychisl. Mat. Mat. Fiz., **12** (2), 334–351 (1972)

[202] Urusov, A. I. "Ob ekvivalentnosti konechnoraznostnykh operatorov pri sovpadenii ikh yader", ("On the Equivalence of Finite-Difference Operators in Case Their Kernels Are Identical") in *Chislenny analiz* (Inst. Teor. Prikl. Mekh. Sib. Otd. Akad. Nauk, Novosibirsk 1978) pp. 94–98

[203] Burstein, S. Z. Finite-difference calculations for hydrodynamic flows containing discontinuities. J. Comput. Phys. **2**, 198–222 (1967)

[204] Rusanov, V. V. Raschet vzaimodeistviya nestatsionarnykh udarnykh voln s prepyadtstviyami. (The calculation of the interaction of nonstationary shock waves with obstacles) Zh. Vychisl. Mat. Mat. Fiz **1** (2), 267–279 (1961)

[205] Lax, P. D., Wendroff, B. Difference schemes for hyperbolic equations with high order of accuracy. Commun. Pure Appl. Math. **17**, 387–398 (1964)

[206] Eilon, B., Gottlieb, D., Zwas, G. Numerical stabilizers and computing time for second-order accurate schemes. J. Comput. Phys. **9**, 387–397, 1972

[207] MacCormack, R. W., Paullay, A. J. The influence of the computational mesh on accuracy for initial value problems with discontinuous and nonunique solutions. Comput. Fluids **2** (3/4), 339–361 (1974)

[208] Lax, P. D. Weak solutions of nonlinear hyperbolic equations and their numerical computation. Commun. Pure Appl. Math. **7** (1), 159–193 (1954)

[209] MacCormack, R. W. The effect of viscosity in hyperbolicity impact cratering. AIAA Pap. 69–354 (Cincinnati, Ohio 1969)

[210] Gourlay, A. R., Morris, J. Ll. Finite difference methods for nonlinear hyperbolic systems. Math. Comput. **22**, 28–39, (1968)

[211] Rubin, E. L., Burstein, S. Z. Difference methods for the inviscid and viscous equations of compressible gas. J. Comput. Phys. **2**, 178–196 (1967)

[212] Yanenko, N. N., Shokin, Yu. I., Tusheva, L. A., Fedotova, Z. I. Klassifikatsiya raznostnykh skhem odnomernoi gazovoi dinamiki metodom differentsial'nogo priblizheniya. (Classification of difference schemes of one-dimensional gas dynamics by the method of the differential approximation) Chislennye Metody Mekh. Sploshnoi Sredy **11** (2), 123–159 (1980)

[213]* Lerat, A. "Numerical shock structure and nonlinear corrections for difference schemes

---

*) The references were added to the manuscript after the translation of the book

in conservation form", in Lecture Notes in Physics, Vol. 90, ed. by H. Cabannes, M. Holt, V. V. Rusanov (Springer, Berlin, Heidelberg, New York 1979) pp. 345–351

[214] Lerat, A. Une classe de schémas aux différences implicites pour les systèmes hyperboliques de lois de conservation. C. R. Acad. Sc. Paris, **288**, série A., 1033–1036 (1979)

[215] Lerat, A., Sidés, J. "Numerical simulation of unsteady transonic flows using the Euler equations in integral form", in *La 21-e Confér. Annuelle sur l'Aviation et l'Astronautique* (1979) pp. 1–8

[216] Shokin, Yu. I. "Analysis of conservative properties of the difference schemes by the method of differential approximation", in Lecture Notes in Physics, Vol. 141, ed. by W. C. Reynolds, R. W. MacCormack (Springer, Berlin, Heidelberg, New York 1981) pp. 383–386

[217] Gour-Tsyn Yeh Numerical solutions of Navier-Stokes equations with an integrated compartment method (ICM). Int. J. Num. Meth. Fluids **1** (3), 207–223 (1981)

[218] Shokin Yu. I., Urusov, A. I. Ob odnom metode postroeniya podvizhnikh setok dlya resheniya uravnenii giperbolicheskogo tipa. (A method of construction of moving grids for the solution of the equations of hyperbolic type) (Inst. Teor. Prikl. Mekh. Sib. Otd. Akad. Nauk SSSR, Novosibirsk 1981)

[219] Kuropatenko, V. V. O tochnosti vychisleniya entropii v raznostnykh skhemakh dlya uravnenii gazovoi dinamiki. (On the accuracy of the computational entropy in difference schemes for the equations of gas dynamics) Chislennye Metody Mekh. Sploshnoi Sredy **9** (7), 49–59 (1978)

[220] Sielecki, A., Wurtele, M. G. The numerical integration of the nonlinear shallow-water equations with sloping boundaries. J. Comput. Phys. **6**, 219–236 (1970)

[221] Kolgan, V. P. Primenenie printsipa minimal'nykh znachenii proizvodnoi k postroeniyu konechnoraznostnykh skhem dlya rascheta razryvnykh reshenii gazovoi dinamiki. (Application of the principle of minimizing the values of the derivative for the construction of finite difference schemes for the calculation of unsteady solutions of gas dynamics) Uchen. Zap. TsAGI **3** (6), 68–77 (1972)

[222] Abarbanel, S., Zwas, G. An iterative finite difference method for hyperbolic systems. Math. Comput. **23** (107), 549–566 (1969)

[223] Abarbanel, S., Goldberg, M. Numerical solutions of quasi-conservative hyperbolic systems for the cylindrical shock problem. J. Comput. Phys. **10**, 1–21 (1972)

[224] Boris, J. P., Book, D. L. Flux-Corrected Transport. III. Minimal-error FCT algorithms. J. Comput. Phys. **20** (4), 397–431 (1976)

[225] Burstein, S. Z., Mirin, A. Third order difference methods for hyperbolic equations. J. Comput. Phys. **5**, 547–557 (1970)

[226] Balakin, V. B. O metodakh tipa Runge-Kutta dlya uravnenii gazovoi dinamiki. (On methods of Runge-Kutta type for the equations of gas dynamics) Zh. Vychisl. Math. Fiz. **10** (6), 1512–1519 (1970)

[227] Strang, G. Trigonometric polynomials and difference method of maximum accuracy. J. Math. Phys. **16** (2), 147–154 (1962)

[228] Paasonen, V. I. Absolyutno ustoichivye raznostnye skhemy povyshennoi tochnosti dlya sistem giperbolicheskogo tipa. (Absolutely stable difference schemes of higher accuracy for systems of hyperbolic type) Chislennye Metody Mekh. Sploshnoi Sredy **3** (3), 82–91 (1972)

[229] Yanenko, N. N., Yaushev, I. K. Ob odnoi absolyutno ustoichivoi skheme integrirovaniya uravnenii gidrodinamiki. (An absolutely stable scheme for the integration of the hydrodynamic equations) Trud. Math. Inst. AN SSSR **74**, 141–146 (1966)

[230] Kuropatenko, V. F. O raznostnykh metodakh dlya uravnenii gidrodinamiki. (On difference methods for the equations of hydrodynamics) Trud. Math. Inst. AN SSSR **74**, 107–137 (1966)

[231] von Neumann, J., Richtmyer, R. A method for numerical calculation of hydrodynamic shocks. J. Appl. Phys. **21**, 232–237 (1950)

[232] Samarskii, A. A., Arsenin, V. Ya. O chislennom reshenii uravnenii gazodinamiki s razlichnymi tipami vyazkosti. (On numerical solutions of the equations of gas dynamics with different types of viscosity) Zh. Vych. Math. Math. Fiz. **1** (2), 357–360 (1961)

[233] Popov, Yu. P., Samarskii, A. A. Pol'nost'yu konservativnye raznostnye skhemy. (Totally conservative difference schemes) Zh. Vych. Math. Math. Fiz. **9** (4), 953–958 (1979)

[234] Latter, R. Similarity solution for a spherical shock wave. J. Appl. Phys. **26**, (8), 955–960 (1955)

Subject Index

Applied Mathematical Sciences

Editors: F. John, L. Sirovich, J. P. LaSalle

Volume 33: **U. Grenander**

Regular Structures

Lectures in Pattern Theory
Volume III
1981. VIII, 569 pages. ISBN 3-540-90560-X

Contents: Introduction. - Patterns: From Chaos to Order. - A Pattern Formalism. - Algebra of Regular Structures. - Some Topology of Image Algebras. - Metric Pattern Theory. - Patterns of Scientific Hypotheses. - Synthesis of Social Patterns of Domination. - Taxonomic Patterns. - Patterns in Mathematical Semantics. - Outlook. - Appendix. - Notes. - Bibliography. - Index.

Volume 34: **J. Kevorkian, J. D. Cole**

Perturbation Methods in Applied Mathematics

1981. 79 figures. X, 558 pages. ISBN 3-540-90507-3

Contents: Introduction. - Limit Process Expansions Applied to Ordinary Differential Equations. - Multiple-Variable Expansion Procedures. - Applications to Partial Differential Equations. - Examples from Fluid Mechanics. - Bibliography. - Author Index. - Subject Index.

Volume 35: **J. Carr**

Applications of Centre Manifold Theory

1981. XII, 142 pages. ISBN 3-540-90577-4

Contents: Introduction to Centre Manifold Theory. - Proofs of Theorems. - Examples. - Bifurcations with Two Parameters in Two Dimensions. - Application to a Panel Flutter Problem. - Infinite Dimensional Problems. - References. - Index.

Volume 36

Dynamic Meteorology: Data Assimilation Methods

Editors: L. Bengtsson, M. Ghil, E. Källén
1981. IX, 330 pages. ISBN 3-540-90632-0

Contents: An Overview of Meteorological Data Assimilation. - A Review of Methods for Objective Analysis. - Normal Mode Initialization. - Assimilation of Asynoptic Data and the Initialization Problem. - Applications of Estimation Theory to Numerical Weather Prediction. - Convergence of Assimilation Procedures. - Some Climatological and Energy Budget Calculations Using the FGGE III-b Analyses During January 1979. - Appendix: Provisional Report on Calculation of Spatial Covariance and Autocorrelation of the Pressure Field.

Volume 37: **S. H. Saperstone**

Semidynamical Systems in Infinite Dimensional Spaces

1981. XIII, 474 pages. ISBN 3-540-90643-6

Contents: Basic Definitions and Properties. - Invariance, Limit Sets, and Stability. - Motions in Metric Space. - Nonautonomous Ordinary Differential Equations. - Semidynamical Systems in Banach Space. - Functional Differential Equations. - Stochastic Dynamical Systems. - Weak Semidynamical Systems and Processes. - Appendix A. - Appendix B. - References. - Index of Terms. - Index of Symbols.

Volume 38: **A. J. Lichtenberg, M. A. Lieberman**

Regular and Stochastic Motion

1983. 140 figures. XXI, 499 pages
ISBN 3-540-90707-6

Contents: Overview and Basic Concepts. - Canonical Perturbation Theory. - Mappings and Linear Stability. - Transition to Global Stochasticity. - Stochastic Motion and Diffusion. - Three or More Degrees of Freedom. - Dissipative Systems. - Appendix A: Applications. - Appendix B: Hamiltonian Bifurcation Theory. - Bibliography. - Author Index. - Subject Index.

Volume 40: **A. W. Naylor, G. R. Gall**

Linear Operator Theory in Engineering and Science

Reprint. 1982. 120 figures. XV, 624 pages
ISBN 3-540-90748-3. (Originally published by Holt, Rinehart and Winston, Inc., 1971)

Contents: Introduction. - Set-Theoretic Structure. - Topological Structure: Introduction to Metric Spaces. Some Deeper Metric Space Concepts. - Algebraic Structure: Introduction to Linear Spaces. - Further Topics. - Combined Topological and Algebraic Structure: Banach Spaces. Hilbert Spaces. Special Operators. - Analysis of Linear Operators (Compact Case): An Illustrative Example. The Spectrum. Spectral Analysis. - Analysis of Unbounded Operators. - Appendix A: The Hölder, Schwarz, and Minkowski Inequalities. - Appendix B: Cardinality. - Appendix C: Zorn's Lemma. - Appendix D: Integration and Measure Theory. - Appendix E: Probability Spaces and Stochastic Processes. - Index of Symbols. - Index.

Springer-Verlag
Berlin
Heidelberg
New York
Tokyo

Springer Series in Computational Physics

Editors: H. Cabannes, M. Holt, H. B. Keller, J. Killeen, S. A. Orszag

F. Bauer, O. Betancourt, P. Garabedian

A Computational Method in Plasma Physics

1978. 22 figures. VIII, 144 pages
ISBN 3-540-08833-4

Contents: Introduction. – The Variational Principle. – The Discrete Equations. – Description of the Computer Code. – Applications. – References. – Listing of the Code with Comment Cards. – Index.

Finite-Difference Techniques for Vectorized Fluid Dynamics Calculations

Editor: **D. L. Book**
With contributions by numerous experts
1981. 60 figures. VIII, 226 pages
ISBN 3-540-10482-8

Contents: Introduction. – Computational Techniques for Solution of Convective Equations. – Flux-Corrected Transport. – Efficient Time Integration Schemes for Atmosphere and Ocean Models. – A One-Dimensional Lagrangian Code for Nearly Incompressible Flow. – Two-Dimensional Lagrangian Fluid Dynamics Using Triangular Grids. – Solution of Elliptic Equations. – Vectorization of Fluid Codes. – Appendices A–E. – References. – Index.

D. P. Telionis

Unsteady Viscous Flows

1981. 132 figures. XXIII, 408 pages
ISBN 3-540-10481-X

Contents: Introduction. – Basic Concepts. – Numerical Analysis. – Impulsive Motion. – Oscillations with Zero Mean. – Oscillating Flows with Non-Vanishing Mean. – Unsteady Turbulent Flows. – Unsteady Separation. – Index.

F. Thomasset

Implementation of Finite Element Methods for Navier-Stokes Equations

1981. 86 figures. VII, 161 pages
ISBN 3-540-10771-1

Contents: Introduction. – Notations. – Elliptic Equations of Order 2: Some Standard Finite Element Methods. – Upwind Finite Element Schemes. – Numerical Solution of Stokes Equations. – Navier-Stokes Equations: Accuracy Assessments and Numerical Results. – Computational Problems and Bookkeeping. – Appendix 1: The Patch Test of the $P1$ Nonconforming Triangle: Sketchy Proof of Convergence. – Appendix 2: Numerical Illustration. – Appendix 3: The Zero Divergence Basis for 2-D $P1$ Nonconforming Elements. – References. – Index.

R. Peyret, T. D. Taylor

Computational Methods for Fluid Flow

1983. 125 figures. X, 358 pages
ISBN 3-540-11147-6

Contents: Numerical Approaches: Introduction and General Equations. Finite-Difference Methods. Integral and Spectral Methods. Relationship Between Numerical Approaches. Specialized Methods. – Incompressible Flows: Finite-Difference Solutions of the Navier-Stokes Equations. Finite-Element Methods Applied to Incompressible Flows. Spectral Method Solutions for Incompressible Flows. Turbulent-Flow Models and Calculations. – Compressible Flows. – Inviscid Compressible Flows. Viscous Compressible Flows. – Concluding Remarks. – Appendix A: Stability. – Appendix B: Multiple-Grid Method. – Appendix C: Conjugate-Gradient Method. – Index.

Springer-Verlag
Berlin
Heidelberg
NewYork
Tokyo